D1667263

Handbook of
Therapeutic Antibodies

Edited by
Stefan Dübel

1807–2007 Knowledge for Generations

Each generation has its unique needs and aspirations. When Charles Wiley first opened his small printing shop in lower Manhattan in 1807, it was a generation of boundless potential searching for an identity. And we were there, helping to define a new American literary tradition. Over half a century later, in the midst of the Second Industrial Revolution, it was a generation focused on building the future. Once again, we were there, supplying the critical scientific, technical, and engineering knowledge that helped frame the world. Throughout the 20th Century, and into the new millennium, nations began to reach out beyond their own borders and a new international community was born. Wiley was there, expanding its operations around the world to enable a global exchange of ideas, opinions, and know-how.

For 200 years, Wiley has been an integral part of each generation's journey, enabling the flow of information and understanding necessary to meet their needs and fulfill their aspirations. Today, bold new technologies are changing the way we live and learn. Wiley will be there, providing you the must-have knowledge you need to imagine new worlds, new possibilities, and new opportunities.

Generations come and go, but you can always count on Wiley to provide you the knowledge you need, when and where you need it!

William J. Pesce
President and Chief Executive Officer

Peter Booth Wiley
Chairman of the Board

Handbook of Therapeutic Antibodies

Volume I

Edited by
Stefan Dübel

WILEY-VCH Verlag GmbH & Co. KGaA

The Editor

Prof. Dr. Stefan Dübel
Technical University of Braunschweig
Institute of Biochemistry and Biotechnology
Spielmannstr. 7
38106 Braunschweig
Germany

Library of Congress Card No.:
applied for

British Library Cataloguing-in-Publication Data
A catalogue record for this book is available from the British Library.

Bibliographic information published by the Deutsche Nationalbibliothek
Die Deutsche Nationalbibliothek lists this publication in the Deutsche Nationalbibliografie; detailed bibliographic data are available in the Internet at <http://dnb.d-nb.de>.

Cover Design: Schulz Grafik-Design, Fussgönheim

Wiley Bicentennial Logo: Richard J. Pacifico

Composition SNP Best-set Typesetter Ltd., Hong Kong

Printing betz-druck GmbH, Darmstadt

Bookbinding Litges & Dopf GmbH, Heppenheim

Printed in the Federal Republic of Germany
Printed on acid-free paper

ISBN 978-3-527-31453-9

Dedication

To Inge, Hans, Ulrike and Tasso Dübel – the four best things in my life

Stefan Dübel

Contents

Volume I

Handbook of Therapeutic Antibodies. Edited by Stefan Dübel

ISBN 978-3-527-31453-9

Volume III Approved Therapeutics

Overview of Therapeutic Antibodies

Handbook of Therapeutic Antibodies. Edited by Stefan Dübel

ISBN 978-3-527-31453-9

A Greeting by the Editor

Today, therapeutic antibodies are essential assets for physicians fighting cancer, inflammation, and infections. These new therapeutic tools are a result of an immense explosion of research sparked by novel methods in gene technology which became available between 1985 and 1995.

This handbook endeavors to present the fascinating story of the tremendous achievements that have been made in strengthening humanity's weapons arsenal against widespread diseases. This story not only includes the scientific and clinical basics, but covers the entire chain of therapeutic antibody production – from downstream processing to Food and Drug Administration approval, galenics – and even critical intellectual property issues.

A significant part is devoted to emerging developments of all aspects of this process, including an article showing that antibodies may only be the first generation of clinically used targeting molecules, making the IgG obsolete in future developments, and novel ideas for alternative therapeutic paradigms.

Finally, approved antibody therapeutics are presented in detail in separate chapters, allowing the clinicians to quickly gain a comprehensive understanding of individual therapeutics.

In such a fast-developing area, it is difficult to keep pace with the rapidly growing information. For example, a PubMed search with "Herceptin" yields more than 1500 citations. Consequently, we have tried to extract the essentials from this vast resource, offering a comprehensive basis of knowledge on all relevant aspects of antibody therapeutics for the researcher, the company expert, and the bedside clinician.

At this point, I express my deep gratitude to all the colleagues who wrote for these books. Without their enthusiasm this project would never have materialized. I would also like to thank Dr Pauly from the publisher's office, who paved the way for this three-volume endeavor, and the biologist Ulrike Dübel – my wife. Both played essential roles in keeping the project on track throughout the organizational labyrinth of its production. The hard work and continuous suggestions of all of these colleagues were crucial in allowing the idea of a comprehensive handbook on therapeutic antibodies to finally become a reality.

Braunschweig, December 2006 Stefan Dübel

Handbook of Therapeutic Antibodies. Edited by Stefan Dübel

ISBN 978-3-527-31453-9

Foreword

The most characterized class of proteins are the antibodies. After more than a century of intense analysis, antibodies continue to amaze and inspire. This *Handbook of Therapeutic Antibodies* is not just an assembly of articles but rather a state-of-the-art comprehensive compendium, which will appeal to all those interested in antibodies, whether from academia, industry, or the clinic. It is an unrivaled resource which shows how mature the antibody field has become and how precisely the antibody molecule can be manipulated and utilized.

From humble beginnings when the classic monoclonal antibody paper by Kohler and Milstein ended with the line, "such cultures could be valuable for medical and industrial use" to the current Handbook you hold in your hand, the field is still in its relative infancy. As information obtained from clinical studies becomes better understood then further applications will become more streamlined and predictable. This Handbook will go a long way to achieving that goal. With the application of reproducible recombinant DNA methods the antibody molecule has become as plastic and varied as provided by nature. This then takes the focus away from the antibody, which can be easily manipulated, to what the antibody recognizes. Since any type, style, shape, affinity, and form of antibody can be generated, then what the antibody recognizes now becomes important.

All antibodies have one focus, namely, its antigen or more precisely, its epitope. In the realm of antibody applications antigen means "target." The generation of any sort of antibody and/or fragment is now a relatively simple procedure so the focus of this work has shifted to the target, and rightly so. Once a target has been identifed then any type of antibody can be generated to that molecule. Many of the currently US Food and Drug Administration (FDA) approved antibodies were obtained in this manner. If the target is unknown then the focus is on the specificity of the antibody and ultimately the antigen it recognizes.

As the field continues to mature the applications of antibodies will essentially mimic as much of the natural human immune response as possible. In this respect immunotherapy may become immunomanipulation, where the immune system is being manipulated by antibodies. With the success of antibody monotherapy the next phase of clinical applications is the use of antibodies with standard chemotherapy, and preliminary studies suggest the combination of

Handbook of Therapeutic Antibodies. Edited by Stefan Dübel

ISBN 978-3-527-31453-9

these two modalities is showing a benefit to the patient. When enough antibodies become available then cocktails of antibodies will be formulated for medical use. Since the natural antibody response is an oligoclonal response then cocktails of antibodies can be created by use of various *in vitro* methods to duplicate this in a therapeutic setting. In essence, this will be oligotherapy with a few antibodies. After all, this is what nature does and duplicating this natural immune response may be effective immunotherapy.

And all of this brings us back full circle to where it all starts and ends, the antibody molecule. No matter what version, isotype, form, or combination used the antibody molecule must first be made and shown to be biologically active. Currently, many of the steps and procedures to generate antibodies can be obtained in kit form and therefore are highly reproducible, making the creation of antibodies a straightfoward process. Once the antibody molecule has been generated it must be produced in large scale for clinical and industrial applications. More often than not this means inserting the antibody genes into an expression system compatible with the end use of the antibody (or fragment). Since many of the steps in generating clinically useful antibodies are labor intensive and costly, care must be used to select antibodies with the specificity and activity of interest before they are mass produced. For commercial applications the FDA will be involved so their guidelines must be followed.

Stating the obvious, it would have been nice to have this Handbook series in the late 1970s when I entered the antibody field. It certainly would have made the work a lot easier! And here it is, about 30 years later, and the generation of antibodies has become "handbook easy." In this respect I am envious of those starting out in this field. The recipies are now readily available so the real challenge now is not in making antibodies but rather in the applications of antibodies. It is hoped that this Handbook will provide a bright beacon where others may easily follow and generate antibodies which will improve our health. The immune system works and works well; those using this Handbook will continue to amaze and inspire.

Mark Glassy
Chairman & Professor, The Rajko Medenica Research Foundation, San Diego, CA, USA
Chief Executive Officer, Shantha West, Inc., San Diego, CA, USA
December 2006

List of Authors

James Allen
University College London
Department of Biochemistry and Molecular Biology
Gower Street
Darwin Building
London WC1E 6BT
UK

Christian Antoni
Schering-Plough
Clinical Research Allergy/Respiratory/Immunology
2015 Galloping Hill Rd
Kenilworth, NJ 07033
USA

Rosalin Arends
Pfizer Inc.
MS 8220-3323
Eastern Point Road
Groten, CT 06339
USA

Michaela A.E. Arndt
Department of Medical Oncology and Cancer Research
University of Essen
Hufelandstr. 55
45122 Essen
Germany

Kenneth D. Bagshawe
Department of Oncology
Charing Cross Campus
Imperial College London
Fulham Palace Road
London W6 8RF
UK

Harald Becker
Wetzbach 26 D
64673 Zwingenberg
Germany

Klaus Bergemann
Boehringer Ingelheim Pharma GmbH & Co. KG
BioPharmaceuticals
Birkendorfer Str. 65
88397 Biberach a.d. Riss
Germany

Thomas Beyer
University Schleswig-Holstein
Campus Kiel
Division of Nephrology
Schittenhelmstr. 12
24105 Kiel
Germany

Handbook of Therapeutic Antibodies. Edited by Stefan Dübel

ISBN 978-3-527-31453-9

Michael Braunagel
Affitech AS
Gaustadalléen 21
0349 Oslo
Norway

Marianne Brüggemann
The Babraham Institute
Protein Technologies Laboratory
Babraham
Cambridge CB22 3AT
UK

Kerry A. Chester
CR UK Targeting & Imaging Group
Department of Oncology
Hampstead Campus
UCL, Rowland Hill Street
London NW3 2PF
UK

Daan J.A. Crommelin
Utrecht University
Utrecht Institute for Pharmaceutical Sciences (UIPS)
Sorbonnelaan 16
3584 CA Utrecht
The Netherlands

Rathin C. Das
Affitech USA, Inc.
1945 Arsol Grande
Walnut Creek, CA 94595
USA

Jason L.J. Dearling
Royal Free and University College Medical School
University College London
Cancer Research UK Targeting & Imaging Group
Department of Oncology
Rowland Hill Street
Hampstead Campus
London NW3 2PF
UK

Michael Dechant
University Schleswig-Holstein
Campus Kiel
Division of Nephrology
Schittenhelmstr. 12
24105 Kiel
Germany

Peter Markus Deckert
Charité Universitätsmedizin Berlin
Medical Clinic III – Haematology, Oncology and Transfusion Medicine
Campus Benjamin Franklin
Hindenburgdamm 30
12200 Berlin
Germany

Eduardo Díaz-Rubio
Hospital Clínico San Carlos
Medical Oncology Department
28040 Madrid
Spain

Dimiter S. Dimitrov
Protein Interactions Group
Center for Cancer Research Nanobiology Program
CCR, NCI-Frederick, NIH
Frederick, MD 21702
USA

Stefan Dübel
Technical University of Braunschweig
Institute of Biochemistry and Biotechnology
Spielmannstr. 7
38106 Braunschweig
Germany

Christian Eckermann
Boehringer Ingelheim Pharma GmbH & Co. KG
BioPharmaceuticals
Birkendorfer Str. 65
88397 Biberach a.d. Riss
Germany

Paul Ellis
Department of Medical Oncology
Guy's Hospital
Thomas Guy House
St. Thomas Street
London SE1 9RT
UK

Thomas Elter
Department of Hematology and Oncology
University of Cologne
Kerpener Str. 62
50937 Köln
Germany

Andreas Engert
Department of Hematology and Oncology
University of Cologne
Kerpener Str. 62
50937 Köln
Germany

Yang Feng
Protein Interactions Group
Center for Cancer Research Nanobiology Program
CCR, NCI-Frederick, NIH
Frederick, MD 21702
USA

Markus Fiedler
Scil Proteins GmbH
Affilin Discovery
Heinrich-Damerow-Str. 1
06120 Halle an der Saale
Germany

Peter Fischer
Boehringer Ingelheim Pharma GmbH & Co. KG
Department of R&D Licensing & Information Management
Birkendorfer Str. 65, K41-00-01
88397 Biberach a.d. Riss
Germany

Martin Foerster
Friedrich-Schiller-University
Department of Pneumology and Allergy
Medical Clinics I
Erlanger Allee 101
07740 Jena
Germany

Patrick Garidel
Boehringer Ingelheim Pharma GmbH & Co. KG
BioPharmaceuticals
Birkendorfer Str. 65
88397 Biberach a.d. Riss
Germany

Guiseppe Giaccone
Vrije Universiteit Medical Center
Department of Medical Oncology
De Boelelaan 1117
1081 HV Amsterdam
The Netherlands

Ralf Gold
Department of Neurology
St. Josef Hospital
Ruhr University Bochum
Gudrunstr. 56
44791 Bochum
Germany

Martin Gramatzki
University of Schleswig-Holstein
Campus Kiel
Division of Stem Cell Transplantation and Immunotherapy
Schittenhelmstr. 12
24105 Kiel
Germany

Stefanos Grammatikos
Boehringer Ingelheim Pharma GmbH & Co. KG
BioPharmaceuticals
Birkendorfer Str. 65
88397 Biberach a.d. Riss
Germany

Larry Green
Abgenix, Inc.
6701 Kaiser Drive
Fremont, CA 94555
USA

Michael Hallek
Department of Hematology and Oncology
University of Cologne
Kerpener Str. 62
50937 Köln
Germany

Melanie Hartsough
Center for Drug Evaluation and Research
Food and Drug Administration
Division of Biological Oncology Products
10903 New Hampshire Ave.
Silver Spring, MD 20993
USA

Mingyue He
The Babraham Institute
Technology Research Group
Cambridge CB2 4AT
UK

Martin H. Holtmann
Johannes-Gutenberg-University
1st Department of Medicine
Rangenbeckstr. 1
55131 Mainz
Germany

Alexandra Huhalov
Royal Free and University College Medical School
University College London
Cancer Research UK Targeting & Imaging Group
Department of Oncology
Rowland Hill Street
Hampstead Campus
London NW3 2PF
UK

Michael Hust
Technical University of Braunschweig
Institute of Biochemistry and Biotechnology
Spielmannstr. 7
38106 Braunschweig
Germany

Alexander Jacobi
Boehringer Ingelheim Pharma GmbH & Co. KG
BioPharmaceuticals
Birkendorfer Str. 65
88397 Biberach a. d. Riss
Germany

Sigbert Jahn
Serono GmbH
Freisinger Str. 5
85716 Unterschleissheim
Germany

Wim Jiskoot
Gorlaeus Laboratories
Leiden/Amsterdam Center for Drug Research (LACDR)
Division of Drug Delivery Technology
P.O. Box 9502
2300 RA Leiden
The Netherlands

Thomas Jostock
Novartis Pharma AG
Biotechnology Development
Cell and Process R & D
CH-4002 Basel
Switzerland

Joachim R. Kalden
University of Erlangen-Nürnberg
Medical Clinic III
Rheumatology, Immunology & Oncology
Krankenhausstrasse 12
91052 Erlangen
Germany

Hitto Kaufmann
Boehringer Ingelheim Pharma GmbH & Co. KG
BioPharmaceuticals
Birkendorfer Str. 65
88397 Biberach a.d. Riss
Germany

Ralph Kempken
Boehringer Ingelheim Pharma GmbH & Co. KG
BioPharmaceuticals
Birkendorfer Str. 65
88397 Biberach a.d. Riss
Germany

Scott Klakamp
AstraZeneca Pharmaceuticals LP
24500 Clawiter Road
Hayward, CA 94545
USA

Roland E. Kontermann
University Stuttgart
Institute for Cell Biology and Immunology
Allmandring 31
70569 Stuttgart
Germany

Zoltán Konthur
Max Planck Institute for Molecular Genetics
Department of Vertebrate Genomics
Ihnestr. 63–73
14195 Berlin
Germany

Jürgen Krauss
Department of Medical Oncology and Cancer Research
University of Essen
Hufelandstr. 55
45122 Essen
Germany

Claus Kroegel
Friedrich-Schiller-University
Department of Pneumology and Allergy
Medical Clinics I
Erlanger Allee 101
07740 Jena
Germany

Hartmut Kupper
Abbott GmbH & Co. KG
Knollstr. 50
67061 Ludwigshafen
Germany

Meina Liang
AstraZeneca Pharmaceuticals LP
24500 Clawiter Road
Hayward, CA 94545
USA

Charito Love
Long Island Jewish Medical Center
Division of Nuclear Medicine
New Hyde Park, New York
USA

Andrew C.R. Martin
University College London
Department of Biochemistry and Molecular Biology
Darwin Building
Gower Street
London WC1E 6BT
UK

Christian Menzel
Technical University of Braunschweig
Institute of Biochemistry and Biotechnology
Spielmannstr. 7
38106 Braunschweig
Germany

Gerhard Moldenhauer
German Cancer Research Center
Department of Molecular Immunology
Tumor Immunology Program
Im Neuenheimer Feld 280
69120 Heidelberg
Germany

Dafne Müller
University Stuttgart
Institute for Cell Biology and Immunology
Allmandring 31
70569 Stuttgart
Germany

Markus F. Neurath
Johannes-Gutenberg-University
1st Department of Medicine
Langenbeckstr. 1
55131 Mainz
Germany

Dianne L. Newton
SAIC Frederick, Inc.
Developmental Therapeutics Program
National Cancer Institute at Frederick
Frederick, MD 21702
USA

Michael J. Osborn
The Babraham Institute
Protein Technologies Laboratory
Babraham
Cambridge CB22 3AT
UK

Christopher J. Palestro
Albert Einstein College of Medicine
Bronx, New York
USA
and:
Long Island Jewish Medical Center
New Hyde Park, New York
USA

Gabriele Pecher
Humboldt University Berlin
Medical Clinic for Oncology and Hematology
Charité Campus Mitte
Charitéplatz 1
10117 Berlin
Germany

Matthias Peipp
University Schleswig-Holstein
Campus Kiel
Division of Stem Cell Transplantation and Immunotherapy
Schittenhelmstr. 12
24105 Kiel
Germany

Edith A. Perez
Mayo Clinic Jacksonville
Division of Hematology/Oncology
4500 San Pablo Road
Jacksonville, FL 32224
USA

Sandra Pisch-Heberle
Boehringer Ingelheim Pharma GmbH & Co. KG
BioPharmaceuticals
Birkendorfer Str. 65
88397 Biberach a.d. Riss
Germany

Ponraj Prabakaran
Protein Interactions Group
Center for Cancer Research Nanobiology Program
CCR, NCI-Frederick
Frederick, MD 21702
USA

Gabriele Reich
Ruprecht-Karls-University
Institute of Pharmacy and Molecular Biotechnology (IPMB)
Department of Pharmaceutical Technology and Pharmacology
Im Neuenheimer Feld 366
69120 Heidelberg
Germany

Josephine N. Rini
Albert Einstein College of Medicine
Bronx, New York
USA
And:
Long Island Jewish Medical Center
Division of Nuclear Medicine
New Hyde Park, New York
USA

Lorin Roskos
AstraZeneca Pharmaceuticals LP
24500 Clawiter Road
Hayward, CA 94545
USA

Susanna M. Rybak
Bionamomics LLC
411 Walnut Street, #3036
Green Cove Springs, FL 32043
USA

José W. Saldanha
National Institute for Medical Research
Division of Mathematical Biology
The Ridgeway
Mill Hill
London NW7 1AA
UK

Jochen Salfeld
Abbott Bioresearch Center
100 Research Drive
Worcester, MA 01605
USA

Huub Schellekens
Utrecht University
Department of Pharmaceutics Sciences
Department of Innovation Studies
Sorbonnelaan 16
3584 CA Utrecht
The Netherlands

Sebastian Schimrigk
Ruhr University Bochum
St. Josef Hospital
Department of Neurology
Gudrunstr. 56
44791 Bochum
Germany

Thomas Schirrmann
Technical University Braunschweig
Institute of Biochemistry and Biotechnology
Department of Biotechnology
Spielmannstr. 7
38106 Braunschweig
Germany

Norbert Schleucher
Hematolgy and Medical Oncology
Marienkrankenhaus Hamburg
Alfredstr. 9
22087 Hamburg
Germany

Alexander C. Schmidt
National Institutes of Health
National Institute of Allergy and Infectious Diseases
Laboratory of Infectious Diseases
50 South Drive, Room 6130
Bethesda, MD 20892
USA
And:
Charité Medical Center at Free University and Humboldt University Berlin
Center for Perinatal Medicine and Pediatrics
Schumannstr 20/21
13353 Berlin
Germany

Karlheinz Schmitt-Rau
Serono GmbH
Freisinger Str. 5
85716 Unterschleissheim
Germany

Marjorie A. Shapiro
Center for Drugs Evaluation and Research
Food and Drug Administration
Division of Monoclonal Antibodies
HFD-123
5600 Fishers Lane
Rockville, MD 20872
USA

Surinder K Sharma
CR UK Targeting & Imaging Group
Department of Oncology
Hampstead Campus
UCL, Rowland Hill Street
London NW3 2PF
UK

Igor A. Sidorov
Center for Cancer Research Nanobiology Program
CCR, NCI-Frederick, NIH
Protein Interactions Group
P.O. Box B, Miller Drive
Frederick, MD 21702-1201
USA

Arne Skerra
Technical University Munich
Chair for Biological Chemistry
An der Saatzucht 5
85350 Freising-Weihenstephan
Germany

Jennifer A. Smith
The Babraham Institute
Protein Technologies Laboratory
Babraham
Cambridge CB22 3AT
UK

Patrick G. Swann
Center for Drug Evaluation and Research
Food and Drug Administration
Division of Monoclonal Antibodies
HFD-123
5600 Fishers Lane
Rockville, MD 20872
USA

Michael J. Taussig
The Babraham Institute
Technology Research Group
Cambridge CB2 4AT
UK

Lars Toleikis
RZPD, Deutsches Ressourcenzentrum für Genomforschung GmbH
Im Neuenheimer Feld 515
69120 Heidelberg
Germany

Daniel Tracey
Abbott Bioresearch Center
100 Research Drive
Worcester, MA 01605
USA

Martina M. Uttenreuther-Fischer
Boehringer Ingelheim Pharma GmbH & Co. KG
Department of Medicine
Clinical Research Oncology
Birkendorfer Str. 65
88397 Biberach a.d. Riss

Thomas Valerius
University Schleswig-Holstein
Campus Kiel
Division of Nephrology
Schittenhelmstr. 12
24105 Kiel
Germany

Udo Vanhoefer
Department of Medicine
Hematology and Medical Oncology, Gastroenterology and Infectious Diseases
Marienkrankenhaus Hamburg
Alfredstr. 9
22087 Hamburg
Germany

Michael Wenger
International Medical Leader
F. Hoffmann-La Roche Ltd.
Bldg. 74/4W
CH-4070 Basel
Switzerland

Maria Wiekowski
Schering-Plough
Clinical Research Allergy/Respiratory/ Immunology
2015 Galloping Hill Road
Kenilworth, NJ 07033
USA

Xiangang Zou
The Babraham Institute
Protein Technologies Laboratory
Cambridge CB22 3AT
UK

Introduction

1
Therapeutic Antibodies – From Past to Future

Stefan Dübel

1.1
An Exciting Start – and a Long Trek

In the late nineteenth century, the German army doctor Emil von Behring (1854–1979), later first Nobel Laureate for Medicine, pioneered the therapeutic application of antibodies. He used blood serum for the treatment of tetanus and diphtheria ("Blutserumtherapie"). When his data were published in 1890 (Behring and Kitasato 1890), very little was known about the factors or mechanisms involved in immune defense. Despite this, his smart conclusion was that a human body needs some defense mechanism to fight foreign toxic substances and that these substances should be present in the blood – and therefore can be prepared from serum and used for therapy against the toxins or infections. It worked, and the success allowed him to found the first "biotech" company devoted to antibody-based therapy in 1904 – using his Nobel Prize money as "venture capital." The company is still active in the business today as part of ChironBehring.

In 1908, Paul Ehrlich, the father of hematology (Ehrlich, 1880) and the first consistent concept of immunology ("lateral chain theory," Fig. 1.1d, Ehrlich, 1908), got the second Nobel Prize related to antibody therapeutics for his groundbreaking work on serum, "particularly to the valency determination of sera preparations." Ehrlich laid the foundations of antibody generation by performing systematic research on immunization schedules and their efficiency, and he was the first to describe different immunoglobulin subclasses. He also coined the phrases "passive vaccination" and "active vaccination." His lateral chain theory ("Seitenketten," sometimes misleadingly translated to "side chain theory") postulated chemical receptors produced by blood cells that fitted intruding toxins (antigens). Through these chemical receptors, cells combine with antigens and the receptors are eventually released as circulating antitoxins (antibodies). Without any knowledge of molecular structure or biochemical binding mechanisms, Paul Ehrlich anticipated much of today's knowledge on immunoglobulin generation and antibody–antigen interaction, even class switching (Fig. 1.1d).

Handbook of Therapeutic Antibodies. Edited by Stefan Dübel

ISBN 978-3-527-31453-9

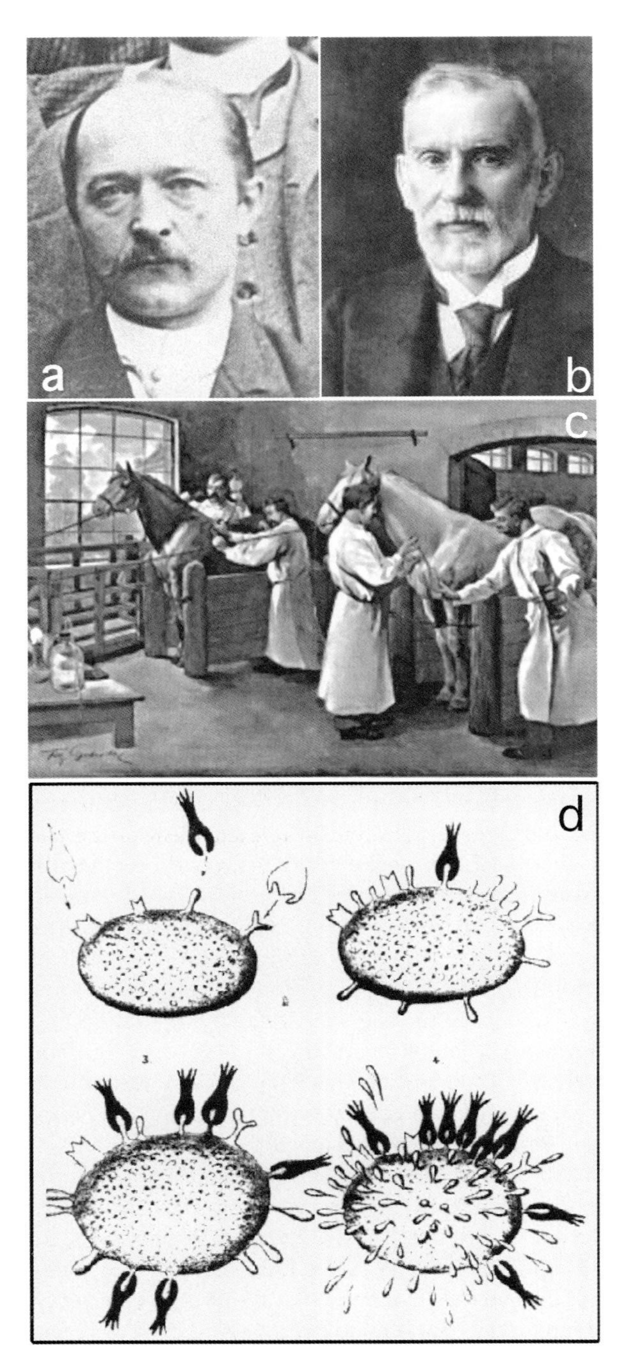

Fig. 1.1 We have come a long way since the first methods for the generation of antibody based therapeutics were established (c), pioneered by (a) Emil von Behring and (b) Paul Ehrlich in the last decade of the nineteenth century. (d) Drawing from Paul Ehrlich on the lateral chain theory (lateral chains = antibodies). He anticipated principles confirmed on a molecular basis many decades later, like the binding of antigens by different specific antibodies (the "lock and key" principle), the differentiation and maturation of B cells and the class switch, allowing the initially cell-bound antibodies to be released in large amounts. Photos: Deutsches Historisches Museum, Berlin.

Passive and active vaccines were developed in rapid succession at the beginning of the twentieth century, and were successful in saving many lives. Snake and insect bites could be treated specifically, and beneficial effects were even observed with human serum immunoglobulin G (IgG) preparations without prior specific immunization (e.g. protection against hepatitis A).

The enormous succes of all these blood products for the prevention and treatment of infections and intoxications, however, could not be expanded to other substantial disease areas, in particular cancer and autoimmunity. Here, the understanding of molecular processes in their etiology or at least the ability to identify molecules strongly correlated to their onset stimulated the desire to produce antibodies targeting these for therapeutic intervention. Unlike a snake bite, however, cancer and autoimmune diseases are chronic, and it was rapidly understood from animal models that antibodies have to be applied to the patient more than once. Immunologists knew very well at that time that repeated application of antigen during antiserum preparation is a good strategy to "booster" the immune response. When using animal serum antibodies to treat chronically ill humans, this must induce an immune response to the therapeutic agent. Further, a drug containing solely an IgG of defined specificity would be tremendously helpful to limit side effects and reach sufficient concentrations at the target site. These prerequisites could by no means be met by the well-established methods of serum antibody preparation.

Much has been learned about the antibody structure (Fig. 1.2) and its function since then (see Chapter 2, Vol I). Hopes were high when Cesar Milstein and Georg Köhler demonstrated that monoclonal antibodies could be produced in mouse cell culture (see Chapter 2, Vol I). However, the excitement of the late 1970s cooled rapidly when almost all first-generation monoclonal therapeutics failed during clinical evaluation. Only one of these products made it through to US Food and Drug Administration (FDA) approval in 1984 – Orthoclone (see Chapter 9, Vol III). However, this was a special case as the typical patient recieving this CD3 antibody was already immune suppressed by disease, a setup not commonly present in cancers or autoimmune diseases. Even more important, it was realized that simple binding to a target (inducing its neutralization) is in most cases not sufficient to make an antibody a good therapeutic agent to treat cancer or immune diseases. Effector functions, like complement activation or cellular responses triggered by Fc receptor binding, are obviously needed, but are not fully provided by the mouse monoclonals.

As a consequence, huge efforts were undertaken to exchange the antibody's effector domains (constant regions) for human ones – thereby also removing the most immunogenic parts from the mouse IgG. By using the then available methods of molecular cloning and recombinant expression, most mouse antibodies were humanized prior to clinical testing. Various methods (chimerization and CDR-exchange-based humanization being the most widely used) were employed (see Chapter 6, Vol I). When tinkering with the antibody sequence, most candidate antibodies were also affinity matured to typical nanomolar and subnanomolar affinities (see Chapter 7, Vol I). The technology for humanization

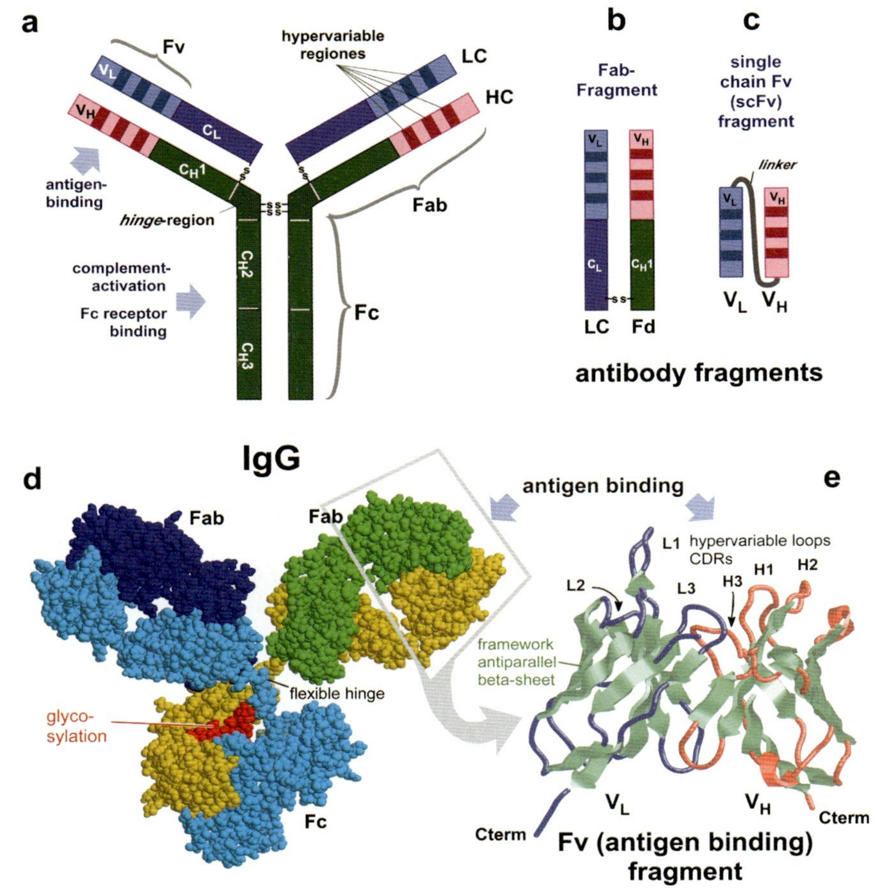

Fig. 1.2 Introduction to antibody structure. (a, d) IgG is a heterotetrameric protein assembled from two identical heavy and light chains (HC, LC), assembled by disulfide bonds. (b) Fab fragments contain the antigen-binding region, and can be generated by proteolysis or recombinant production. (c) In single-chain Fv fragments (scFv), the two antigen-contacting domains (variable regions of the heavy and light chains, V_H, V_L) are connected by an oligopeptide linker to form a single polypeptide. They can be produced in *E. coli* and are typically employed for the selection of human antibodies by phage display and other display systems. (d) Space fill model based on X-ray crystallographic data of an antibody. The typical Y shape is only one of the many conformations the Fab fragments of an antibody can assume relative to each other. T-shaped structures can be assumed, and the hinge region to the Fc part can also bend significantly relative to the Fab fragments. (e) Alpha carbon backbone of an Fv fragment, the antigen-binding fragment of an antibody located at the two tips of the Y- or T-shaped complex, emphasizing the typical antiparallel beta sheet framework structure which holds together the hypervariable loops (L1–3, H1–3) composing the antigen-binding surface (CDRs).

and affinity maturation became available from the mid-1980s. Taking into account that drug development, testing and approval needs about 10 years to reach the market, a growing number of therapeutic antibodies were approved starting from the mid-1990s. These "first generation" recombinant antibodies – cloned from rodent hybridoma cells and improved by gene engineering – dominate the current list of approved antibody therapeutics.

Early in the 1990s, two novel enabling technologies were developed that revolutionized the generation of therapeutic antibodies, as for the first time they provided a robust and reliable method to prepare specific antibodies of human origin. Phage display (see Chapter 3, Vol I) and transgenic mice (see Chapter 4, Vol I) allowed the production of antibodies that are genetically 100% identical to human immunoglobulins. Recently, even transgenic rabbits producing human antibodies have been generated (Buelow et al. 2006). Many other approaches for the *in vitro* selection of antibodies, such as yeast display or ribosomal display, followed and are still in various stages of maturation. These are all based on the selection of antibodies from a large antibody gene repertoire in a heterologous expression system (Fig. 1.3).

The experience with other recombinant human protein drugs (e.g. insulin) raised great hopes that these human antibodies are completely stealthy in respect of an anti-drug immune reaction. However, it was soon realized that immunoge-

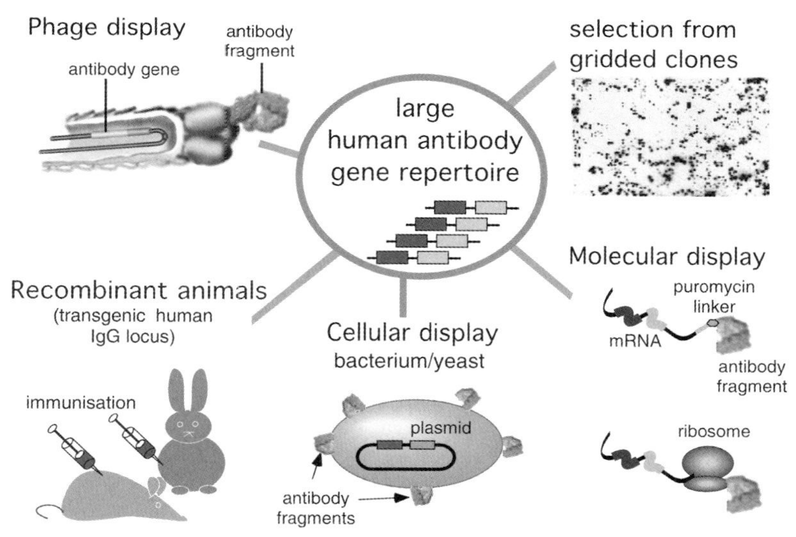

Fig. 1.3 Systems employed for the generation of human antibodies. They all include a heterologous expression of a repertoire of human immunoglobulin genes. Selection is achieved either *in vivo* by immunization (in case of the recombinant animals) or *in vitro* by binding to the antigen, allowing clonal selection of the gene host.

nicity was not switched off completely in most cases – Fc-glycosylation patterns resulting from recombinant production can be very diffferent from endogenous human IgG, and a variable human antibody region with its "lottery"-derived CDRs (complementarity determining regions) generated outside of the context of the human immune system can be quite immunogenic. However, these problems were in most cases minor compared with the effects observed previously with animal-produced immunoglobulins. Today, the first antibody with a completely "human" sequence origin reached the pharmacy shelves: Humira (see Chapter 1, Vol III). It was genetically assembled entirely *in vitro*, with an antigen-binding region selected from an *E. coli*-hosted gene repertoire by phage display (see Chapter 3, Vol I). Many more antibodies derived from human gene repertoires, selected by phage display or generated in transgenic mice carrying the human immunoglobulin locus, have entered clinical testing. The *in vitro* technologies, like phage display, offer an additional advantage when antibody generation in animals is difficult, for example due to the high homology (resulting in low immunogenicity) of the human antigen used for immunization to a mouse protein, or in case of highly toxic or deadly pathogenic antigens. A few animal or even human serum-derived antibody products are still available (Rohrbach et al. 2003) but recombinant human or humanized products vastly dominate the current clinical studies.

Approval time was also getting increasingly short for antibodies when compared with a typical small molecule drug, and success rates of clinical studies improved, mainly due to the more predictable pharmacokinetics and lower risk of toxicity and other side effects when using molecules almost identical to the IgG in our veins – of course with the usual unavoidable exceptions, usually due to effects caused by the antigen binding itself (e.g. seen with TGN1412).

It is a surprising side note of history that in contrast to so many other biomedical achievements, almost all of the enabling antibody technologies (polyclonal sera, monoclonal antibodies, production of functional antibodies in *E. coli* and phage display, humanization, among others) were pioneered in Europe, mainly in the UK and Germany.

1.2
The Gold Rush

A gold rush for new therapeutic agents started when it was realized that all enabling technologies are in place to develop and produce monospecific, nearly human antibodies that are only mildly immunogenic but provide high-affinity target binding and human effector functions, long serum half-life and other pharmacologic advances. Many promising new concepts for the treatment of a huge variety of different diseases were envisaged. In fact, there are only few theoretical restrictions for antibody treatments. First comes the necessity to find a molecular target (antigen) accessible from the bloodstream (i.e. typically a target at the cell surface that is located solely or in a higher concentration on the cell compartment to be effected). Second, the antibody in most cases needs to activate some immune reaction at the binding site (e.g. to kill a tumor cell). Exceptions are antibodies that act by neutralization of an infectious agent or an overexpressed factor, which can be achieved by simple sterical inhibition of the binding of the agent to its natural receptor. Affinity is no practical limitation anymore, as with existing methods antibodies usually can be engineered to provide affinities better than those needed for a maximal therapeutic effect (see Chapter 7, Vol I). Specificity is always an issue, of course, as no antibody is a priori unsusceptible to a cross-reaction, but many strategies have been developed to tackle this problem. Most simply, large numbers of different human antibody clones can now easily be evaluated in parallel using high-throughput assays (see Chapter 4, Vol II).

Some commentators are pessimistic about the shortage of manufacturing capacity for antibodies, and their high cost of production in comparison with a small molecule drug. They have calculated that health systems could not afford all of these new, expensive drugs even if they were made available. However, a scale-up of capacity and novel alternative production systems (e.g. microbial, eukaryotic or plant-derived) may allow much cheaper production of antibodies for many applications, and may even allow the "expensive" antibodies to enter new, low-margin therapeutic markets (see Chapter 6, Vol II). An example already paving the way in this direction is a plant-produced anti-caries antibody which is in phase II clinical testing today.

1.3 Success and Disappointment

All these exciting opportunities offered by the new technologies have resulted in an explosion in the number of clinical studies and approvals (Fig. 1.4). After the slow start in the 1980s and no substantial increase between 1985 and 2000 (Reichert 2001), the number of clinical studies has changed dramatically. Today, more than 400 studies are ongoing, targeting a broad range of diseases, from various cancers to infection and autoimmunity (Tables 1.1 and 1.2). Cancer therapeutics clearly dominate the field, but infection/immunity applications are catching up considering the low number of already approved antibodies, with about a third of the ongoing studies now. The nature of the antigens is just as diverse, with the approved antibodies targeting both cell surface markers (e.g. CD11a, CD20, CD25, CD33, CD52, EGFr, Her2) and soluble molecules (e.g. TNF-α, RSV, VEGF), with three different molecular formats: IgG, Fab fragments, and radioconjugates. A range of new targets and conjugates will follow. This is perfectly illustrated by the antibody variants made to block the tumor necrosis factor (TNF)/TNFR interaction to downregulate overshooting inflammations (Fig. 1.5). Starting with a clinically failed mouse hybridoma antibody, the next steps were chimeric and then humanized antibodies. Finally, a completely human antibody

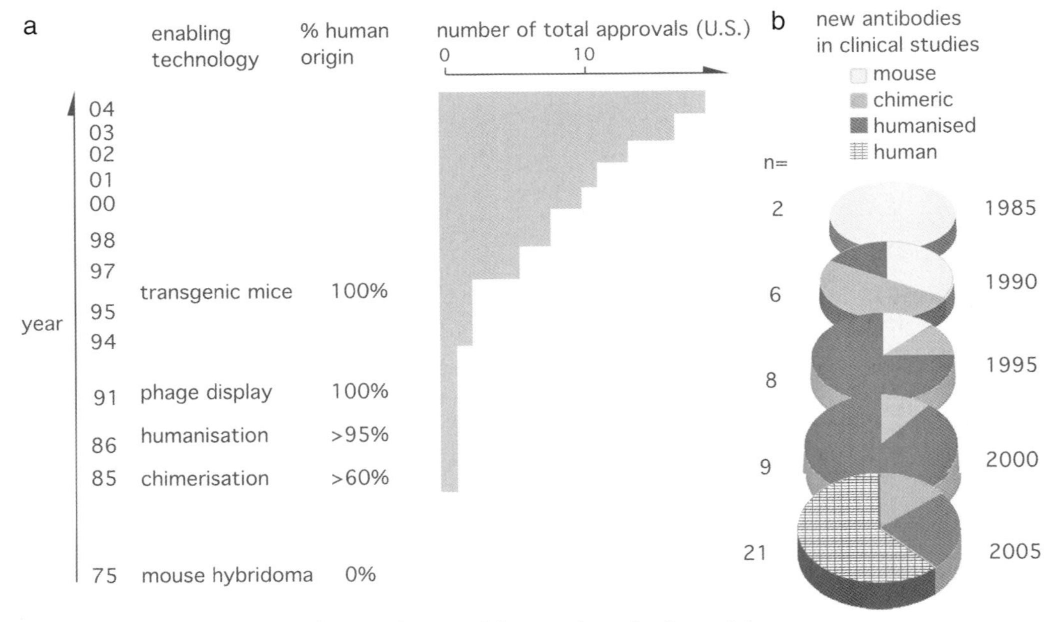

Fig. 1.4 Development history of therapeutic antibodies and the technologies for their generation. Data assembled from Reichert (2001), Reichert et al. (2005), Fabrizi (2005) and public domain sources of the companies.

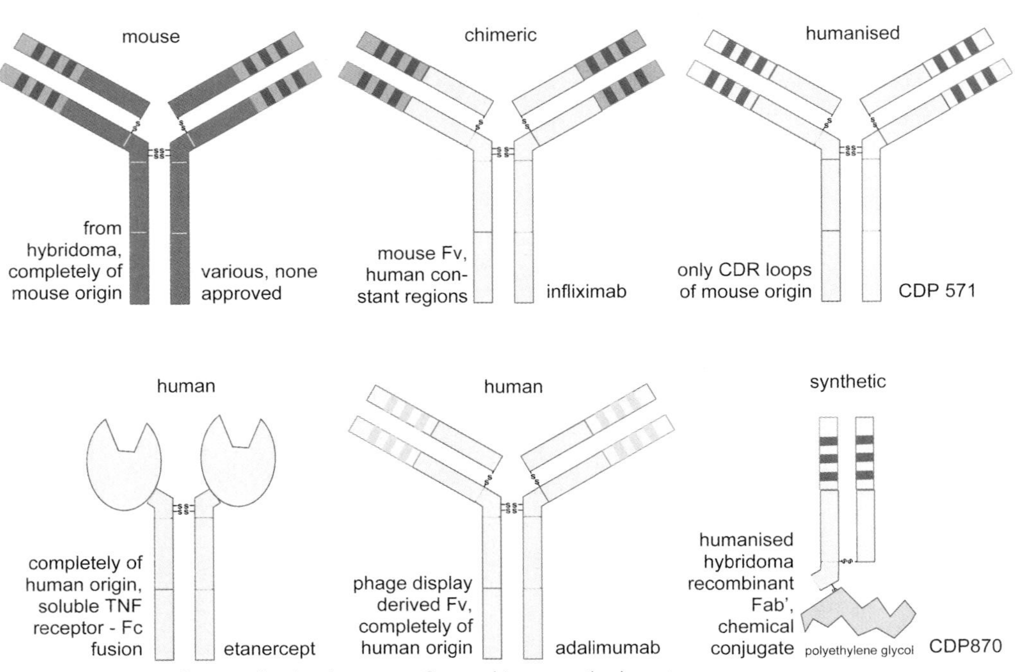

Fig. 1.5 The development of recombinant antibody variant TNF antagonists. Inspired by Rutgeerts et al. (2004).

Table 1.1 Clinical studies with therapeutic antibodies 2005.

Disease	*Clinical phase*			
	I	*II including I/II*	*III including II/III*	*Total*
Cancer				283
Systemic treatment	120	118	18	256
Ex-vivo purging				27
Infection and immunity	40	56	24	120
Total	160	174	42	403

Assembled from Bayes et al. (2005), Reichert et al. (2005), data presented on http://www.clinicaltrials.gov, http://www.centerwatch.com/, company homepages, other public domain sources and the Monoclonal Antibody Index.

and a human soluble receptor–Fc fusion were the next clinically tested and approved iterations of the idea. Now, a Fab fragment of human origin fused to PEG (polyethylene glycol) to improve its pharmacokinetics marks the sixth generation of TNF antagonist antibody constructs no longer resembling a molecule present in nature. Recombinant antibody technology may even reach its limits here, and a seventh-generation TNF antagonist could well be a novel synthetic binder of nonantibody structure (see Chapter 7, Vol II). A very good introduction to the clinical success story of recombinant antibodies is given by the review series by Reichert and colleagues (Reichert 2001, 2002; Reichert et al. 2005).

In February 2005, marketing of the therapeutic antibody natalizumab (marketed as Tysabri), a promising drug for the treatment of multiple sclerosis, was voluntarily suspended after only a few months of approval, based on reports of serious adverse events that have occurred in patients treated in combination with interferon beta-1a in clinical trials (Kleinschmidt-DeMasters and Tyler 2005). It seems, however, that this suspension will be released at time of print due to the reported positive effect of this drug. In December 2005, a voluntary market suspension due to serious and life-threatening cardiopulmonary events following the administration of NeutroSpec (technetium (^{99m}Tc) fanolesomab) was published. In March 2006, dramatic adverse effects not predicted from the animal studies were observed in clinical phase I for the anti-CD28 antibody TGN1412. These unfortunate stories of agents, some of which have even passed clinical studies for their approval, although far from final conclusion, as seen by the lift of the clinical hold for natalizumab after one year of evaluation, show that despite of all the theoretical advantages, there are still risks. These mainly originate from our still very incomplete understanding of the molecular and immunological processes, in particular in combination therapies. Here, great hopes are put on the

Table 1.2

Drug name*	***FDA name ***** ***(original name)***	***Antigen***	***Ab origin ******	***Trageted diseases***	***Used for***
Avastin	bevacizumab (rhuMAb-VEGF)	VEGF	humanised	colon, breast, kidney, lung cancer, myeloma, mesothelioma	combin. therapy
Bexxar	tositumomab (Bl)	CD20 (B cells)	murine Iodine 131I conjug.	lymphoma (NHL)	therapy
CEA-Scan	arcitumomab (IMMU-4)	CEA	Fab' fragment, 111In conjug.	colorectal, lung, breast cancer	imaging
Cotara	– (TNT-1)	histone H1, DNA	chimeric	lung, uterine cancer, glioma, sarcoma	therapy (China)
Erbitux	cetuximab (C225)	EGFR	chimeric	breast, head and neck, pancreatic, colorectal, lung cancer, melanoma	therapy
Herceptin	trastuzumab (rhuMAb-4D5-8)	p185/HER-2	humanised	breast, lung, ovarian, pancreatic cancer	therapy
HumaSpect-Tc	votumumab (88BV59H21-2V67-66)	CTAA 16.88	human BV-transformed cells 99mTc conjug.	Colorectal, breast, prostatic, lung, ovarian, pancreatic cancer	imaging
Humira	adalimumab (D2E7)	TNF-alpha	phage display, human library	rheumatoid arthritis, Crohn's disease	therapy
Indimacis-125	igovomab (OC125)	CA125	F(ab')2 fragment	Gastrointestinal, ovarian cancer	imaging
LeukoScan	sulesomab (MN-3)	NCA-90 (granulocytes)	Fab' fragment, 99Tcm conjug.	inflammation, infection, Crohn's disesase, Meckel's diverticulum, IBD	imaging
Leukosite	alemtuzumab (CAMPATH-1H)	CD52	humanised from rat	leukemia (CLL), lymphoma	therapy
Mylotarg	gemtuzumab ozogamicin (hP67.6)	CD33	humanised, conjugated to calicheamicin (cytostatic)	leukemia (AML)	therapy
Myoscint	imciromab pentetate RllDl0	myosin heavy chain	murine Fab, 111In conjug.	myocardial infarction	imaging
Neutrospec	fanolesomab (RB5)	CD15 granulocytes	murine IgM, 99Tcm conjug.	Inflammation, infection	imaging
OncoScint Oncorad	satumomab (B72.3n)	TAG-72 (CA72-4) / sialyl Tn	murine, 111In conjugate	breast, pancreatic, colorectal, ovarian, lung, endometrial cancer mesothelioma	imaging

Orthoclone	muromonab (OKT3)	CD3 (T cells)	murine	transplantation, rheumatoid arthritis	therapy
Panorex	edrecolomab (17-1A)	Ep-CAM	humanised	colon cancer	therapy (Germany)
ProstaScint	capromab (7Ell-C5.3)	PSA, PSMA	Murine, glycyl-tyrosyl-lysyl-DTPA konjugated	prostatic cancer	imaging
Raptiva	efalizumab (hu-1124)	CD2a	humanised	psoriasis, rheumatoid arthritis transplantation, autoimmunity	therapy
Remicade	infliximab (cA2)	TNF-alpha	chimeric	Crohn's disease, psoriasis Rheumatoid arthritis, asthma, Ankylosing spondylitis Wegener's granulomatosis,	therapy
ReoPro	abciximab (7E3)	gp llb/IIIa Receptor complex (platelets)	chimeric Fab	Restenosis, heart ischemic complication	therapy
Rituxan	rituximab	CD20 (B cells)	chimeric	Non-hodgkin Lymphoma Rheumatoid arthritis,	therapy
Simulect	basiliximab (CHB-201)	CD25	chimeric	transplantation	therapy
Synagis	palivizumab (1129)	F protein of RS virus	humanised	RSV infection in infants	therapy
Tysabri	natalizumab (HP2/1)	Integrin a4-ß (T cells)	humanised	multiple sclerosis	therapy
Verluma	nofetumomab merpentan NR-LU-l0	ca. 40kDa glycoprotein	murine Fab, 99mTc conjug.	lung, colon, pancreatic, breast, ovarian, renal, gastric cancer	imaging
Xolair	omalizumab (rhuMAb-E25)	human IgE Fc	humanised	asthma, autoimmunity, peanut allergy	therapy
Zenapax	daclizumab (anti-Tac-H)	CD25	humanised	transplantation, asthma, autoimmunity, inflammation, multiple sclerosis	therapy
Zevalin	imbritumomab (C2B8)	CD20	chimeric	lymphoma (NHL), autoimmunity, transplantation, rheumatoid arthritis	therapy

Table 2 Approved Antibody Therapeutics. Please note that a listing does not indicate that the antibodies are still available, approved or in use at date of print! Quite a number of products, inclusing most of the radiolabelled mouse antibodies, have been withdrawn from the market or suspended. Some were approved by other bodies than the U.S. FDA (e.g. Panorex or Cotara).

* Trade names are Copyright of distributing companies.

** please see addendum for nomenclature.

*** chimeric / humanised indicates murine origin of V regions /CDRs, respectively, if not stated otherwise.

intense research going on worldwide into "-omics" and systems biology, which is intended to lead to a mathematical interaction model for all involved factors. Only then, and with quite a number of years to go, will there be a chance to better predict adverse effects of novel drugs and combination therapies on a truly rational basis.

1.4
The Gleaming Horizon

Despite all of the recent success stories of recombinant human-like IgGs, they do not mark the end of the development, but just the start. As we understand more and more of the complex molecular interactions between immune cells or in cancer tissues, and in expectation of a significant speed-up of knowledge gain from the "-omics" and systems biology approaches, we can endeavor to expand the design limits of an antibody drug. All approved drugs are based on IgG molecules close to the naive structure of the antibodies in our bloodstream (a few on the IgG–fragment Fab), sometimes conjugated to an effector or label. But this can only be the first step of engineering applied to this fascinating molecule. For example, by engineering the Fc glycosylation, dramatic improvements in efficiency can be obtained (Jefferis 2005).

Further, we should learn from nature by looking at the modular design it has used to create the highest diversity group of proteins from repeats of slightly changed domains with a single common basic structure (immunoglobulin fold). We can be inspired to utilize this modular approach for completely novel molecular designs. This has already successfully been attempted since the early 1990s, and led to a plethora of novel molecular designs. It allowed the creation of bispecific antibodies (see Chapter 2, Vol II), the adjustment of the size for optimal pharmacokinetics (Hu et al. 1996), and the addition of functions that nature does not provide with an IgG at all (see Chapters 1 to 3, Vol II). Unfortunately, clinical results with these new designs are sparse and frequently disappointing, although this may simply reflect the fact that the molecular design of today is still rather a result of trial and error than of an understanding of the underlying mechanisms, or is even dictated mainly by the developer's patent portfolio. Neverthless, this will change as well; an example is immunotoxins, which failed in many clinical studies for more than three decades, before novel concepts and acknowledgement of the vast body of knowledge collected thoughout this time brought optimization and new ideas (see Chapter 3, Vol II). Fascinating concepts are under evaluation in hundreds of labs. There are so many ideas and so many parameters affecting therapeutic efficiency to be learned about that the development of antibody therapeutics will not reach saturation for any foreseeable time. Furthermore, major technology patents which have blocked some developments in the past will expire in the foreseeable future (see Chapter 13). Gíven that only the first generation of antibody drugs just has been approved in significant numbers, given the availability of recombinant human antibodies and the number of targets in the developer's pipelines plus the advent of completely novel therapeutic strategies using antibody fusion proteins, therapeutic antibodies look forward to a golden future.

Further Reading

Benny, K.C. (ed.) (2003) *Antibody Engineering: Methods and Protocols (Methods in Molecular Biology)*. Totowa, NJ: Humana Press (collection of relevant protocols).

Breitling, F., Dübel, S. (1999) *Recombinant Antibodies*. New York: John Wiley and Sons (introduction to basic principles and applications).

Brekke, O.H., Sandlie, I. (2003) Therapeutic antibodies for human diseases at the dawn of the twenty-first century. *Nat Rev Drug Discov* 2: 52–62.

Carter, P.J. (2006) Potent antibody therapeutics by design. *Nat Rev Immunol* 6: 343–357.

Chowdhury, P.S., Wu, H. (2005) Tailor-made antibody therapeutics. *Methods* 36: 11–24.

Gavilondo, J.V., Larrick, J.W. (2000) Antibody engineering at the millennium. *Biotechniques* 29: 128–132.

Groner, B., Hartmann, C., Wels, W. (2004) Therapeutic antibodies. *Curr Mol Med* 4: 539–547.

Haurum, J., Bregenholt, S. (2005) Recombinant polyclonal antibodies: therapeutic antibody technologies come full circle. *IDrugs* 8: 404–409

Hudson, P., Souriau, C. (2003) Engineered antibodies. *Nature Med* 9: 129–134.

Huston, J.S., George, A.J. (2001) Engineered antibodies take center stage. *Hum Antibodies* 10: 127–142.

Kontermann, R., Dübel, S. (ed.) (2001) *Antibody Engineering*. Heidelberg, New York: Springer (collection of relevant protocols).

Monoclonal Antibodies and Therapies. Supplement of *Nat Biotechnol*, November 2004 (collection of reviews).

Monoclonal Antibody Index (MAI), Gallart Biotech SL, 2005 (collection of data on therapeutic antibodies, updated regularily).

Presta, L.G. (2002) Engineering antibodies for therapy. *Curr Pharm Biotechnol* 3: 237–256.

Subramanian, G. (ed.) (2004) *Antibodies: Volume 2: Novel Technologies and Therapeutic Use*. Heidelberg, New York: Springer.

Urch, C.E., George, A.J.T. (eds) (2000) *Diagnostic and Therapeutic Antibodies (Methods in Molecular Medicine)*. Totowa, NJ: Humana Press.

References

Bayes, M., Rabasseda, X., Prous, J.R. (2005) Gateways to clinical trials. *Methods Find Exp Clin Pharmacol* 27: 569–612.

Behring, E., Kitasato, S. (1890) Über das Zustandekommen der Diphterie-Immunität und der Tetanus-Immunität bei Thieren. *Dt Med Wschr* 16: 1113–1114.

Buelow, R., Puels, J., Hansen-Wester, I., Thorey, I., Schueler, N., Siewe, B., Niersbach, H., van Schooten, W., Platzer, J. (2006) Expression of a human antibody repertoire in transgenic rabbits. *Human Antibodies* 15: 19–23.

Ehrlich, P. (1880) Methodologische Beiträge zur Physiologie und Pathologie der verschiedenen Formen der Leukocyten. *Z Klin Med* 1: 553–560.

Ehrlich, P. (1908) Partial cell functions. Nobel Lecture, December 11: 1908.

Hu S.Z., Shively, L.E., Raubitschek, A., Sherman, M., Williams, L.E., Wong, J.Y.C., Shively J.E., Wu A.M. (1996) Minibody: a novel engineered anti-carcinoembryonic antigen antibody fragment (single-chain Fv-C_H3) which exhibits rapid, high-level targeting of xenografts. *Cancer Res* 56: 3055–3061.

Jefferis, R. (2005) Glycosylation of recombinant antibody therapeutics. *Biotechnol Prog* 21: 11–16.

Kleinschmidt-DeMasters, B.K., Tyler, K.L. (2005) Progressive multifocal leukoencephalopathy complicating treatment with natalizumab and interferon beta-1a for multiple sclerosis. *N Engl J Med* 353: 369–374.

Rutgeerts, P.J., Targan, S.R, Hanauer, S.B., Sandborn, W.J. (2004) Challenges in Crohn's disease: the role for current and future TNF antagonists. www.medscape.com/ viewprogram/3261_pnt

Reichert, J.M. (2001) Monoclonal antibodies in the clinic. *Nat Biotechnol* 19: 819–822.

Reichert, J.M. (2002) Therapeutic monoclonal antibodies: trends in development and approval in the US. *Curr Opin Mol Ther* 4: 110–118.

Reichert, J.M., Rosenzweig, C.J., Faben, L. B., Dewitz, M.C. (2005) Monoclonal antibody successes in the clinic. *Nat Biotechnol* 23: 1073–1078.

Rohrbach, P., Broders, O., Toleikis, L., Dübel, S. (2003) Therapeutic antibodies and antibody fusion proteins. *Biotechnol/ Genet Eng Rev* 20: 129–155.

Part I
Selecting and Shaping the Antibody Molecule

2
Selection Strategies I: Monoclonal Antibodies

Gerhard Moldenhauer

2.1
Introduction

Since the late nineteenth century and the days of Paul Ehrlich, who considered antibodies as "magic bullets" (Ehrlich, 1900), immunologists have been attracted by the idea of destroying tumor cells with antibody molecules alone or conjugates made up of such molecules. Emil von Behring and Shibasaburo Kitasato were the first to demonstrate the efficacy of a heterologous polyclonal antiserum directed against the exotoxin produced by *Corynebacterium diphtheria* ("Diphtherie-Heilserum"), thereby saving the lives of many children (von Behring and Kitasato 1890). Although similar attempts were subsequently made to employ antisera for tumor treatment, the outcomes were less successful. With the advent of monoclonal antibody (mAb) technology 30 years ago, there was renewed enthusiasm for the development of a modern immunotherapy for cancer. This promise was fulfilled, however, only in some rare cases of non-Hodgkin's lymphomas (Levy and Miller 1990). The application of mouse mAbs for therapy has raised several problems. One major drawback was, for instance, that mouse antibodies are not usually able to activate human immunological effector functions as antibody-dependent cellular cytotoxicity (ADCC) and complement-dependent cytotoxicity (CDC). The formation of a human anti-mouse antibody (HAMA) response after repeated injection constitutes another reason for the low response rates observed (Khazaeli et al. 1994; DeNardo et al. 2003).

A breakthrough was achieved in 1997 with the approval of the chimeric (mouse/human) rituximab antibody for the treatment of relapsed/refractory low-grade non-Hodgkin's lymphomas by the US Food and Drug Administration (FDA) (McLaughlin et al. 1998). During the past decade molecular biology has provided the means to create chimeric, humanized or fully human antibodies for therapy. To date, eight antibody-based cancer therapeutics have been approved and are on the market. They comprise unmodified antibodies as well as conjugates with toxins or radionuclides. Thus, antibody engineering finally has led to a renaissance of antibody-guided tumor therapy. The new reagents can interact with

Handbook of Therapeutic Antibodies. Edited by Stefan Dübel

ISBN 978-3-527-31453-9

human effector molecules and have been shown to synergize with or even substitute for conventional chemotherapeutic regimens (reviewed by Glennie and van de Winkel 2003; Brekke and Sandlie 2003; Ross et al. 2004; Stern and Herrmann 2005; Adams and Weiner 2005).

In this chapter I describe the principles of mAb creation by somatic cell hybridization, give an overview on frequently used screening procedures and modifications of the antibody molecule and finally outline the mechanisms by which antibodies exert their effector functions. For the sake of clarity, I have restricted my viewpoint mainly to the field of cancer research and oncology/hematology.

2.2 Historical Remarks

The immune system is capable of generating about 10^{11} to 10^{12} different antibody molecules. An individual B lymphocyte, however, synthesizes only the one distinct antibody it is genetically programmed for. How this enormous antibody repertoire is generated was a central question of immunology over the past 50 years. Starting from the natural clonal selection theory of antibody formation (Jerne 1955) it became obvious that two mechanisms affecting the immunoglobulin genes are of paramount importance: rearrangement of gene segments and somatic hypermutation. To study somatic mutations of antibody genes in more detail, an antibody-secreting cell line recognizing the same antigen was urgently needed. This was the aim of experiments performed by Georges Köhler and César Milstein, leading to the discovery of hybridoma technology for the production of mAbs (Köhler and Milstein 1975). One of the major methodological advances in biology and medicine, this work was honored by a Nobel Price in 1984, was achieved by answering an academic question (Milstein 1999). Recently, the immunologist Klaus Eichmann has published an interesting book on the scientific and historical background leading to the invention of hybridoma technology (Eichmann 2005).

Hybridoma production basically relies on the fusion of immunized lymphocytes from an experimental animal with immortal myeloma cells. The resulting cell hybrid contains the genetic material of both parents. From the tumor cell the hybrid acquires the capacity for indefinite growth while the B lymphocyte confers the ability to synthesize a specific antibody. After stabilization by repeated cell cloning, hybridomas produce fairly large quantities of identical mAb for years (Fig. 2.1). The early development of the method is documented in a couple of scientific anthologies (Melchers et al. 1978; Kennett et al. 1980; Hämmerling et al. 1981). Importantly, the first patient suffering from non-Hodgkin's lymphoma was treated in 1979 with a mouse mAb at the Dana Farber Cancer Center in Boston (Nadler et al. 1980). Taken together, mAbs have led to a revolution in basic sciences, medicine and industry during the past 30 years.

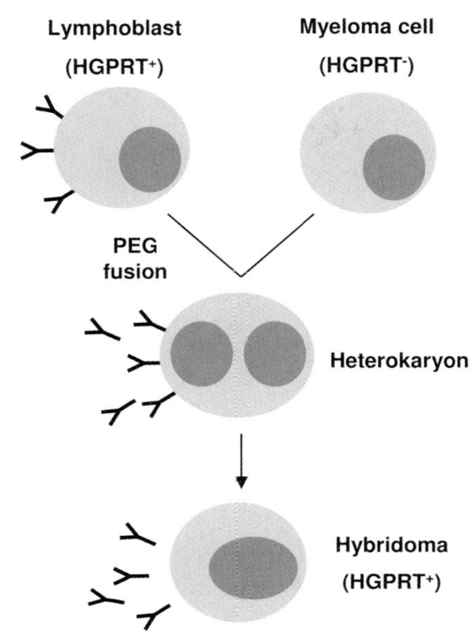

Fig. 2.1 Principle of hybridoma production by cell fusion. Both parents confer their key functions to the hybrid: production of an individual antibody (lymphoblast) and indefinite growth (myeloma cell). Introduction of a selectable marker (HGPRT) allows only hybridoma cells to proliferate.

2.3 Antibody Structure and Function

2.3.1 Membrane-bound and Secreted Forms of Antibodies

The basic structure of an IgG antibody has been elucidated as a symmetric monomer consisting of two identical heavy chains and two identical light chains, connected via disulfide bonds (Cohen and Porter 1964; Edelman et al. 1969). Five classes of immunoglobulins can be distinguished according to their distinct heavy chains: IgG, IgM, IgD, IgA, and IgE. Furthermore, each antibody contains one type of light chain, kappa or lambda. Both, heavy and light chains harbor a variable region of 110 amino acids at the N-terminus with three hypervariable segments called complementarity determining regions (CDRs). The hypervariable loops form the two antigen binding (or antigen-combining) sites of an IgG molecule and determine its specificity. In contrast, the constant part of the immunoglobulin (named the Fc portion) is responsible for secondary effects like activation of the complement system or binding to cellular Fc receptors (Fig. 2.2).

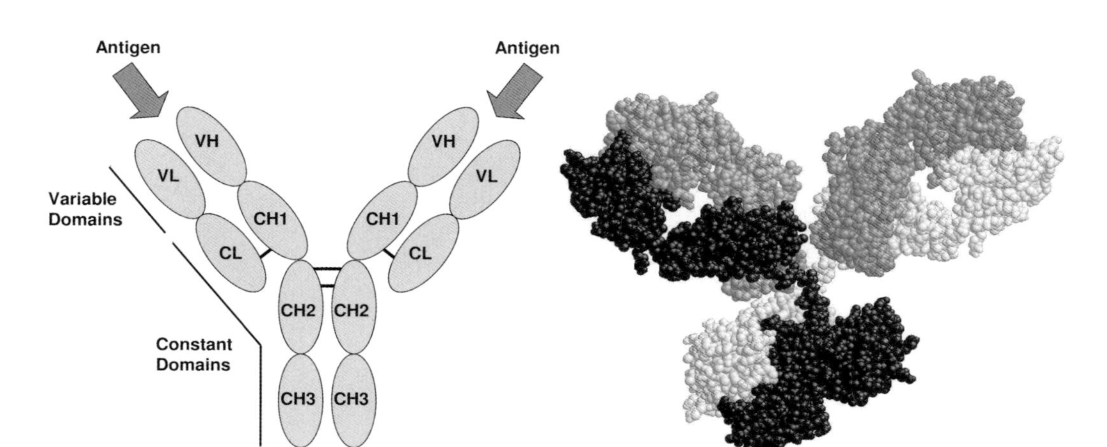

Fig. 2.2 Structure of an IgG molecule. The Y-shaped simplified representation in (a) shows the functional domains as well as the two antigen-binding sites. A more realistic space-fill model is depicted in (b) (courtesy of Dr C.W. von der Lieth, DKFZ Heidelberg).

Importantly, immunoglobulins display a dual function since they are exposed on the surface membrane of B lymphocyte as antigen receptors and are also secreted by plasma cells. They circulate in the blood and other body fluids and are able to bind, neutralize and eliminate foreign antigens, such as viruses, bacteria or toxins. Antibodies represent the effector molecules of the humoral immune system. Transmembrane and secreted forms of an antibody are generated by differential splicing of a primary transcript RNA. In the course of an ongoing immune response specific B lymphocytes undergo isotype switching, that is the transition of the IgM^+ and IgD^+ phenotype to IgG surface and secreted immunoglobulin. In addition, point mutations of the rearranged heavy and light chain variable genes occur in response to T lymphocyte signaling, giving rise to somatic hypermutation necessary to increase antibody affinity. These processes take place in the germinal centers of secondary lymphoid tissues such as lymph node and spleen. The enzyme activation-induced cytidine deaminase (AID) has been identified as a key player for both class switching and affinity maturation (Barreto et al. 2005). Detailed information on how an effective antibody response to antigen develops at a molecular level can be looked up in one of several excellent textbooks of immunobiology (Roitt et al. 2001; Abbas and Lichtman 2003; Janeway et al. 2005).

Immunoglobulin G (IgG, 150kDa) is the most abundant immunoglobulin in serum, accounting for up to 80% of all secreted antibodies. There are four different IgG isotypes in humans (IgG1, IgG2, IgG3, and IgG4) and mouse (IgG1, IgG2a, IgG2b, and IgG3). Human IgG1 and IgG3 antibodies are potent activators of the complement system and also bind with high affinity to Fc receptors on phagocytic cells, resulting in ADCC. Immunoglobulin M (IgM, 900 kDa) accounts for approximately 10% of serum antibodies. It is expressed as monomer on B lymphocyte as antigen receptor, whereas the secreted form consists of a pentamer hold together by a J (joining) chain. Immunoglobulin D (IgD, 180 kDa) is only found in fairly low amounts in the serum but is, together with IgM, the major membrane-bound form expressed on mature B cells. Immunoglobulin A (IgA, 160 kDa) constitutes about 10–15% of serum antibodies. It represents the major antibody class, being secreted into tears, saliva, and mucus of the bronchial, genitourinary, and digestive tracts. While the prevalent form in the serum is a monomer, the secretory IgA usually consists of dimers covalently connected via J (joining) chains together with an additional polypeptide called secretory component. Immunoglobulin E (IgE, 180 kDa) is found in the serum only in trace amounts. IgE antibodies are responsible for immediate hypersensitivity and cause the symptoms of hay fever, asthma and anaphylactic shock. Mast cells and basophils bind IgE via Fc receptors and subsequent contact with an allergen will cause degranulation and release of histamine and other mediators.

2.3.2
Monoclonal Antibodies

The key feature of an mAb is its unique specificity. It recognizes only one particular antigenic determinant (called epitope) on a given molecule – that means it is monospecific. All antibodies secreted by an individual hybridoma represent identical immunoglobulin molecules that display identical binding strengths to its antigen (referred to as affinity) and have identical physicochemical properties (isotype, stability). This homogeneity will give rise to the same immunological effector functions. In principle, mAbs can be produced in unlimited quantities, the hybridoma cell itself survives after cryopreservation at least for decades.

When compared with polyclonal antisera (for instance from rabbit) the affinity of mAbs might sometimes be inferior. Because mAbs consist of homogeneous molecules of the same isotype they may not elicit certain biological responses. The majority of monoclonals are directed against conformational epitopes of an antigen and may lose reactivity when tested on denatured samples, for instances by Western blotting or by immunohistology on paraffin sections. In common with polyclonal reagents, mAbs may show unexpected cross-reactivity with antigens being expressed in unrelated tissues.

Being glycoproteins, antibodies are potent immunogens when injected into another species. There are three different types of antigenic determinants against which an immune response can be induced: anti-isotype, anti-allotype, and anti-idiotype. If a human being is injected several times with a mouse mAb, it is very likely a HAMA response will develop (Khazaeli et al. 1994; DeNardo et al. 2003).

2.4 Production of Monoclonal Antibodies

2.4.1 Immunization

The aim of an immunization is to elicit a strong immune response against a certain antigen.

For mAb production, most commonly mice and rats and less frequently hamsters and rabbits are immunized with antigen by distinct routes of administration. Antigen may consist of cellular components, purified proteins, peptides, carbohydrates, lipids, or nucleic acids and for each of these specific immunization protocols are available (Harlow and Lane 1988; Goding 1996; Coligan et al. 2004). The purity of the antigen used for immunization plays a major role in the outcome of antibody response. If rather impure preparations are used, problems may arise from the possible immunodominance of contaminants. This might occur when complete cells are employed as immunogen; on the other hand intact cells are highly immunogenic. Molecular biology allows the expression of fusion proteins in eukaryotic cells, a method that has largely improved the preparation of immunizing agents.

Especially for mounting an immune response against soluble antigens, the use of a strong adjuvant is highly recommended. Adjuvants are nonspecific stimulators of the immune system, the most famous representative being Freund's complete adjuvant (Freund 1956). This consists of mineral oil and inactivated *Mycobacterium tuberculosis* particles. When mixed with immunogen, a water-in-oil emulsion is prepared that allows the release of antigen over a long period of time. The mycobacteria give rise to an inflammatory response with the production of numerous cytokines. In general, for the first immunization it is appropriate to use a strong adjuvant (complete Freund's or *Bordetella pertussis*), whereas the second injection and following challenges can be given with incomplete Freund's (mineral oil only) or without any adjuvant. By repeated immunization the response of the animal is shifted against high-affinity antibodies of IgG isotype. Since previously activated lymphoblasts show preferential fusion with myeloma cells, the final booster immunization should be given 3 days prior to fusion to maximize the yield of hybrids.

If the amount of chosen protein antigen is short or if the antigen is not available at all, intrasplenic immunization with minute amounts of antigen (Spitz et al. 1984; Grohmann et al. 1991) or DNA immunization (Barry et al. 1994) might offer alternative approaches. During the past years a couple of useful antibodies were produced by immunizing with synthetic peptides coupled to immunogenic carriers like keyhole limpet hemacyanin (KLH) or bovine serum albumin (BSA). For this, only the amino acid sequence of the protein has to be known, which can be found in several databases. Coupling of peptides carrying an N- or C-terminal cysteine residue to the carrier is achieved by *m*-maleimidobenzoyl-*N*-hydroxysuccinimide ester or another heterobifunctional crosslinker (Green et al. 1982). One drawback of the method is that such anti-peptide antibodies often exclusively react with the denatured but not with the native protein. *In vitro* immunization, invented to prime naive lymphocytes in cell culture, did not fulfill the expectations because in most instances solely an IgM response was induced (Borrebaeck 1983).

2.4.2 Myeloma Cell Lines

Multiple myeloma or plasmacytoma represents a malignancy of plasma cells in which large numbers of antibody-secreting cells residing in bone marrow are produced. They secrete monoclonal immunoglobulin, the specificity of which is usually not known, therefore they are regarded as "antibody without antigen." All available mouse myeloma lines for fusion are derived from the MOPC-21 tumor that has been induced in BALB/c mice by mineral oil injection into the peritoneal cavity (mineral oil-induced plasmacytoma) and was then adapted to growth in tissue culture. While the very first hybridomas were made with myeloma fusion partners that endogenously secreted complete antibody, later on loss variants were selected producing solely kappa light chains (e.g. P3-NS1-Ag4-1; Köhler et al. 1976) or no immunoglobulin. Such nonproducer lines are mostly used for cell fusion these days, prominent examples being X63-Ag8.653 (Kearney et al. 1979), Sp2/0-Ag-14 (Köhler and Milstein 1976) and F0 (Fazekas de St. Groth and Scheidegger 1980).

Similarly, for the production of mAbs against mouse antigens some rat myeloma lines have been established from the LOU/C strain. Frequently used are Y3-Ag1.2.3 (secreting kappa light chains; Galfrè et al. 1979) and line IR983F (nonproducer; Bazin et al. 1990). Since rats are not that much easier to handle than mice and since rat hybridomas are sometimes dependent on growth factors, making cell culture more complicated, interspecies hybrids have been constructed. For this, immune rat spleen cells were fused with a murine nonsecretor myeloma cell line. These rat/mouse hybrids turned out to be stable and secreted amounts of mAbs comparable to mouse/mouse hybridomas (Ledbetter and Herzenberg 1979). Mouse interspecies hybridomas have also been created with hamster and rabbit lymphoblasts to obtain respective mAbs (Sanchez-Madrid et al. 1983; Raybould and Takahashi 1988). A more advanced method for the production of mouse antibodies against mouse antigens is the use of knockout mice for immunization. Since they lack expression of the target antigen they are not tolerant and are able to mount a normal immune response.

2.4.3 Cell Fusion

In early experiments cell fusion was facilitated by means of agglutinating viruses like Sendai. The introduction of polyethylene glycol (PEG) as a fusing agent (Pontecorvo 1975) has simplified the procedure drastically and is used throughout the field today. PEG renders the membrane of cells to be fused gluey, so that they stick together. Subsequently plasma membrane fusion occurs, giving rise to a cell with two (or more) nuclei called a heterokaryon. During cell division the nuclear membranes are degraded and the chromosomes are distributed into the daughter cell. These hybrid cells contain only one nucleus but the genetic material of both parents and are named synkaryons. The double set of chromosomes in hybrids causes genetic instability during further mitoses, leading to improper segregation or loss of chromosomes. If chromosomes coding for immunoglobulin heavy or light chain genes (in the mouse chromosomes 6, 12, and 16) are affected, antibody secretion of this hybrid will ultimately stop. To prevent overgrowth of early hybridoma culture by nonproducing variants, immediate cloning of cultures is mandatory. Following many cell divisions the hybrid line is stabilized in its chromosomal inventory. However, prior to mass production of hybridoma cells and in case the antibody titer of culture supernatant declines recloning should be performed (Fig. 2.3).

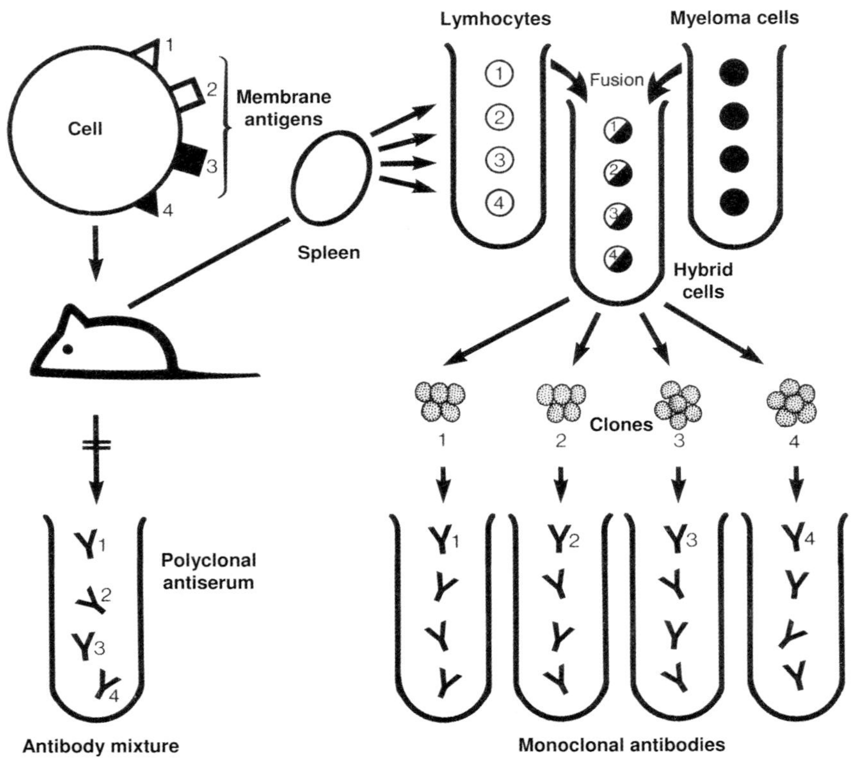

Fig. 2.3 Generation of monoclonal antibodies. Spleen cells from an immunized mouse are fused with myeloma cells to obtain hybridomas. After reduction to individual cell clones they secrete monoclonal antibodies of defined specificity.

Factors critically influencing the outcome of a fusion experiment are the choice of fetal calf serum (FCS) and the health status of the immunized mice. Depending on the content of the growth-promoting constituents, some FCS batches evolve better suited for hybridoma growth than others and it is worth testing for an optimal one. Bacterial and viral infections of rodents can cause severe immune suppression and will lead to low yields of hybridomas. Some protocols recommend the use of feeder cells as a source of growth factors as well as lysis of erythrocytes present in spleen cell preparations from immunized animal prior to fusion. In addition, repeated medium changes to remove potentially harmful substances derived from dying cells have been suggested. In my experience a simplified method works best, leaving the fused cells as untouched as possible. If a spleen from an immunized mouse (containing approximately 1–1.2 × 10^8 lymphocytes) is fused with an equal number of myeloma cells and the fusion mixture is distributed into 15–20 96-well microtiter plates (1440–1920 individual cultures), cell density is high enough to support hybridoma growth even without feeder cells and medium exchange (Fig. 2.4).

As an alternative to the PEG procedure described, electrically induced cell fusion has been developed. It is based on the delivery of high-voltage electrical field pulses to physically fuse lymphoblasts and myeloma cells (Ohnishi et al. 1987; Schmitt and Zimmermann 1989). More recently, a method was published

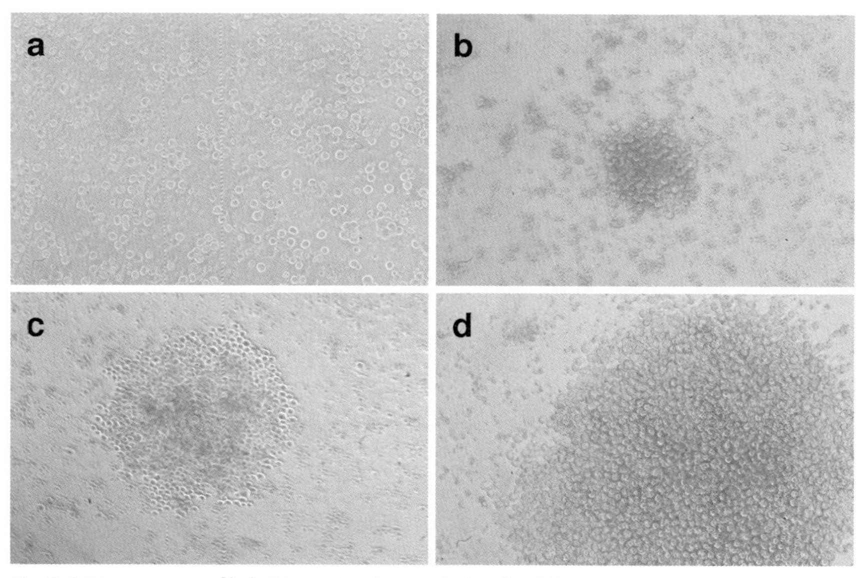

Fig. 2.4 Time course of hybridoma development visualized by phase-contrast light microscopy. Fusion mixture is plated on day 0 (a) and early hybridomas became visible after 4–6 days (b). At day 6–10 hybridomas grow vigorously (c) and are ready for screening at day 10–14 (d).

allowing the production of mAbs without hybridomas by using transgenic mice harboring a mutant temperature-sensitive simian virus-40 large tumor antigen (Pasqualini and Arap 2004).

2.4.4 Drug Selection of Hybridomas

During somatic cell hybridization only a small number of cells will actually fuse and only a minor proportion of them will develop into hybridomas (in the range of 1 in 10^5). Consequently, the culture is rapidly overgrown by myeloma cells or myeloma–myeloma fusions that also occur. This makes the introduction of a drug selection system indispensable. To accomplish this, enzyme-deficient myeloma lines are employed for cell fusion. In the system described by Littlefield (1964) the myeloma lacks the enzyme hypoxanthine guanine phosphoribosyl transferase (HGPRT). If the main biosynthetic pathway for purine and pyrimidine nucleotides required for both DNA and RNA synthesis is blocked by folic acid antagonists like aminopterin, cells can survive using an alternative "salvage pathway" that requires the enzymes HGPRT and thymidine kinase (Fig. 2.5). If the fusion mixture is cultured in medium containing hypoxanthine, aminopterin, and thymidine (HAT medium) only hybrids can actively grow. Normal spleen cells from the immunized mouse will die spontaneously *in vitro*. HGPRT-deficient myeloma cells undergo cell death in the presence of aminopterin since they cannot use the

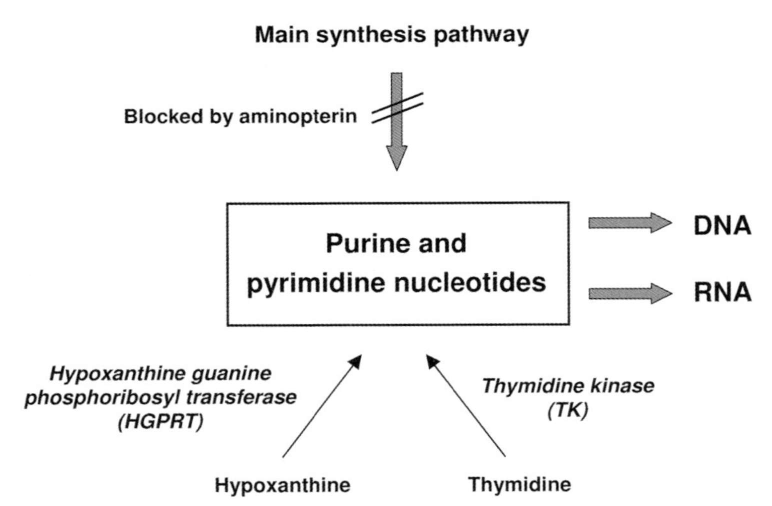

Fig. 2.5 Chemical selection of hybridomas. While the main pathway for synthesis of purine and pyrimidine nucleotides is blocked by the folic acid antagonist aminopterin, cells are forced to use the salvage pathway that requires the enzymes hypoxanthine guanine phosphoribosyl transferase (HGPRT) and thymidine kinase (TK).

salvage pathway. Hybridomas having acquired HGPRT from immune lymphoblast represent the only cell type that continues to proliferate.

HGPRT-negative myeloma mutants can be selected by culturing the cells in the presence of toxic purine analogs like 8-azaguanine or 6-thioguanine. Cells harboring the enzyme are killed after incorporation of the toxic nucleotide. Spontaneously arising mutants can be simply established because the enzyme is encoded on the X chromosome and only one gene locus has to be targeted.

2.4.5 Screening Hybridoma Cultures for Specific Antibody

Establishment of a reliable, sensitive, and fast screening assay for the detection of desired mAb is the most important prerequisite for successful hybridoma production. It is not recommended to start a fusion experiment before an appropriate screening assay has been set up. There are numerous different test types available for the initial screening that all are based on the measurement of antigen–antibody binding. Taking blood from the immunized animal for instance by puncture of the retro-orbital venous plexus will provide a serum sample that is very useful to establish a sophisticated assay system for screening.

During the decade following invention of hybridoma technology many laboratories have successfully applied solid-phase and cellular radioimmunoassay using ^{125}I-iodine-labeled second step antibodies or protein A (Bjork et al. 1972). This method as well as laborious rosette techniques employing sheep red blood cells as indicator for antibody binding are only of historical interest today. The most frequently performed tests for early hybridoma screening are discussed below. Of course much more tailor-made assays are now established to identify mAbs with special features, such as antibodies reacting with sugar or glycolipid epitopes, antibodies working in Western blotting or neutralizing antibodies against a bacterial toxin or a virus.

2.4.5.1 Enzyme-linked Immunosorbent Assay (ELISA)

Solid-phase ELISA (Engvall and Perlman 1971), where the antigen is immobilized on the well of a microtiter plate, represents a universal test system that can easily be customized and allows rapid analysis of many samples in parallel. Special ELISA plates (the surfaces of which are specifically treated) are coated first with (semi-) purified antigen or peptide that attaches to the plastic surface by adhesive forces. To prevent nonspecific binding plates are then blocked by incubation with a gelatin or bovine albumin solution. Plates can thus be stored in the cold for month. Hybridoma supernatants are allowed to react with immobilized antigen for a while. If the supernatant contains specific antibody it will be strongly retained by its corresponding antigen whereas all other contaminating proteins are subsequently washed away. Next, bound mAb is detected by an enzyme-labeled second step reagent, usually anti-mouse immunoglobulin linked to horseradish peroxidase or alkaline phosphatase, that attaches to the already formed

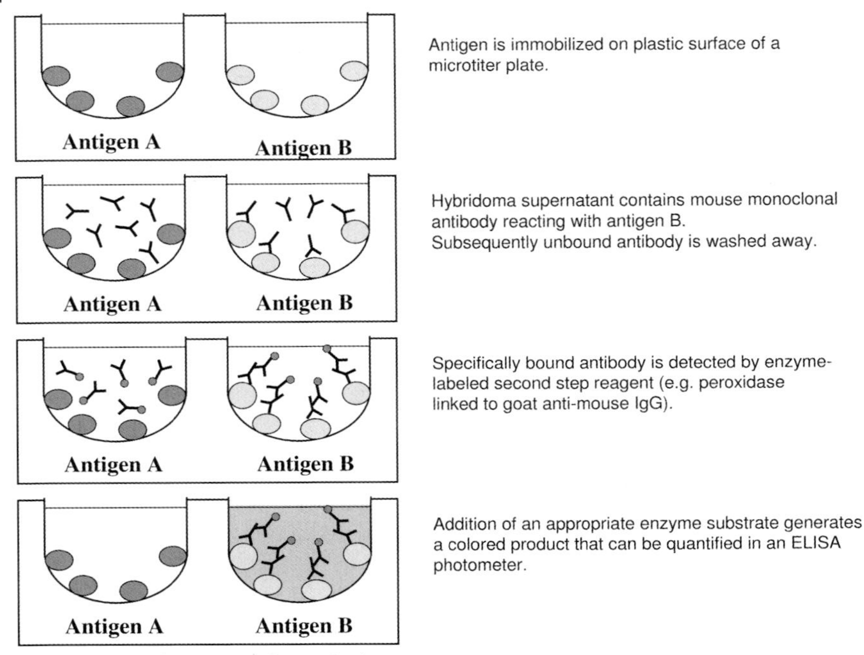

Fig. 2.6 Solid-phase ELISA for antibody screening.

antigen–antibody complex. Finally, the bound conjugate is visualized by a substrate reaction in which a colorless substrate is enzymatically converted into a dye. The reaction can be quantified by an ELISA photometer measuring the appropriate optical density in each well of the plate (Fig. 2.6).

ELISA is also the method of choice to determine the heavy and light chain subclasses of hybridoma antibodies. For this purpose, isotype-specific, enzyme-labeled secondary antibodies are supplied by commercial sources. Isotype-specific enzyme immunoassay was also engaged to select for isotype switch variants of hybridomas. The idea behind it implies that hybridoma clones at rather low frequency spontaneously perform class switching to the subsequent isotype encoded by the heavy chain immunoglobulin gene locus. These rare variants can be traced, enriched, and established by limiting dilution in combination with isotype-specific ELISA (Spira et al. 1984). Many variations of standard indirect ELISA (with immobilized antigen), such as sandwich and competitive ELISA, were suggested, allowing an increase in sensitivity that under normal conditions is not necessary. In addition, for the detection of antibodies directed against cell membrane-exposed antigens cellular ELISA protocols have been developed (Feit et al. 1983). These are especially suitable if a homogeneous cell population, as for instance a gene transfected cell line, is available for screening.

2.4.5.2 **Flow Cytometry**

The invention of the fluorescence-activated cell sorter (FACS) was a hallmark in the analysis of cell membrane antigens detected by polyclonal antibodies in the beginning and subsequently by mAbs (Bonner et al. 1972, Parks and Herzenberg 1984). First a cell suspension (for instance white blood cells) is incubated with hybridoma supernatant. Depending on the nature of the antigen recognized, a certain proportion of cells will bind the respective antibody to which a second layer, a fluorescently labeled anti-immunoglobulin (e.g. goat anti-mouse IgG coupled to fluorescein or phycoerythrin), is added. Single cells are then passed through a laser beam leading to excitation of the fluorochrome. Taking advantage of highly sophisticated optical and electronic devices, emitted fluorescence from stained cells is measured by a photomultiplier. Usually forward and side scatter signals providing information on the size and granularity of cellular subpopulations are also monitored. Dead cells can be discriminated in parallel by propidium iodide, a fluorescent dye that stains DNA and that is only taken up by injured cells.

In essence, fluorescence-activated cell analyzers and cell sorters provide extremely valuable tools for the rapid, reliable and quantitative screening of antibodies interacting with cell surface receptors. In addition, flow cytometry offers a broad spectrum of applications in immunology, cell biology, and other disciplines. Modern instruments equipped with argon and krypton lasers make the simultaneous use of multiple antibody–fluorochrome conjugates possible, thus allowing multi-parameter analysis of cellular subsets.

2.4.5.3 **Immunohistology and Immunocytology**

If morphological aspects play a major role in antibody screening immunohistology and immunocytology may provide appropriate methods. Several techniques have been advised to fix tissue or cell suspensions. In traditional immunohistology formalin-fixed and paraffin-embedded sections are mostly used. Unfortunately, many mAbs raised against protein antigens will not work with such material due to the destruction of native conformation by the fixation process. On the other hand, antibodies raised against protein-derived peptides are likely to react with paraffin sections because they often recognize linear epitopes of the antigen. Frozen sections fixed with acetone are well suited for initial antibody screening. Likewise, cells to be used for immunocytology should be fixed with acetone and subsequently air-dried. For this, adherent cells can be grown on coverslips, whereas cells growing in suspension are immobilized on a glass slide using a cytocentrifuge. The staining procedure is reminiscent of that described for ELISA. Briefly, it consists of a first incubation with the hybridoma-derived antibody followed by a secondary enzyme- or less frequently fluorochrome-labeled reagent. If the hybridoma secrets a mouse mAb, a secondary reagent like goat anti-mouse IgG conjugated to alkaline phosphatase or horseradish peroxidase may be applied. The reaction's sensitivity can be enhanced by employing preformed complexes consisting of enzyme and anti-enzyme antibodies (APAAP, alkaline phosphatase anti-alkaline phosphatase and PAP, peroxidase anti-peroxidase). Another option for signal augmentation is the use of biotinylated secondary reagents that exhibit high-affinity binding to streptavidin linked to the enzyme. In contrast to ELISA, for immunohistology and immunocytology an insoluble substrate is needed that forms an insoluble colored precipitate at the site where the antibody has bound.

2.4.5.4 **Cytotoxicity Assays**

In case antibodies needs to be selected that do not only bind to a particular cell type but in addition should fix complement, a screening assay for cytotoxicity may be performed. This type of test, aimed at tracing CDC, is based on the measurement of cell membrane leakiness following complement attack. Target cells are plated in microtiter wells and mixed with hybridoma supernatants together with a source of complement proteins like rabbit or guinea-pig serum. If the antibody under investigation is able to activate the complement cascade, cell lysis will occur within minutes to hours. Killed target cells can be microscopically visualized by addition of dyes such as trypan blue or acridine orange. For quantitative evaluation either the chromium release test (that requires initial labeling of target cells with radioactive chromium-51) or flow cytometric staining with propidium iodide may be applied.

Similar assays have been established to detect antibodies capable of inducing ADCC. Here, target cells are lysed by Fc receptor-bearing effector cells (mainly natural killer (NK) cells and monocytes) that are attracted by and interact with the Fc portion of the antibody bound to the target cell (Sondel and Hanks 2001).

In very rare instances, antibody binding per se can cause target cell destruction. One prominent example is the crosslinking of certain death receptors such as CD95 exposed on the surface of lymphoid cells leading to apoptosis (Schulze-Bergkamen and Krammer 2004).

2.4.5.5 **Screening for Function**

A great variety of different assays have been set up to identify hybridoma antibodies with special features. To mention only few examples related to oncology here, those screening procedures relying on biological interference with tumor cell growth are of particular interest. A simple method to monitor cell proliferation is incorporation of ^{3}H-thymidine or bromodesoxyuridine into cellular DNA that can be quantified by beta-counting or ELISA, respectively. Antibodies directed against the growth factor interleukin 6 (IL-6) or its cellular receptor, IL-6R, can be recognized by growth inhibition of a sensitive multiple myeloma target cell. Similarly, epidermal growth factor receptor (EGFR) constitutes a growth-promoting receptor on colon cancer cells that can be used for screening. Recently, antibodies interfering with angiogenesis that bind to vascular endothelial growth factor (VEGF) or its receptors on endothelial cells have been successfully identified and brought to the clinics.

2.4.6 Cloning

Rapid cloning of hybridoma cultures is mandatory to select for stable antibody-secreting cell lines. There are mainly two reasons for single-cell cloning: first, as mentioned before, early hybridomas are sequestering chromosomes to stabilize their genetic inventory. Second, the culture of interest may contain two or even more individual antibody-producing hybrids, making the maintenance of the desired clone not an easy task. The method of choice for single-cell cloning is limiting dilution. In principle, hybridoma cells are distributed in 96-well plates so that one well will contain theoretically 0.5 or 1 cell. Based on the individual cloning efficiency of a particular hybridoma culture that actually is not known at time of cloning, few to many of the seeded cells will give rise to cell clones. Because single hybridoma cells are dependent on several poorly characterized growth factors and in addition require "cellular togetherness," feeder cells have to be added. Feeder cell cultures are usually prepared from BALB/c spleen cells, thymocytes, or peritoneal macrophages. After 10–14 days clones are reanalyzed for specificity. Ideally, every growing clone should secrete the mAb of interest. To be on the safe side, hybridoma cultures should be cloned at least twice. Interspecies hybrids like mouse/rat or mouse/human hybrids are often unstable and need repeated recloning to preserve antibody production. It is advisable to reclone any hybridoma before starting mass production.

2.4.7 Expansion and Freezing of Hybridoma Clones

After cloning and reanalysis cultures of interest are slowly expanded. Vigorous dilution can cause sudden death of the culture. At this time point hypoxanthine thymidine (HT) medium is gradually replaced by normal medium. To be on the safe side one should freeze a small cell aliquot as backup as soon as possible. If afterwards hybridoma cells divide rapidly, a couple of samples can be frozen and stored in liquid nitrogen for decades. Freezing medium usually consists of 20–90% fetal calf serum and 10% dimethyl sulfoxide.

One of the most adverse events during hybridoma culture is contamination with mycoplasma (Hay et al. 1989; Rottem and Barile 1993). Since mycoplasma infection can interfere with numerous cellular, biochemical, and molecular biological assays, early detection is essential. It is advisable to check all permanently growing cell lines in a laboratory on a regular basis by a sensitive method such as polymerase chain reaction (PCR) or ELISA (Uphoff et al. 1992). Some antibiotics can eliminate mycoplasma from cell culture but this needs a long lasting treatment procedure without guarantee of success and carries the risk of inducing resistant variants. Prevention of contamination by regular testing and clean cell culture working is of course the best way to solve the problem.

2.5
Purification and Modification of Monoclonal Antibodies

2.5.1
Mass Culture and Purification of Monoclonal Antibody

Once the hybridoma is established, large quantities of antibody can be produced employing modern cell culture devices for long-term propagation. At least two systems are on the market meeting the demands of laboratory-scale production because they can simply be installed in a normal CO_2 incubator and do not require complicated pumping and other sophisticated equipment. Both, the miniPERM modular minifermenter (Falkenberg et al. 1995) and the two-chamber cell culture device CELLine 1000 (Trebak et al. 1999) are easy to handle and allow culturing of hybridoma cells at high density (above 10^7 cells per mL). Harvest of the antibody-enriched product can be performed several times until productivity ceases. The antibody yield throughout is comparable to the formerly favored ascites production in mice that now is prohibited in most Western countries by animal protection laws.

Purification of mouse monoclonal IgG antibodies by affinity chromatography over protein A–Sepharose represents the method of choice (Seppälä et al. 1981). There are suitable protocols available for isolation of all different IgG isotypes. Mouse IgG antibodies are very robust molecules, easy to handle and of high efficiency in many biological assays. By contrast, only the rat IgG2c subtype will bind with sufficient strength to protein A, for the remaining IgG isotypes of rat protein G–Sepharose is the best-suited affinity matrix (Bjorck and Kronvall 1984). Monoclonals of IgM isotype, however, cause problems in purification, storage and handling. Therefore, one should carefully decide whether an IgM reagent is really useful for a certain application. If the epitope of interest is located on the carbohydrate or glycolipid portion of an antigen one has certainly to deal with IgM antibodies. IgM antibodies can be purified by a combination of gel filtration and ion exchange chromatography or alternatively by affinity chromatography on immobilized mannan-binding protein (Nevens et al. 1992).

2.5.2
Fragmentation of Monoclonal IgG Antibodies

There are certain applications where smaller versions of an antibody may perform better (e.g. in immunoscintigraphy). Furthermore, in some settings it is necessary to exploit monovalent binding of the antibody, thereby preventing crosslinking of the antigen. Another reason to use immunoglobulin fragments is to get rid of the antibody Fc portion that may bind in a nonspecific fashion to Fc receptors exposed on the surface of myelomonocytic cells.

Digestion of IgG (MW approx. 150 kDa) with the thiol protease papain results in two monovalent Fab fragments (MW approx. 50 kDa). The nonspecific protease pepsin cuts below the first disulfide bond in the hinge region, giving rise to a

$F(ab')_2$ fragment (MW approx. 100kDa). Fragmentation of mouse mAbs is not an easy task and the protocol has to be adapted for each individual antibody. There is, however, a clear hierarchy with regard to the immunoglobulin subclasses (Parham 1983; Lamoyi and Nisonoff 1983). More recently a method has been reported that cleaves mouse IgG2b antibodies with lysyl endopeptidase to obtain $F(ab')_2$ fragments (Yamaguchi et al. 1995).

2.5.3 Labeling of Monoclonal Antibodies

For many applications, such as multicolor staining of cells, enzyme immunoassay, or affinity determination of antibody by Scatchard plot, directly labeled monoclonals are needed. Conjugation of purified antibody with either fluorescein isothiocyanate (FITC) and biotin or iodine-125 can be easily performed. FITC binds by a hydrolysis reaction to the free amino group of lysines in the immunoglobulin. Biotin connected by a spacer of variable length to succinimide ester is also covalently bound to the antibody via lysine. The biotin–streptavidin system is especially attractive because of its flexibility. In essence, streptavidin binds with such high affinity to biotin that this bond is rapidly formed and irreversible. In addition, one biotin can accommodate four streptavidins, leading to an amplification effect. There are many streptavidin conjugates commercially available containing fluorescent dyes, enzymes, or even particles for electron microscopy. If the antibody of choice contains many lysines in its combining site, conjugation may abolish binding activity. By contrast, sodium ^{125}I-iodide is coupled to immunoglobulin via tyrosine residues, usually by an oxidation reaction (for instance the chloramine-T method, the Iodogen method, or application of Iodobeads). Alternatively, radiolabeling of antibodies can be achieved via lysine by the Bolton–Hunter procedure (Bolton and Hunter 1973). Enzyme conjugation of antibodies is technically more demanding because the labeled products have to be separated from the unlabeled by biochemical means. Many conjugates of high quality are now commercially available and thus individual enzyme labeling of monoclonals is performed only in special cases.

2.6 Monoclonal Antibodies for Tumor Therapy

2.6.1 Leukocyte Differentiation Antigens

Shortly after the invention of mAb technology in laboratories all around the world a huge variety of reagents were raised against white blood cells and normal as well as malignant cells from numerous tissues. This led to a Babylonian confusion with regard to antibody names and designation of detected antigens. The problem was approached by the organization of the well-known Workshops and

Conferences on Human Leukocyte Differentiation Antigens (HLDA), the first of which took place 1982 in Paris, France, and the most recent 8th meeting was held 2004 in Adelaide, Australia (Bernard et al. 1984; Zola et al. 2005). Antibody samples submitted to the Workshop were grouped into panels and simultaneously analyzed by a couple of reference laboratories with expertise for particular methods such as flow cytometry, immunohistology, biochemistry, or molecular genetics. The results were subsequently compared and statistically evaluated. This allowed the identification of distinct "clusters of differentiation" that became the basis of CD antigen nomenclature. To date, 339 CD antigens have been defined and characterized in depth by approaches taking advantage of immunology, cell biology, biochemistry, and molecular biology.

Monoclonal antibodies have proven to be unique reagents for the analysis of surface antigens on lymphocytes that are expressed on certain stages of lymphocyte differentiation and maturation. Using a whole panel of such antibodies has facilitated the phenotyping of functional subpopulations of normal lymphocytes. Likewise, the malignant counterparts derived from the respective stages of differentiation can be classified. Figure 2.7 illustrates how individual B cell antigens show up and vanish during B cell development. The identification and diagnosis of distinct entities among malignant lymphomas is essentially based on immunohistological staining with a set of antibodies recognizing lymphocyte differentiation antigens (Harris et al. 2001). More recently, classical immunophenotyping

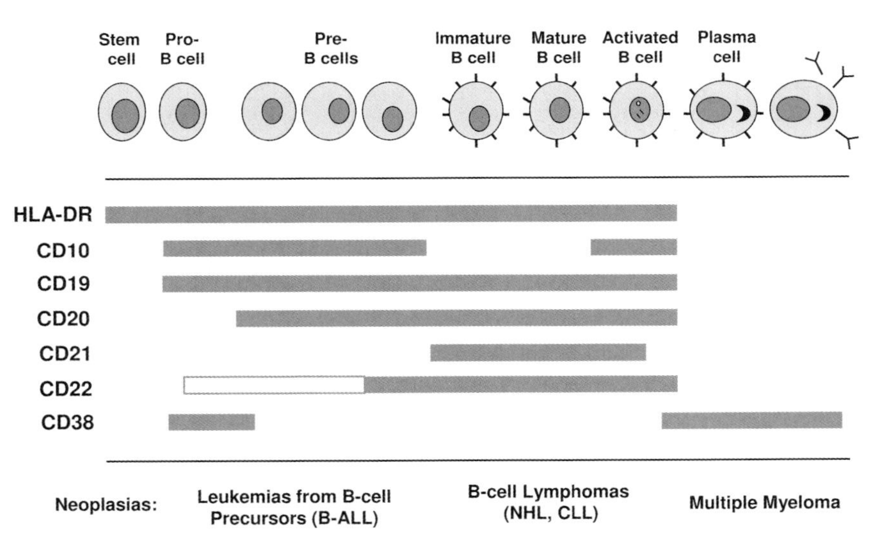

Fig. 2.7 Differentiation antigens of human B lymphocyte. Cluster of differentiation antigens (CD antigens) are expressed on certain maturation stages during B-cell ontogeny (upper panel). Monoclonal antibodies to CD antigens are also reactive with tumor-derived samples of corresponding counterparts (bottom panel), allowing for lymphoma phenotyping.

Table 2.1 Target antigens on lymphocytes for monoclonal antibody therapy.

Antigen	*Expression*	*Structure*	*Function*	*Antibody*	*Indication*
CD3	Mature T cells	Complex of five polypeptides	Signal transduction, TCR coreceptor	Muromonab	Transplant rejection
CD20	B cells	Pp35/37 kDa	Ion channel for Ca^{2+}	Rituximab	B-NHL
CD22	Mature B cells	Gp130/140 kDa heterodimer	Modulation of B cell activation, cell–cell adhesion	Epratuzumab	B-NHL
CD33	Monocytes, myeloid cells	Gp67	Cell–cell adhesion?	Gemtuzumab	AML
CD25	Activated B and T cells	Gp55 kDa	α-chain of interleukin 2 receptor	Daclizumab	Transplant rejection
CD52	Lymphocytes, monocytes	Gp21–28 kDa	Unknown	Alemtuzumab	B-CLL, T-lymphoma

Gp, glycoprotein; Pp, phosphoprotein (number given is the molecular weight under reducing conditions); TCR, T-cell receptor; NHL, Non-Hodgkin's lymphoma; AML, acute myelogenous leukemia; CLL, chronic lymphocytic leukemia.

of lymphomas was complemented by gene expression profiling technology (Staudt and Dave 2005).

Only a small proportion of the 339 CD antigens have yet evolved as valuable targets for antibody therapy not only of malignant lymphomas but also of certain autoimmune diseases and for the prevention of allograft rejection. In Table 2.1 the characteristics of six antigens that serve as targets for therapeutic antibodies already on the market are listed. These antigens differ with respect to their tissue distribution and show different traits concerning stability of surface expression and internalization. The CD22 antigen, for example, has a high internalization rate, making it an exquisite candidate for manufacturing an immunotoxin that has to reach the cytosol of a target cell to become effective (Messmann et al. 2000; Kreitman et al. 2001).

2.6.2
Epithelial Differentiation Antigens

The most frequent tumor type in humans, carcinoma, is derived from epithelial cells. Therefore, tremendous efforts have been made to identify tumor-associated or even tumor-specific membrane antigens on epithelial tumors by means of mAbs. With time it has emerged that all antigens initially regarded as tumor-specific were actually differentiation antigens and are also expressed on certain normal cells. Today it is clear that tumor-specific antigens recognized by antibodies most likely do not exist. The same experience emerged from studies focusing on other tumor types such as melanoma and brain tumors. We have learned, however, that differentiation antigens, although not tumor-specific, are valuable targets for antibody-based tumor therapy.

Members of the epidermal growth factor receptor (EGFR) family, in particular, such as HER2/neu and EGFR, hold great promise as therapeutic targets since they are overexpressed in a variety of solid tumors (Hynes and Lane 2005). The induction of antitumor responses using the antibodies trastuzumab and cetuximab are discussed in Volume III in Chapters 14 and 4. Further good candidates for antibody therapy of solid tumors are the epithelial cell adhesion molecule, Ep-CAM, and the carcinoembryonic antigen, CEA. Ep-CAM represents a very stable marker even in highly de-differentiated adenocarcinomas and its overexpression is associated with poor prognosis in breast cancer (Gastl et al. 2000) whereas CEA appears to be a suitable target for radioimmunotherapy of colorectal and medullary thyroid cancer (Mayer et al. 2000; Sharkey et al. 2005).

2.6.3
Mechanisms of Action of Monoclonal Antibodies

The antitumor effects of antibodies can be induced by direct and indirect mechanisms (Table 2.2). In some instances antibody binding per se will lead to cell death. For instance, if a surface receptor is crosslinked that transmits an apoptosis signal, programmed cell suicide is started. Likewise binding to growth receptors or their ligands might abrogate vital signals required for cell proliferation. Antibodies against EGFR family members are prominent examples of this mode of action, as already mentioned. Recently, reagents interfering with angiogenesis have become increasingly attractive (Ferrara et al. 2003). Antibodies specific for VEGF or its receptors can prevent tumor vessel formation and thus deprive the tumor of nutrients. A rather special case is represented by anti-idiotype antibodies in B cell lymphoma. By mechanisms that are poorly understood, these antibodies are able to facilitate long lasting growth control of tumor cells (Davis et al. 1998).

The classical effector functions of antibody are CDC and ADCC. Depending on the isotype of the therapeutic antibody, complement component C1q is activated and triggers a cascade of enzymatic reactions resulting in recruitment of phagocytes and formation of a membrane-attack complex that finally leads to the

Table 2.2 Therapeutic effects of monoclonal antibodies.

Direct effects	Induction of apoptosis Inhibition of proliferation Blockade of growth factors or growth factor receptors Interference with angiogenesis
Indirect effects	Complement-dependent cytotoxicity (CDC) Antibody-dependent cellular cytotoxicity (ADCC) Vehicle for toxins, radionuclides and cytostatic drugs Anti-idiotype antibody formation Effector cell targeting using bispecific antibodies

lysis of tumor cells (Gelderman et al. 2004). In case of ADCC tumor cell-bound antibodies interact via their Fc portion with Fc receptors expressed at high density on NK cells, neutrophils, and monocytes (Ravetch and Bolland 2001). Upon activation, these effector cells release cytotoxic granules from the cytosol delivering a kiss of death to the tumor target. Unfortunately, many antibodies elicit neither direct nor indirect effects, this holds especially true for murine antibodies. However, these reagents can be successfully used as carriers for toxins, radionuclides, or chemotherapeutic substances. There has been much debate on the issue whether anti-idiotypic networks, forming an internal image of tumor antigens, really contribute to tumor regression. Finally, bispecific antibodies are synthetic molecules that carry two different antigen binding sites. By virtue of their dual specificity they can trigger effector cells via a membrane receptor and at the same time link them to a tumor cell. This interaction leads to the subsequent destruction of the tumor cell.

2.6.4 Human Monoclonal Antibodies

Great efforts have been made to take human myeloma cells in culture suitable for cell fusion in order to raise human mAbs. These attempts were largely hampered by the fact that most of the laboriously established lines later turned out to be Epstein–Barr virus (EBV) transformed lymphoblastoid B cell cultures. Although some human cell lines capable of producing human hybridomas have been described, for instance SK-007 (Olsson and Kaplan 1980), GM1500 (Croce et al. 1980) LICR-LON-Hmy2 (Edwards et al. 1982) and Karpas 707 (Karpas et al. 1982), the overall experience remains disappointing. In addition, for ethical reasons it is not possible to immunize a human volunteer with an experimental antigen. As already mentioned, *in vitro* immunization was not able to solve the problem due to predominant IgM responses.

In an alternative attempt, antigen-specific B lymphocytes were isolated from the peripheral blood of human donors and immortalized by EBV to establish permanent cell lines (Steinitz et al. 1977). Unfortunately, the production rate of the lines was low and decreased with time. It further turned out that the EBV-transformed lines were extremely difficult to clone. To circumvent those problems, the EBV hybridoma technique was developed, which combined EBV-induced immortalization of human antibody-secreting cells with fusion of a variant of the human myeloma line GM1500 to obtain human–human hybrids (Kozbor and Roder 1981). This method, however, is complex and often leads to instable hybridomas that require repeated recloning. Lacking a human non-secretor myeloma cell line with high fusion frequency the production of human mAbs by the hybridoma technique was no longer pursued for many years. Recently, the EBV method has been improved to immortalize memory B cells from a patient with severe acute respiratory syndrome (SARS) coronavirus infection. Neutralizing mAbs of high affinity against the virus, conferring protection in a mouse model, were successfully isolated (Traggiai et al. 2004).

At present, there are at least three alternative core technologies available allowing for the creation of human mAbs. The variable regions or only the CDRs from mouse heavy and light chains can be grafted onto a human IgG scaffold giving rise to chimeric or humanized antibodies, respectively (Carter 2001). Screening of large recombinant antibody libraries is exploited to build human antibodies with high specificity and affinity (Hoogenboom 2005). Transgenic mice carrying human immunoglobulin genes will respond to immunization with the production of entirely human antibodies. After fusion with mouse myeloma cells, these human antibodies are secreted by resulting hybridomas (Lonberg 2005). In addition, recombinant antibodies containing minimal binding fragments can be reconstructed to multivalent high-affinity reagents (Holliger and Hudson 2005).

2.7 Outlook

Monoclonal antibodies secreted by hybidoma cells have led to a revolution in biology, medicine, and many applied sciences due to their excellent specificity. After a first wave of innovation based on mouse monoclonals, molecular biology has provided tools for reshaping the antibody molecule to obtain chimeric, humanized, and fully human antibodies as well as recombinant antibody fragments. Therapeutic antibodies have evolved as effective pharmaceutical compounds not only for the treatment of malignant tumors but also of autoimmune diseases and infections. Currently we are encountering a third wave of scientific advancement by subtle antibody engineering, making it possible to tune the molecule in a way that it can meet special therapeutic demands (Weiner and Carter 2005). In the end there is no doubt that antibody-based therapeutics will play an outstanding role in several fields of modern medicine.

Acknowledgments

I would like to thank Dr Reinhard Schwartz-Albiez for critical reading of the manuscript.

References

Abbas, A.K., Lichtman, A.H. (2003) *Cellular and Molecular Immunology*, 5th edn. Philadelphia: Saunders.

Adams, G.P., Weiner, L.M. (2005) Monoclonal antibody therapy of cancer. *Nat Biotechnol* 23: 1147–1157.

Barreto, V.M., Ramiro, A.R., Nussenzweig, M.C. (2005) Activation-induced deaminase: controversies and open questions. *Trends Immunol* 26: 90–96.

Barry, M.A., Barry, M.E., Johnston, S.A. (1994) Production of monoclonal

antibodies by genetic immunization. *BioTechniques* 16: 616–618.
Bazin, H., Pear, W.S., Klein, G., SÜmegi, J. (1990) Rat immunocytomas (IR) In: *Rat Hybridomas and Rat Monoclonal Antibodies* (Bazin, H., ed.), pp. 53–68. Boca Raton: CRC Press.
Bernard, A., Boumsell, L., Dausset, J., Milstein, C., Schlossman S.F. (1984) *Leucocyte Typing*. Berlin, Heidelberg: Springer-Verlag.
Bolton, A.E., Hunter W.M. (1973) The labeling of proteins to high specific radioactivities by conjugation to a 125I-containing acylating agent. *Biochem J* 133: 529–539.
Bjorck, L., Kronvall, G. (1984) Purification and some properties of streptococcal protein G, a novel IgG-binding reagent. *J Immunol* 133: 969–974.
Bjork, I., Petersson, B.A., Sjoquist, J. (1972) Some physicochemical properties of protein A from Staphylococcus aureus. *Eur J Biochem* 29: 579–584.
Bonner, W.A., Hulett, H.R., Sweet, R.G., Herzenberg, L.A. (1972) Fluorescence activated cell sorting. *Rev Sci Instrum* 43: 404–409.
Borrebaeck, C.A. (1983) In vitro immunization of mouse spleen cells and the production of monoclonal antibodies. *Acta Chem Scand B* 37: 647–648.
Brekke, O.H., Sandlie, I. (2003) Therapeutic antibodies for human diseases at the dawn of the twenty-first century. *Nat Rev Drug Disc* 2: 52–62.
Carter, P. (2001) Improving the efficacy of antibody-based cancer therapies. *Nat Rev Cancer* 1: 118–129.
Cohen, S., Porter, R.R. (1964) Structure and biological activity of immunoglobulins. *Adv Immunol* 3: 287–349.
Coligan, J.E. Bierer, B.E., Margulies, D.H., Shevach, E.M., Strober, W. (2004) *Current Protocols in Immunology*. New York: John Wiley and Sons.
Croce, C.M., Linnenbach, A., Hall, W., Steplewski, Z., Koprowski, H. (1980) Production of human hybridomas secreting antibodies to measles virus. *Nature* 288: 488–489.
Davis, T.A., Maloney, G.G., Cerwinski, D.K., Liles, T.M., Levy, R. (1998) Anti-idiotype antibodies can induce long-term complete remissions in non-Hodgkin's lymphoma without eradicating the malignant clone. *Blood* 15: 1184–1190.
DeNardo, G.L., Bradt, B.M., Mirick, G.R., DeNardo, S.J. (2003) Human antiglobulin responses to forein antibodies: therapeutic benefit? *Cancer Immunol Immunother* 52: 309–316.
Edelman, G.M., Cunningham, B.A., Gall, W.E., Gottlieb, P.D., Rutishauser, U., Waxdal, M.J. (1969) The covalent structure of an entire γG immunoglobulin molecule. *Proc Natl Acad Sci USA* 63: 78–85.
Edwards, P.A.W., Smith, C.M., Neville, A.M., O'Hare, M. (1982) A human-human hybridoma system based on a fast-growing mutant of the ARH-77 plasma cell leukemia-derived line. *Eur J Immunol* 12: 641–648.
Ehrlich, P. (1900) On immunity with special reference to cell life. *Proc R Soc Lond* 66: 424–448.
Eichmann, K. (2005) *Köhler's Invention*. Basel-Boston-Berlin: Birkhäuser-Verlag.
Engvall, E., Perlman, P. (1971) Enzyme-linked immunosorbent assay (ELISA): Quantitative assay of immunoglobulin G. *Immunochemistry* 8: 871–879.
Falkenberg, F.W., Weichert, H., Krane, M., Bartels, I., Palme, M., Nagels, H.-O., Fiebig, H. (1995) In vitro production of monoclonal antibodies in high concentration in a new and easy to handle modular minifermenter. *J Immunol Methods* 179: 13–29.
Fazekas de St. Groth S., Scheidegger, D. (1980) Production of monoclonal antibodies: strategy and tactics. *J Immunol Methods* 47: 129–144.
Feit, C., Bartal, A.H., Tauber, G., Dymbort, G., Hirshaut, Y. (1983) An enzyme-linked immunosorbent assay (ELISA) for the detection of monoclonal antibodies recognizing antigens expressed on viable cells. *J Immunol Methods* 58: 301–308.
Ferrara, N., Gerber, H.-P., LeCouter, J. (2003) The biology of VEGF and its receptors. *Nat Med* 9: 669–676.
Freund, J. (1956) The mode of action of immunologic adjuvants. *Adv Tuberc Res* 7: 130–148.
Galfrè,G., Milstein, C., Wright, B. (1979) Rat x rat hybrid myelomas and a monoclonal

anti-Fd portion of mouse IgG. *Nature* 277: 131–133.

Gastl, G., Spizzo, G., Obrist, P., Dünser, M., Mikuz, G. (2000) Ep-CAM overexpression in breast cancer as a predictor of survival. *Lancet* 356: 1981–1982.

Gelderman, K.A., Tomlinson, S., Ross, G.D., Gorter, A. (2004) Complement function in mAb-mediated cancer immunotherapy. *Trends Immunol* 25: 158–164.

Glennie, M.J., van de Winkel, J.G.J. (2003) Renaissance of cancer therapeutic antibodies. *Drug Disc Today* 8: 503–510.

Goding, J.W. (1996) *Monoclonal Antibodies: Principles and Practice*. 3rd edn. London, San Diego: Academic Press.

Green, N., Alexander, H., Olson, A., Alexander, S., Shinnick, T.M., Sutcliffe, J. G., Lerner, R.A. (1982) Immunogenic structure of the influenza virus hemagglutinin. *Cell* 28: 477–487.

Grohmann, U., Romani, L. Binaglia, L., Fioretti, M.C., Puccetti, P. (1991) Intrasplenic immunization for the induction of humoral and cell-mediated immunity to nitrocellulose-bound antigen. *J Immunol Methods* 137: 9–15.

Hämmerling, G.J., Hämmerling, U., Kearney, J.F. (Eds) (1981) *Monoclonal Antibodies and T-cell Hybridomas.* Amsterdam: Elsevier/North-Holland Biomedical Press.

Harlow, E., Lane, D. (1988) *Antibodies: A Laboratory Manual.* Cold Spring Harbor, NY: Cold Spring Harbor Laboratory.

Harris, N.L., Stein, H., Coupland, S.E., Hummel, M., Favera, R.D., Pasqualucci, L., Chan, W.C. (2001) New approaches to lymphoma diagnosis. *Hematology (Am Soc Hematol Educ Program)*, 194–220.

Hay, R.J., Macy, M.L., Chen, T.R. (1989) Mycoplasma infection of cultured cells. *Nature* 339: 487–488.

Holliger, P., Hudson, P. (2005) Engineered antibody fragments and the rise of single domains. *Nat Biotechnol* 23: 1126–1136.

Hoogenboom, H. (2005) Selecting and screening recombinant antibody libraries. *Nat Biotechnol* 23: 1105–1116.

Hynes, N.E., Lane, H.A. (2005) ERBB receptors and cancer: the complexity of targeted inhibitors. *Nat Rev Cancer* 5: 341–354.

Janeway, C.A., Travers, P., Walport, M., Shlomchik, M.J. (2005) *Immunobiology. The Immune System in Health and Disease,* 6th edn. London: Garland Science.

Jerne, N.K. (1955) The natural selection theory of antibody formation. *Proc Natl Acad Sci USA* 41: 849–857.

Karpas, A., Fischer, P., Swirsky, D. (1982) Human plasmacytoma with an unusual karyotype growing in vitro and producing light-chain immunoglobulin. *Lancet* I: 931–933.

Kearney, J.F., Radbruch, A., Liesengang, B., Rajewsky, K. (1979) A new mouse myeloma cell line that has lost immunoglobulin expression but permits the construction of antibody-secreting hybrid cell lines. *J Immunol* 123: 1548–1550.

Kennett, R.H., McKearn, T.J., Bechtol, K.B. (Eds) (1980) *Monoclonal Antibodies. Hybridomas: A New Dimension in Biological Analyses.* New York: Plenum Press.

Khazaeli, M.B., Conry, R.M., LoBuglio, A.F. (1994) Human immune response to monoclonal antibodies. *J Immunother* 15: 42–52.

Köhler, G., Milstein, C. (1975) Continuous cultures of fused cells secreting antibody of predefined specificity. *Nature* 256: 495–497.

Köhler, G., Milstein, C. (1976) Derivation of specific antibody-producing tissue culture and tumor lines by cell fusion. *Eur J Immunol* 6: 511–519.

Köhler, G., Howe, S.C., Milstein. C. (1976) Fusion between immunoglobulin-secreting and nonsecreting myeloma cell lines. *Eur J Immunol* 6: 292–295.

Kozbor, D., Roder, J.C. (1981) Requirements for the establishment of high-titered human monoclonal antibodies using the Epstein-Barr virus technique. *J Immunol* 127: 1275–1280.

Kreitman, R.J., Wilson, W.H., Bergeron, K., Raggio, M., Stetler-Stevenson, M., FitzGerald, D.J., Pastan, I. (2001) Efficacy of the anti-CD22 recombinant immunotoxin BL22 in chemotherapy-resistant hairy-cell leukemia. *N Engl J Med* 345: 241–247.

Lamoyi, E., Nisonoff, A. (1983) Preparation of $F(ab')_2$ fragments from mouse IgG of various subclasses. *J Immunol Methods* 56: 235–243.

Ledbetter, J.A., Herzenberg, L.A. (1979) Xenogeneic monoclonal antibodies to mouse lymphoid differentiation antigens. *Immunol Rev* 47: 63–90.

Levy, R., Miller, R.A. (1990) Therapy of lymphoma directed at idiotypes. *J Natl Cancer Inst Monogr* 10: 61–68.

Littlefield, J.W. (1964) Selection of hybrids from matings of fibroblasts in vitro and their presumed recombinants. *Science* 145: 709–710.

Lonberg, N. (2005) Human antibodies from transgenic animals. *Nat Biotechnol* 23: 1117–1125.

Mayer, A., Tsiompamou, E., O'Malley, D., Boxer, G.M., Bhatia, J., Flynn, A.A., Cester, K.A., Davidson, B.R., Lewis, A.A., Winslet, M.C., Dhillon, A.P., Hilson, A.J., Begent, R.H. (2000) Radioimmunoguided surgery in colorectal cancer using a genetically engineered anti-CEA single-chain Fv antibody. *Clin Cancer Res* 6: 1711–1719.

McLaughlin, P., Grillo-Lopez, A.J., Link, B. K., Levy, R., Czuczman, M.S., Williams, M.E., Heyman, M.R., Bence-Bruckler, I., White, C.A., Cabanillas, F., Jain, V., Ho, A.D., Lister, J., Wey, K., Shen, D., Dallaire, B. (1998) Rituximab chimeric anti-CD20 monoclonal antibody therapy for relapsed indolent lymphoma: half of patients respond to a four-dose treatment program. *J Clin Oncol* 16: 2825–2833.

Melchers, F., Potter, M., Warner, N.L. (Eds) (1978) Lymphocyte hybridomas. *Current Topics in Microbiology and Immunology*, Vol. 81. Berlin, Heidelberg, New York: Springer-Verlag.

Messmann, R.A., Vitetta, E.S., Headlee, D., Senderowicz, A.M., Figg, W.D., Schindler, J., Michiel, D.F., Creekmore, S., Steinberg, S.M., Kohler, D., Jaffe, E.S., Stetler-Stevenson, M., Chen, H., Ghetie, V., Sausville E.A. (2000) A phase I study of combination therapy with immunotoxins IgG-HD37-deglycosylated ricin A chain (dgA) and IgG-RFB4-dgA (Combotox) in patients with refractory CD19(+), CD22(+) B cell lymphoma. Clin. *Cancer Res* 6: 1302–1313.

Milstein, C. (1999) The hybridoma revolution: an offshoot of basic research. *BioAssays* 21: 966–973.

Nadler, L.M., Stashenko, P., Hardy, R., Kaplan, W.D., Button, L.N., Kufe, D.W., Antman, K.H., Schlossman, S.F. (1980) Serotherapy of a patient with a monoclonal antibody directed against a human lymphoma-associated antigen. *Cancer Res* 40: 3147–3154.

Nevens, J.R., Mallia, A.K., Wendt, M.W., Smith, P.K. (1992) Affinity chromatographic purification of immunoglobulin M antibodies utilizing immobilized mannan binding protein. *J Chromatogr* 597: 247–256.

Ohnishi, K., Chiba, J., Goto, Y., Tokunaga, T. (1987) Improvement in the basic technology of electrofusion for generation of antibody-producing hybridomas. *J Immunol Methods* 26: 181–189.

Olsson, L., Kaplan, H.S. (1980) Human-human hybridomas producing monoclonal antibodies of predefined antigenic specificity. *Proc Natl Acad Sci USA* 77: 5429–5434.

Parham, P. (1983) On the fragmentation of monoclonal IgG1: IgG2a and IgG2b from BALB/c mice. *J Immunol* 131: 2895–2902.

Parks, D.R., Herzenberg, L.A. (1984) Fluorescence-activated cell sorting: theory, experimental optinization, and application in lymphoid cell biology. *Methods Enzymol* 108: 197–241.

Pasqualini, R., Arap, W. (2004) Hybridoma-free generation of monoclonal antibodies. *Proc Natl Acad Sci USA* 101: 257–259.

Pontecorvo, G. (1975) Production of mammalian somatic cell hybrids by means of polyethylene glycol treatment. *Somatic Cell Genet* 1: 397–400.

Ravetch, J.V., Bolland, S. (2001) IgG Fc receptors. *Annu Rev Immunol* 19: 275–290.

Raybould, T.J.G., Takahashi, M. (1988) Production of stable rabbit-mouse hybridomas that secrete rabbit MAb of defined specificity. *Science* 240: 1788–1790.

Roitt, I., Brostoff, J., Male, D. (2001) *Immunology*, 6th edn. St Louis: Mosby.

Ross, J.S., Schenkein, D.P., Pietrusko, R., Rolfe, M., Linette, G.P., Stec, J., Stagliano, N.E., Ginsburg, G.S., Symmans, W.F., Pusztai, L., Hortobagyi, G.N. (2004) Targeted therapies for cancer 2004. *Am J Clin Pathol* 122: 598–609.

Rottem, S., Barile, M.F. (1993) Beware of mycoplasmas. *Trends Biotechnol* 11: 143–151.

Sanchez-Madrid, F., Szklut, P., Springer T.A. (1983) Stable hamster-mouse hybridomas producing IgG and IgM hamster monoclonal antibodies of defined specificity. *J Immunol* 130: 309–317.

Schmitt, J.J., Zimmermann, U. (1989) Enhanced hybridoma production by electrofusion in strongly hypo-osmolar solutions. *Biochim Biophys Acta* 983: 42–50.

Schulze-Bergkamen, H., Krammer, P.H. (2004) Apoptosis in cancer – implications for therapy. *Semin Oncol* 31: 90–119.

Seppälä, I., Sarvas, H., Péterfly, F., Mäkela, O. (1981) The four sub-classes of IgG can be isolated from mouse serum by using protein A-Sepharose. *Scand J Immunol* 14: 335–342.

Sharkey, R.M., Hajjar, G., Yeldell, D., Brenner, A., Burton, J., Rubin, A., Goldenberg, D.M. (2005) A phase I trial combining high-dose 90Y-labeled humanized anti-CEA monoclonal antibody with doxorubicin and peripheral blood stem cell rescue in advanced medullary thyroid cancer. *J Nucl Med* 46: 620–633.

Sondel, P.M., Hank, J.A. (2001) Antibody-directed, effector cell-mediated tumor destruction. *Hematol Oncol Clin North Am* 15: 703–721.

Spira, G., Bargellesi, A., Teillaud, J.L., Sharff, M.D. (1984) *J Immunol Methods* 74: 307–315.

Spitz, M., Spitz, L., Thorpe, R., Eugui, E. (1984) Intrasplenic primary immunization for the production of monoclonal antibodies. *J Immunol Methods* 11: 39–43.

Staudt, L.M., Dave, S. (2005) The biology of human lymphoid malignancies revealed by gene expression profiling. *Adv Immunol* 87: 163–208.

Steinitz, M., Klein, G., Koskimies, S., Mäkela, O. (1977) EB virus-induced B lymphocyte cell lines producing specific antibody. *Nature* 269: 420–422.

Stern, M., Herrmann, R. (2005) overview of monoclonal antibodies in cancer therapy: present and promise. *Crit Rev Oncol/Hematol* 54: 11–29.

Traggiai, E., Becker, S., Subbarao, K., Kolesnikova, L., Uematsu, Y., Gismondo, M.R., Murphy, B.R., Pappuoli, R., Lanzavecchia, A. (2004) An efficient method to make human monoclonal antibodies from memory B cells: potent neutralization of SARS coronavirus. *Nat Med* 10: 871–875.

Trebak, M., Chong, J.M., Herly, D., Speicher, D.W. (1999) Efficient laboratory-scale production of monoclonal antibodies using membrane-based high-density cell culture technology. *J Immunol Methods* 230: 59–70.

Uphoff, C.C., Brauer, S., Grunicke, D., Gignac, S.M., MacLeod, R.A.F., Quentmeier, H., Steube, K., Tummier, M., Voges, M., Wagner, B., Drexler, H.G. (1992) Sensitivity and specificity of five different mycoplasma detection assays. *Leukemia* 6: 335–241.

von Behring, E., Kitasato, S. (1890) Über das Zustandekommen der Diphtherie-Immunität und der Tatanus-Immunität bei Thieren. *Dtsch Med Wochenschr* 16: 1113–1114.

Weiner, L.M., Carter, P. (2005) Tunable antibodies. *Nat Biotechnol* 23: 556–557.

Yamaguchi, Y., Kim, H.H., Kato, K., Masuda, K., Shimida, I., Arata, Y. (1995) Proteolytic fragmentation with high specificity of mouse immunoglobulin G. *J Immunol Methods* 181: 259–267.

Zola, H., Swart, B., Nicholson, I., Aasted, B., Bensussan, A., Boumsell, L., Buckley, C., Clark, G., Drbal, K., Engel, P., Hart, D., Horejsí, V., Isacke, C., Macardle, P., Malavasi, F., Mason, D., Olive, D., Saalmueller, A., Schlossman, S.F., Schwartz-Albiez, R., Simmons, P., Tedder, T.M., Uguccioni, M., Warren, H. (2005) CD molecules 2005: human cell differentiation molecules. *Blood* 106: 3123–3126.

3
Selection Strategies II: Antibody Phage Display

Michael Hust[1], *Lars Toleikis*[2] *and Stefan Dübel*[1]

3.1
Introduction

The production of polyclonal antibodies by immunisation of animals is a method established for more than a century. The first antibody serum was directed against Diphterie and produced in horses (von Behring and Kitasato, 1890). Hybridoma technology was the next milestone, allowing the production of monoclonal antibodies by fusion of an immortal myeloma cell with an antibody producing spleen cell (Köhler and Milstein 1975). However, hybridoma technology has some limitations, the possible instability of the aneuploid cell lines, most of all its inability to produce human antibodies and to provide antibodies against toxic or highly conserved antigens (Winter and Milstein 1991).

To overcome the limitations of hybridoma technology, antibodies or antibody fragments can be generated by recombinant means (Fig. 1). The most common used antibody fragments are the Fragment antigen binding (Fab) and the single chain Fragment variable (scFv). The Fab fragment consists of the fd fragment of the heavy chain and the light chain linked by a disulphide bond. The variable region of the the heavy chain (V_H) and the variable region of the light chain (V_L) are connected by a short peptide linker in the scFv. A major breakthrough in the field of antibody engineering was the generation of antibody fragments as recombinant proteins in the periplasmatic space of *E. coli* (Better et al. 1988, Huston et al. 1988, Skerra and Plückthun 1988). To circumvent the instability of hybridoma cell lines, the genes encoding V_H and V_L of a monoclonal antibody can be cloned into an *E. coli* expression vector in order to produce antibody fragments in the periplasmatic space of *E. coli* which preserve the binding specificity of the parental hybridoma antibody (Toleikis et al. 2004).

1) Technische Universität Braunschweig, Institut für Biochemie und Biotechnologie, Abteilung Biotechnologie, Spielmannstr.7, 38106 Braunschweig

2) RZPD, Deutsches Ressourcenzentrum für Genomforschung GmbH, Im Neuenheimer Feld 515, 69120 Heidelberg

Handbook of Therapeutic Antibodies. Edited by Stefan Dübel

ISBN 978-3-527-31453-9

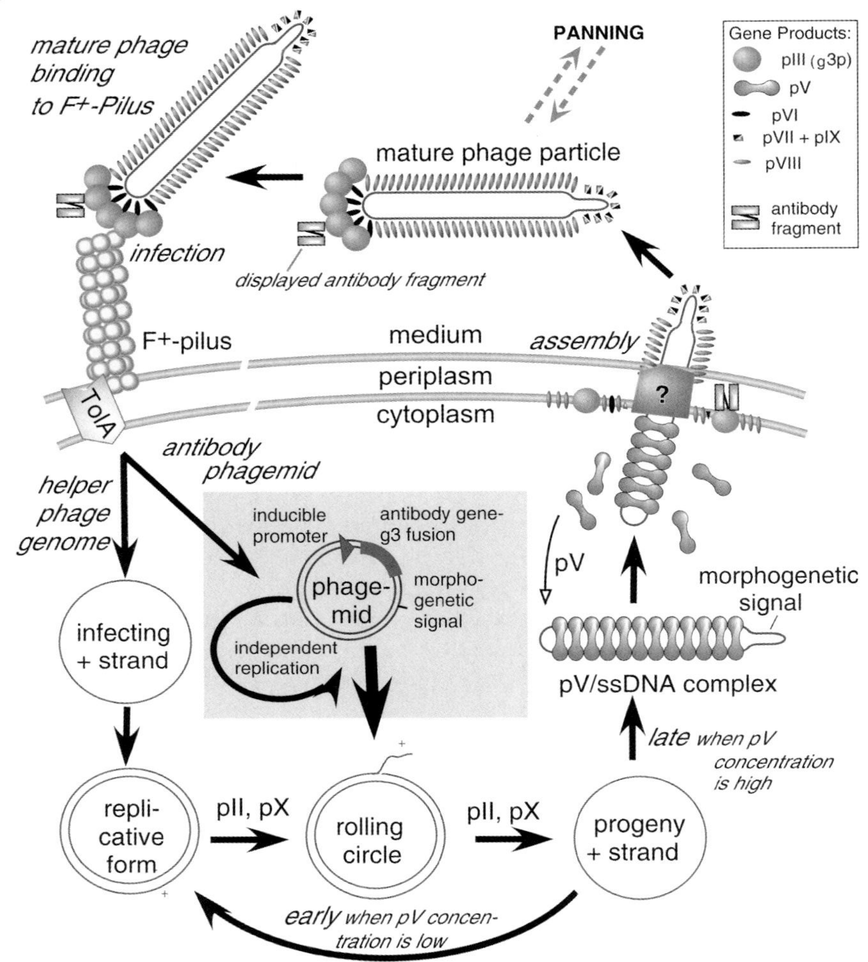

Fig. 3.1 Filamentous phage life cycle and antibody phage display. The plasmid encoding a fusion of an antibody fragment with pIII (as shown in the grey box) typically replicates and produces the fusion protein independently) from the phage genome. It carries the morphogenetic signal of the phage (which make it a "phagemid"). This allows it to be packaged into phage particles by the assembly machinery, typically up to 20 fold more effective than the phage genomes when replication deficient M13KO7 or its derivatives (like Hyperphage) are used. This limits the contamination of the panned antibody phage with "meaningless" helper phage genomes to less than 5%.

The production of mouse derived monoclonal antibody fragments in *E. coli* did not remove the major barrier for the broad application of antibodies in therapy as repeated administration of mouse derived antibodies causes a human anti-mouse antibody (HAMA) response (Courtenay-Luck et al. 1986). This problem can be overcome by two approaches: By humanisation of mouse antibodies (Studnicka et al. 1994) or by employing repertoires of human antibody genes. The second approach was achieved in two ways. First, human antibody gene repertoires were inserted into the genomes of IgG-knockout-mice, allowing to generate hybridoma cell lines which produce human immunoglobulins (Jakobovits 1995, Lonberg und Huszar 1995, Fishwild et al. 1996). However, this method still requires immunisation and has limitations in respect of toxic and conserved antigens.

These restrictions do not apply for the more rational second approach: the complete *in vitro* generation of specific antibodies from human antibody gene repertoires. There, despite of the constant suggestion of novel methods like bacterial surface display (Fuchs et al. 1991, for review see Jostock and Dübel 2005), ribosomal display (Hanes and Plückthun 1997), puromycin display (Roberts and Szostak 1997) or yeast surface display (Boder and Wittrup et al. 1997), phage display has become the most widely used selection method (Table 3.1, chapter "An exciting start- and a long trek" Fig. 1.3), which is based on the groundbreaking work of Smith (1985). The genotype and phenotype of a polypeptide were linked by fusing short gene fragments to the minor coat protein III gene of the filamentous bacteriophage M13. This resulted in the expression of this fusion

Table 3.1 Comparison of recombinant antibody selection systems (modified from Hust et al. 2005).

Selection system	*Advantages*	*Disadvantages*
Transgenic mice	Somatic hypermutation	Immunisation requiered, not freely avaible
Cellular display		
Bacteria	N- and C-terminal and sandwich fusion	Not matured, requires individual sorting
Yeast	Display of larger proteins, N- and C-terminal and sandwich fusion	Requires individual sorting
Intracellular display		
Yeast two hybrid	Screening library versus library possible	Cytoplasm not optimal for antibody folding
Molecular display		
Puromycin/ribosomal	Largest achievable library size *in vitro*	Finicky method
Phage display		
Filamentous		
Genomic	*In vitro*, robust, multivalent display	Prone to mutation, only C-terminal fusion
Phagemid	*In vitro*, robust	Only C-terminal fusion
T7	Well suited for peptide display	No display of antibody fragments
Arrays		
Gridded clones	Robust, simple	Small library sizes

protein on the surface of phage, allowing affinity purification of the gene of interest by the polypeptide binding. The first antibody gene repertoires in phage were generated and screened by using the lytic phage Lambda (Huse et al. 1989, Persson 1991) with limited success. Consequently, antibody fragments were presented on the surface of M13, fused to pIII (McCafferty 1990, Barbas et al. 1991, Breitling et al. 1991, Clackson et al. 1991, Hogenboom et al. 1991, Marks et al. 1991). By uncoupling antibody gene replication and expression from the phage life cycle by locating them on a separate plasmid (phagemid), genetic stability, propagation and screening of antibody libraries was greatly facilitated (Barbas et al. 1991, Breitling et al. 1991, Hoogenboom 1991, Marks et al. 1991). To date, "single-pot" (see below) antibody libraries with a theoretical diversity of up to 10^{11} independent clones were assembled (Sblattero and Bradbury 2000) to serve as a molecular repertoire for phage display selections.

3.2 The Phage Display System

Due to its robustness and straightforwardness, phage display has been the selection method most widely used in the past decade. Display systems employing insertion of antibody genes into the phage genome have been developed for phage T7 (Danner and Balesco 2001), phage Lambda (Huse et al. 1989, Mullinax et al. 1990, Kang et al. 1991a) and the Ff class (genus *inovirus*) of the filamentous phages f1, fd and M13 (McCafferty 1990). Being well established for peptide display, the phage T7 is not well suited for antibody phage display because it is assembled in the reducing enviroment of the cytoplasm, thus leaving most antibodies unfolded (Danner and Balesco 2001). In contrast, the oxidizing milieu of the bacterial periplasm allows antibody fragments to be folded and assembled properly (Skerra and Plückthun 1988). The Ff class non-lytic bacteriophages are assembled in this cell compartment and allow the production of phages without killing the host cell (Fig. 1). This is a major advantage compared to the lytic phages Lambda (Huse et al. 1989). In addition, filamentous phages allow the production of soluble proteins by introducing an amber stop codon between the antibody gene and gene III when using phagemid vectors. In an *E. coli* supE suppressor strain, the fusion proteins will be produced, whereas soluble antibodies are made in a non suppressor strain (Marks et al. 1992a, Griffiths et al. 1994), but expression in suppressor strains is also possible (Kirsch et al. 2005). Therefore, the members of the Ff class are the phages of choice for antibody phage display.

To achieve surface display, five of the M13 coat proteins (Fig. 1) have been used in fusion to foreign proteins, protein fragments or peptides. In the commonly used system the antibody fragment is coupled to the N-terminus or second domain of the minor coatprotein pIII (Barbas et al. 1991, Breitling et al. 1991, Hoogenboom et al. 1991). The function of the 3–5 copies of pIII, in particular their N-terminal domain, is to provide interaction of the phage with the F pili expressed on the surface of *E. coli* (Crisman and Smith 1984). The major coat protein pVIII has been considered as an alternative fusion partner, with only very few success reports in the past decade (Kang et al. 1991b). pVIII fusions are obviously more useful for the display of short peptides (Cwirla et al. 1990, Felici et al. 1991). Fusions to pVI have also been tried, but not yet with antibody fragments (Jespers et al. 1995). pVII and pIX were used in combination, by fusing the V_L domain to pIX and the V_H domain to pVII, allowing the presentation of a Fv fragment on the phage surface. Thus this format offers the potential for heterodimeric display (Gao et al. 1999). However, the fusion with pIII remains the most widely used system for antibody phage display and is still the only system of practical relevance.

Two different systems have been developed for the expression of the antibody:: pIII fusion proteins. First, the fusion gene can be inserted directly into the phage genome substituting the wildtype (wt) pIII (McCafferty et al. 1990). Second, the fusion gene encoding the antibody fusion protein can be provided on a separate plasmid with an autonomous replication signal, a promoter, a resistance marker and a phage morphogenetic signal, allowing this "phagemid" to be packaged into assembled phage particles. A helperphage, usually M13KO7, is necessary for the production of the antibody phage to complement the phage genes not encoded on the plasmid. Due to its mutated origin, the M13KO7 helperphage genome is not efficiently packaged during antibody phage assembly when compared to the phagemid (Vieira and Messing 1987).

In the system using direct insertion into the phage genome, every pIII protein on a phage is fused to an antibody fragment. This is of particular advantage in the first round of panning, where the desired binder is diluted in millions to billions of phages with unwanted specificity. The oligovalency of these phages improves the chances of a specific binder to be enriched due to the improvement of apparent affinity by the avidity effect. This advantage, however, has to be weighted to a number of disadvantages. The transformation efficiency of phagemids is two to three orders of magnitude better than the efficiency of phage vectors, thus facilitating the generation of large libraries. Second, the additional protein domains fused to pIII may reduce the function of pIII during reinfection. In a phagemid system, the vast majority of the pIII assembled into phage are wt proteins, thus providing normal pilus interaction. This may explain why only two "single-pot" antibody libraries (Griffiths et al. 1994, O'Connel et al. 2002) were made using a phage vector. In contrast, in phagemid systems both replication and expression of foreign fusion proteins are independent from the phage genome. There is no selection pressure, as propagation of the phagemid occurs in the absence of helperphage. The fusion proteins can be produced in adjustable quantities and allowing to use the amber/suppressor system for switching to soluble expression of antibody fragments without a pIII domain. Finally, despite usually not derived from highest copy plasmids, the dsDNA of phagemids is more easy to handle than phage DNA, facilitating cloning and analysis. Therefore, most pIII display systems use the phagemid approach. There is, however, a disadvantage which originates from the two independent sources for the pIII during phage packaging. During assembly, the wt pIII of the phage are inserted into the phage particles with much higher rate than the pIII fusion protein. As a result, the vast majority of resulting phage particles carry no antibody fragments at all. The few antibody phages in these mixtures are mainly monovalent, with phage carrying two or more antibody fragments being extremely rare. This allows to select for antibodies with a high monovalent affinity, since avidity effects decreasing the dissociation rate from the panning antigen can be avoided. In the first panning round, however, when a few binders have to be fished out of a huge excess of unwanted antibody phages, the fact that only a few percent of the phages carry antibodies hampers the efficiency of the system (Barbas et al. 1991, Breitling et al. 1991, Hoogenboom et al. 1991, Garrard et al. 1991, Lowman et al. 1991, O'Connel et al. 2002). This problem can be overcome by using a newly developed helperphage "Hyperphage". Hyperphage does not have a functional pIII gene and therefore the phagemid encoded pIII antibody fusion is the sole source of pIII in phage assembly offering multivalent display for phagemid vectors. This method improves antibody phage display by two orders of magnitude and vastly improves panning efficiency (Rondot et al. 2001).

Multivalent display can be also achieved by the integration of two amber stop codons into the gIII gene of the helper phage genome, offering the production of a functional helper phage "Ex-phage" in an *E. coli* suppressor strain. In the associated phagemid pIGT3, the antibody::pIII fusion is made without an amber stop codon and the antibody phage is produced in an *E. coli* non suppressor strain (Baek et al. 2002). However, the necessary deletion of the amber stop codon in the phagemid makes it imperative to subclone the antibody gene or to use a protease to produce soluble antibodies in contrast to the Hyperphage system where the amber/suppressor system can be used for switching to soluble expression of antibody fragments without a pIII domain.

A niche application of phage display is the selective infective phage (SIP) technology. Here, antibody fragments are fused to the N-terminal domain of pIII by cloning into the phage genome, therefore every pIII carries an antibody and the deletion of the pIII N-terminal region made the phage non-infective. In turn, the antigen is fused to the C-terminal end of seperately produced soluble pIII N-terminal domain. The functional, F pili binding pIII is reconstituted when the antibody phage binds to the antigen, allowing only the correct antibody phage to infect *E. coli* and to be propagated (Spada and Plückthun 1997). However, due to the fast kinetics of pIII/pilin interactions and very low concentrations of the three reaction partners if not coexpressed in the same cell, this method is not applicable for the convenient panning of larger libraries.

3.3
Selection and Evaluation of Binders

The novel procedure for isolating antibody fragments by their binding activity *in vitro* was called "panning", referring to the gold washers tool (Parmley and Smith

1988). The antigen is immobilized to a solid surface, such as nitrocellulose (Hawlisch et al. 2001), magnetic beads (Moghaddam 2003), a column matrix (Breitling et al. 1991) or, most widely used, plastic surfaces as polystyrol tubes (Hust et al. 2002) or 96 well microtitre plates (Barbas et al. 1991). The antibody phages are incubated with the surface-bound antigen, followed by thorough washing to remove the vast excess of non-binding antibody phages. The bound antibody phages can subsequently be eluted and reamplified by infection of *E. coli*. This amplification allows detection of a single molecular interaction during panning, as a single antibody phage, by its resistance marker, can give rise to a bacterial colony after elution. This illustrates the tremendous sensitivity of the method. The selection cycle can be repeated by infection of the phagemid bearing *E. coli* colonies from the former panning round with a helperphage to produce new antibody phages, which can be used for further rounds of panning until a significant enrichment is achieved. The number of antigen specific antibody clones should increase with every panning round. Usually 2–6 panning rounds are necessary to select specifically binding antibody fragments (Fig. 3.2). For an

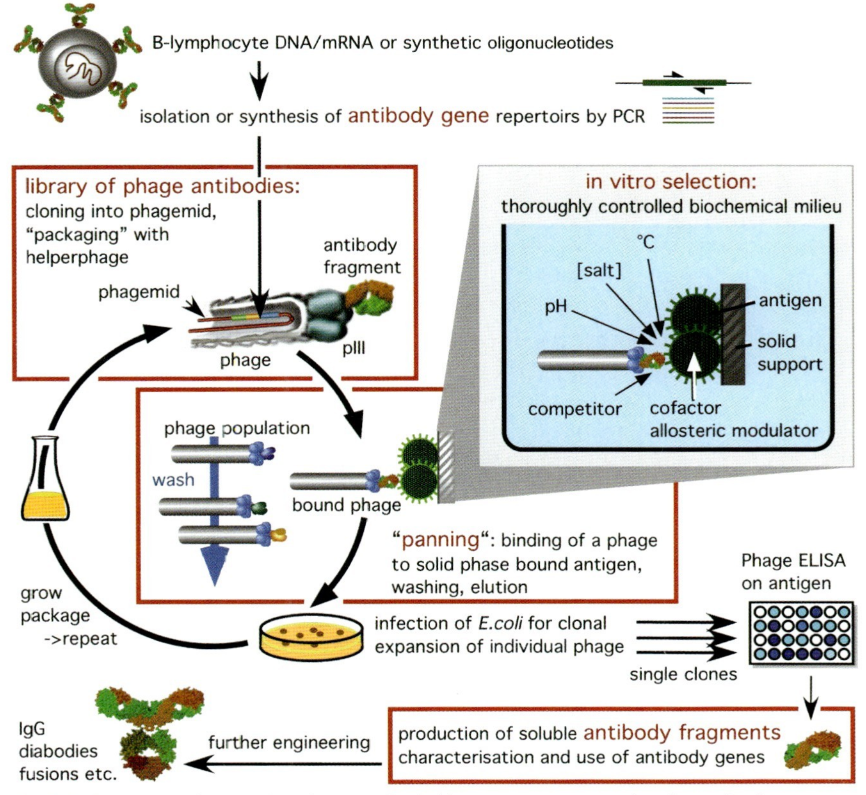

Fig. 3.2 Selection of antibodies from antibody libraries ("panning") by phage display.

overview of available stratagies and protocols, refer to McCafferty et al. (1996) and Kontermann and Dübel (2001). High throughput methods using microtitreplates and robotics can facilitate and enhance the panning procedure (for review see Konthur et al. 2005).

In most cases, the first step in the evaluation process of potential binders is an ELISA with polyclonal phage preparations from each panning round on coated target antigen and a control protein, e.g. BSA. The next step is the production of soluble monoclonal antibody fragments – from the panningrounds showing a significant enrichment of specific binders in polyclonal phage ELISA – in microtitreplates, followed by ELISA on coated antigen and in parallel on control protein. Soluble Fab fragments can be detected by their constant domains, wheras soluble scFvs can be detected by their engineered tags, e.g. his- or cmyc-tag. The clones producing specific antibody fragments will be further analysed by sequencing. Here, specific binders in duplicate can be rejected. A subcloning into *E. coli* expression vectors, like pOPE101 (Schmiedl et al. 2000) offers high scale production of antibody fragments for further analysis, e.g. analysis by flow cytometry (Schirrmann and Pecher 2005). Another important feature of an antibody is its affinity, which is analysed by surface plasmon resonance (BIAcore) (Lauer et al. 2005). After analysis of specificity and affinity the selected antibody fragment can be subcloned into other formats like IgG or scfv::Fc-fusion in order to achieve avidity and immunological effector functions (Jostock et al. 2004).

3.4 Phage Display Vectors

A large number of different phage display vectors have been constructed. Table 3.2 lists a selection of phage display vectors, without pretending to be complete. Some of them have not been used for the construction of a library up to now, but have been included since they offer ideas and alternatives, e.g. a system which allows to control the success of antibody gene cloning by green fluorescent protein (GFP) expression (Pascke et al. 2001).

A variety of different promoters have been employed for the expression of antibody fragments on the surface of phages. Widely used is the lac Z promoter (lacZ) derived from the lactose operon (Jacob and Monod 1961). The gIII promoter (gIII) from the bacteriophage M13 (Smith 1985), the tetracycline promoter (1x $tet^{o/P}$) (Zahn et al. 1999) and the phoA promotor of the *E. coli* alkaline phosphatase (Garrard et al. 1991) were also successfully used. It seems that very strong promoters, e.g. the synthetic promoter PAI/04/03 (Bujard et al. 1987), are rather a disadvantage (Dübel personnel communication). To our knowledge, a systematic comparison of the different promotors has not been done.

The targeting of the antibody fragments to the periplasmatic space of *E. coli* requires the use of signal peptides. The pelB leader of the pectate lyase gene of *Erwinia caratovora* (Lei et al. 1987) is commonly used. The gIII leader (Smith 1985), the phoA leader of the *E. coli* alkaline phosphatase and the ompA leader

Table 3.2 Antibody Phage display vectors in alphabetical order (modified from Hust and Dübel 2005).[1] Cre/lox recombination[2] λ recombination[3] Construction of the HuCAL library is described, but the pMorph vectorsystem is unpublished.

Phage display vector	*Promoter*	*Secretion*	*Antibody format used by reference*	*C-domains in vector*	*Sites heavy chain*	*Sites light chain*	*Tags*	*gIII*	*Expression of soluble Ab*	*Reference*
pAALFab	1x lac Z, 2x RBS	2x pelB	Fab	no	*EcoRI* – *BstPI*	*SpeI* – *XhoI*	–	truncated	subcloning	Iba ct al. (1997)
pAALFv	1x lac Z, 2x RBS	2x pelB	Fv	no	*EcoRI* – *BstPI*	*SpeI* – *XhoI*	–	truncated	subcloning	Iba et al. (1997)
pAALSC	1x lac Z, 1x RBS	1x pelB	scFv	no	*EcoRI* – *BstPI*	*SpeI* – *XhoI*	–	truncated	subcloning	Iba et al. (1997)
pAK100	1x lac Z, 1x RBS	1x pelB	scFv	no	*SfiI* (tet resistance will be removed)		FLAG, myc	truncated	amber, supE strain	Krebber et al. (1997)
pAPIII$_6$ scFv	1x phoA, 1x RBS	1x OmpA	scFv	no	*Hind*III – *SalI*		FLAG, His	truncated	*SalI* – KI digest, deletion of gIII	Haidaris et al. (2001)
pCANTAB3his	1x lac Z, 1x RBS	1x g3p	scFv	no	*NcoI*/*SfiI* – *NotI*		His, myc	full	amber, supE strain	McCafferty et al. (1994)
pCANTAB5his/ pCANTAB 6	1x lac Z, 1x RBS	1x cat	scFv	no	*NcoI*/*SfiI* – *NotI*		His, myc	full	amber, supE strain	
pCANTAB 5 E	1x lac Z, 1x RBS	1x g3p	scFv	no	*SfiI* – *NotI*		E tag	full	amber, supE strain	www.amershambiosciences.com
pCES1	1x lac Z, 2x RBS	1x gIII (L) 1x pelB (H)	Fab	yes	*SfiI* – *PstI*/*BstEII* (VH)	*ApaL1* – *AscI* (L chain), *ApaL1* – *XhoI* (VL)	His, myc	full	amber, supE strain	Haardt et al. (1999)
pComb3	2x lac Z, 2x RBS	2x pelB	Fab	no	*XhoI* – *SpeI*	*SacI* – *XbaI*	–	truncated	*NheI* – *SpeI* digest, deletion of gIII	Barbas et al. (1991)
pComb3H	1x lac Z, 2x RBS	ompA (LC) pelB (HC)	Fab, scFv	yes	*XhoI* – *SpeI*	*SacI* – *XbaI*	–	truncated	*NheI* – *SpeI* digest, deletion of gIII	Barbas et al. (2001)
pComb3X	1x lac Z, 2x RBS	ompA (LC) pelB (HC)	Fab, scFv	yes	*XhoI* – *SpeI*	*SacI* – *XbaI*	His, HA	truncated	amber, supE strain	
pCW93/H, pCW99/L[1]	1x lac Z, 1x RBS	1x pelB	scFv	no	*NcoI* – *NheI*	*SacI* – *BglII*	myc	truncated	amber, supE strain	Tsurushita et al. (1996)
pDAN5[1]	1x lac Z, 1x RBS	undiscribed leader	scFv	no	*XhoI* – *NheI*	*BssHII* – *SalI*	SV5, his	full	amber, supE strain	Sblattero and Bradbury (2000)

Table 3.2 *Continued*

Phage display vector	*Promoter*	*Secretion*	*Antibody format used by reference*	*C-domains in vector*	*Sites heavy chain*	*Sites light chain*	*Tags*	*gIII*	*Expression of soluble Ab*	*Reference*
pDH188	2x phoA, 2x RBS	2x stII	Fab	no	n.d.	n.d.	–	truncated	subcloning	Garrard et al. (1991)
pDN322	1x lac Z, 1x RBS	1x pelB	scFv	no	*Nco*I – *Not*I		FLAG, His	full	amber, supE strain	Pini et al. (1998)
pDNEK	1x lac Z, 1x RBS	1x pelB	scFv	no	*Nco*I – *Not*I		FLAG, His	full	amber, supE strain	Viti et al. (2000)
pEXmide3	1x lac Z, 2x RBS	2x pelB	Fab	yes	*Sfi*I/*Nco*I – *Kpn*I/*Apa*I	*Eag*I/*Not*I – *Nhe*I/*Spe*I	–	full	amber, supE strain	Söderlind et al. (1993)
pEXmide4	1x lac Z, 1x RBS	1x pelB	scFv	CH1	*Nco*I – *Sal*I		–	full	amber, supE strain	Kobayashi et al. (1997)
pEXmide5	1x lac Z, 1x RBS	1x pelB	scFv	?	*Nco*I – *Sal*I		–	full	amber, supE strain	Jirholt et al. (1998)
pFAB4	2x lac Z, 2x RBS	2x pelB	Fab	no	*Sfi*lI – *Not*I		–	truncated	amber, supE strain	Ørum et al. (1993)
pFAB4H	1x lac Z, 2x RBS	2x pelB	Fab	CH1	*Sfi*lI – *Not*I		–	truncated	subcloning	Dziegel et al. (1995)
pFAB5c	2x lac Z, 2x RBS	2x pelB	Fab	no	*Sfi*I – *Not*I		–	truncated	amber, supE strain	Örum et al. (1993)
pFAB5c-His	1x lac Z, 2x RBS	2x pelB	scFv	no	*Sfi*I – *Not*I		his	truncated	amber, supE strain	Söderlind et al. (2000)
pFAB60	1x lac Z, 1x RBS	2x pelB	Fab	CH1	*Sfi*I – *Spe*I (VH) *Sfi*I – *Not*I (Fd)	*Nhe*I – *Asc*I (L chain)	his	truncated	*Eag*I digest, deletion of gIII	Johansen et al. (1995)
pFAB73H	1x lac Z, 1x RBS	2x pelB	Fab	CH1	*Nhe*I – *Apa*I (VH)	*Sfi*I – *Asc*I (L chain)	his	truncated	*Eag*I digest, deletion of gIII	Engberg et al. (1996)
pGP-F100	1x tet$^{o/p}$, 1x RBS	1x pelB	scFv	no	*Sfi*I (GFPuv will be removed)		myc	truncated	TEV protease site	Pascke et al. (2001)
pGZ1	1x tet$^{o/p}$, 1x RBS	1x pelB	scFv	no	*Sfi*I – *Not*I		myc	full	amber, supE strain	Zahn et al. (1999)
pHAL14	1x lac Z, 1x RBS	1x pelB	scFv	no	*Nco*I – *Hind*III	*Mlu*I – *Not*I	Yol1/34, his, myc	full	amber, supE strain	Hust et al. (unpublished)

pHEN1	1x lac Z, 1x RBS	1x pelB	scFv, Fab, Fd, LC	no	*Sfi*I – *Not*I		myc	full	amber, supE strain	Hoogenboom et al. (1991)
pHEN1-Vλ3	1x lac Z, 1x RBS	1x pelB	scFv	no	*Nco*I – *Xho*I	Vλ3 anti-BSA Ab chain	myc	full	amber, supE strain	Hoogenboom and Winter (1992)
pHEN2	1x lac Z, 1x RBS	1x pelB	scFv	no	*Nco*I – *Xho*I	*ApaLI* – *Not*I	his, myc	full	amber, supE strain	http://www.mrc-cpe.cam.ac.uk
pHENIX	1x lac Z, 1x RBS	1x pelB	scFv	no	*Sfi*I/*Nco*I – *Sal*I/*Xho*I	*ApaL*1 – *Not*I	myc	full	amber, supE strain	Finnern et al. (1997)
pHG-1m/A27Jκ1	1x lac Z, 1x RBS	1x pelB	scFv	no	*ApaLI* – *Sfi*I	A27Jκ1 (VL)	his, myc	full	amber, supE strain	Rojas et al. 2002
phh3mu-γ1	2x lac Z, 2x RBS	2x pelB	Fab bidirectional	yes	*Xho*I – *EcoRI*	*Sac*I – *Hind*III	–	truncated	subcloning	Den et al. (1999)
pIG10	1x lac Z, 1x RBS	1x OmpA	scFv	no	*EcoRV* – *EcoRI*		myc	full	amber, supE strain	Ge et al. (1995)
pIGT2 (vector)	1x lac Z, 1x RBS	1x g3p	scFv	no	*Sfi*I – *Not*I		myc	full	amber, supE strain	Baek et al. (2002)
pIGT3 (vector)	1x lac Z, 1x RBS	1x g3p	scFv	no	*Sfi*I – *Sfi*I		myc	full	subcloning	
pIT2	1x lac Z, 1x RBS	1x pelB	scFv	no	SfiI/*Nco*I – *Xho*I	*SalLI* – *Not*I	His, myc	full	amber, supE strain	Goletz et al. (2002)
pLG18	1x phoA, 2x RBS	2x stII	Fab	yes	*BssHII* – *Nco*I (CDR2-3)	*BstEII* – *Asp*718 (CDR1-3)	–	truncated	subcloning	Garrard and Henner (1993)
pM834, pM827[2]	2x lac Z, 2x RBS	2x pelB	Fab	no	*Xho*I – *Spe*I	*Sac*I – *Xba*I	–	full	amber, supE strain	Geoffroy et al. (1994)
pMorph series[3]	1x lac Z ?	1x phoA	scFv	no	*Xba*I – *EcoRI*		FLAG ?	?	subcloning	Knappik et al. (2000)
pScUAGΔcp3	1x lac Z, 1x RBS	1x pelB	scFv with Cκ	Cκ	*Xho*I – *Nhe*I	*Sst*I – *Bgl*II	–	truncated	amber, supE strain	Akamatsu et al. (1993)
pSEX	1x PA1/04/03, 1x RBS	1x pelB	scFv	no	–		Yol1/34	full	subcloning	Breitling et al. (1991)
pSEX20	1x PA1/04/03, 1x RBS	1x pelB	scFv	no	–		Yol1/34	full	subcloning	Dübel et al. (1993)
pSEX81	1x lac Z, 1x RBS	1x pelB	scFv	no	*Nco*I – *Hind*III	*Mlu*I – *Not*I	Yol1/34	full	subcloning	Welschof et al. (1997)

of *E. coli* outer membrane protein OmpA have also been used, being common to many protein expression vectors (Skerra et al. 1993, Skerra and Schmidt 1999). Further examples are the heat-stable enterotoxin II (stII) signal sequence (Garrard et al. 1991) and the bacterial chloramphenicol acetyltransferase (cat) leader (McCafferty et al. 1994).

Due to the inability of *E. coli* to assemble complete IgG, with one exception (Simmons et al. 2002), smaller antibody fragments are used for phage display. In particular, Fabs and scFvs have been shown to be the antibody formats of choice. As aforementioned, in Fabs, the fd fragment and light chain are connected by a disulphide bond. In scFvs, the V_H and V_L are connected by a 15–25 amino acid linker (Bird et al. 1988, Bird and Walker 1991, Huston et al. 1988). Soluble scFvs tend to form dimers, in particular when the peptide linker is reduced to three to twelve amino acid residues. Diabodies or tetrabodies are produced if the linker between V_H and V_L is reduced to a few amino acids (Kortt et al. 1997, Arndt et al. 1998, Le Gall et al. 1999). The dimerization agravates the determination of the affinity, due to the possible avidity effect of the antibody complex (Marks et al. 1992b). Furthermore, some scFvs show a reduced affinity up to one order of magnitude compared to the corresponding Fabs (Bird and Walker 1991). ScFvs with a higher affinity than the corresponding Fabs were rarely found (Iliades et al. 1998). Small antibody fragments like Fv and scFv can easily be produced in *E. coli*. The yield of functional Fvs expressed in *E. coli* is higher than the yield of the corresponding Fabs, due to a lower folding rate of the Fabs (Plückthun 1990, Plückthun 1991). In one example, the stability in long-term storage was much higher for Fabs than for scFvs. After 6 month the functionality of scFvs stored at 4°C was reduced by 50 %, Fabs, however, showed no significant loss of functionality after one year (Kramer et al. 2002). The overall yield of Fvs expressed in *E. coli* vary from 0.5 to 10 mg/l culture compared to 2 to 5 mg/l culture for Fabs (Ward 1993), but very high yields of 1–2 mg/l of soluble and functional F(ab′)2 have been reported (Carter et al. 1992). Therefore, the choice of the antibody format, scFv or Fab, depends on the desired application.

For the expression of Fabs in *E. coli* two polypeptide chains have to be assembled. In the monocistronic systems, e.g. pComb3, the antibody genes are under control of two promoters and each has its own leader peptide (Barbas et al. 1991), whereas in plasmids like pCES1 with a bicistronic Fab operon, both chains are under control of a single promoter, leading to a mRNA with two ribosomal binding sites (De Haard et al. 1999). The bicistronic system is more sufficient for the expression of Fabs (Kirsch et al. 2005).

Two variants of the antibody::pIII-fusion have been made. Either full size pIII or truncated version of pIII were used. The truncated version was made by deleting the pIII N-terminal domain. This domain mediated the interaction with the F pili of *E. coli*. Infection is provided by wt pIII, as only a small percentage of phage in phagemid-based systems are carrying an antibody. These truncated vectors are therefore not compatible with the use of Hyperphage or Ex-phage, as the fullsize pIII is necessary for infection (Rondot et al. 2001, Baek et al. 2002). Some phagemids, e.g. pSEX81 (Welschof et al. 1997) allow the elution of antibody phages during panning by protease digestion instead of pH shift. This is possible due to a protease cleavage site between pIII and the antibody fragment. Therefore complete recovery of specifically antigen bound antibody phages is possible, even in case of very strong antigen binding.

Most of the described phagemids have an amber stop codon between the antibody gene and gIII. This allows the production of soluble antibody fragments after transformation of the phagemid to a non suppressor bacterial strain like HB2151 (Griffiths et al. 1994). For phagemids like pComb, it is necessary to delete the gIII by digestion and religate the vector before tranformation into *E. coli* (Barbas et al. 1991). In the case of phagmids like pSEX81, the selected antibody genes have to be subcloned into a separate *E. coli* expression vector like the pOPE series (Breitling et al. 1991, Dübel et al. 1993, Schmiedl et al. 2000).

3.5 Phage Display Libraries

Various types of phage display libraries have been constructed. Immune libraries are generated by amplification of V genes isolated from IgG secreting plasma cells of immunised donors (Clackson et al. 1991). From immune libraries antibody fragments with monovalent dissociations constants in the nM range can be isolated. Immune libraries are typically created and used in medical research to select an antibody fragment against one particular antigen, e.g. an infectious pathogen, and therefore would not be the source of choice for the selection of a large number of different specificities. Naive, semi-synthetic and synthetic libraries have been subsumed as "single-pot" libraries, as they are designed to isolate antibody fragments binding to every possible antigen. A correlation is seen between the size of the repertoire and the affinities of the isolated antibodies. Antibody fragments with a µM affinity have been isolated from a "single-pot" library consisting of approximately 10^7 clones, whereas antibody fragments with nM affinities were obtained from a library consisting of 10^9 independent clones (Hoogenboom 1997). It is evident that the chance to isolate an antibody with a high affinity for a particular antigen increases almost linearly to the size of the library. According to the source of antibody genes, "single pot" libraries (Table 3.3) can be naive libraries, semisynthetic libraries or fully synthetic libraries. Naive libraries are constructed from rearranged V genes from B cells (IgM) of non-immunized donors. An example for this library type is the naive human Fab library constructed by de Haardt et al. (1999), yielding antibodies with affinities up to 2.7×10^{-9} M. Semi-synthetic libraries are derived from unrearranged V genes from pre B cells (germline cells) or from one antibody framework with genetically randomized complementary determining region (CDR) 3 regions, as described by Pini et al. (1998). The antibody fragments obtained from this library show affinities between 10^{-8} M and 10^{-9} M, with one scFv having a dissociation constant of 5×10^{-11} M. A combination of naive and synthetic repertoire was used by Hoet et al. (2005). They combined light chains from autoimmune patients

Table 3.3 Human "single-pot" phage display libraries (modified from Hust and Dübel 2004). [1]Sheets et al. 1998, [2]Persson et al. 1991, [3]Griffiths et al. 1994.

Library vector	*Library type*	*Antibody type*	*Library cloning stratagy*	*Library size*	*reference*
DY3F63	synthetic and naive repertoire	Fab	ONCL, 4 step cloning with integration of naive CDRH3 in synthetic HC	3.5×10^{10}	Hoet et al. 2005
fdDOG-2lox, pUC19-2lox	semi-synthetic	Fab	PCR with random CDR3 Primers, Cre-lox	6.5×10^{10}	Griffiths et al. 1994
fdTet	naive	scFv	recloning of a naive library[1]	5×10^{8}	O'Connel et al. 2002
pAALFab	semi-synthetic (anti-hen egg white lysozyme Ab framework)	Fab	PCR with random CDR Primers, assembly PCR	2×10^{8}	Iba et al. 1997
pAP-III_6 scFv	naive	scFv	assembly PCR	n.d.	Haidaris et al. 2001
pCANTAB 6	naive	scFv	assembly PCR	1.4×10^{10}	Vaughan et al. 1996
pCES1	naive	Fab	3 step cloning (L chain, VH)	3.7×10^{10}	Haardt et al. 1999
pComb3	semi-synthetic (anti-tetanus Ab framework[2])	Fab	PCR with random CDR H3 Primers	5×10^{7}	Barbas et al. 1992
pComb3	semi-synthetic (anti-tetanus Ab framework[2])	Fab	PCR with random CDR H3 Primers	$>10^{8}$	Barbas et al. 1993
pDAN5	naive	scFv	Cre-lox	3×10^{11}	Sblattero and Bradbury 2000
pDN322	semi-synthetic (VH DP47 and VL DPK22 V-genes)	scFv	random CDR3 Primer, assembly PCR	3×10^{8}	Pini et al. 1998
pDN322	semi-synthetic (anti-AMCV CP ab framework)	scFv	random CDR3 Primer, assembly PCR	3.75×10^{7}	Desiderio et al. 2001

pDNEK (ETH2 library)	semi-synthetic (VH DP47, Vλ DPL16 and Vκ DPK 22 V-genes)	scFv	random CDR3 Primer, assembly PCR	5×10^8	Viti et al. 2000
pEXmide5	semi-synthetic (germline VH-DP47 and VL-DPL3 framework)	scFv	assembly PCR, CDR shuffling	9×10^6	Söderlind et al. 1993
pFAB5c-His (n-CoDeR library)	semi-synthetic (germline VH-DP47 and VL-DPL3 framework)	scFv	assembly PCR, CDR shuffling	2×10^9	Söderlind et al. 2000
pHAL14 (HAL4)	naive, kappa	scFv	2 step cloning	2.2×10^9	Hust et al. unpublished
pHAL14 (HAL7)	naive, lambda	scFv	2 step cloning	2.8×10^9	Hust et al. unpublished
pHEN1	naive	scFv	assembly PCR	10^7–10^8	Marks et al. 1991
pHEN1	naive	scFv	assembly PCR	$2 \times 10^5/2 \times 10^6$	Marks et al. 1992a
pHEN1	naive	scFv	assembly PCR	6.7×10^9	Sheets et al. 1998
pHEN1-Vλ3	semi-synthetic (Vλ3 anti-BSA Ab light chain)	scFv	PCR with random CDR H3 Primers	10^7	Hoogenboom and Winter 1992
pHEN1-Vλ3	semi-synthetic (Vλ3 anti-BSA Ab light chain)	scFV	PCR with random CDR H3 Primers	$>10^8$	Nissim et al. 1994
pHEN1-Vκ3	semi-synthetic (VH)/naive (VL)	scFv	3 step cloning, PCR with random CDR H3 Primers	3.6×10^8	de Kruif et al. 1995
pHEN2 (Griffin 1. library)	semi-synthetic	scFv	recloning of the lox library in scFv format[3]	1.2×10^9	www.mrc-cpe.cam.ac.uk
pIT2 (Tom I/J library)	semi-synthetic (3x VH and 4x Vκ V-genes)	scFv	PCR with random CDR2 and CDR3 Primers	1.47×10^8/ 1.37×10^8	Goletz et al. 2002
pLG18	semi-synthetic (anti-HER2 Ab framework)	Fab	PCR with random CDR Primers, 2 step cloning	2–3×10^8	Garrard and Henner 1993

Table 3.3 *Continued*

Library vector	*Library type*	*Antibody type*	*library cloning stratagy*	*library size*	*reference*
pMorph series (HuCAL library)	synthetic	scFv	2 step cloning, CDR3 replacement	2×10^9	Knappik et al. 2000
pMorph series (HuCAL GOLD library)	synthetic	Fab	2 step cloning, all CDR replacement	1.6×10^{10}	www.morphosys.com
pMID21	synthetic and naïve repertoire	Fab	ONCL, 4 step cloning with integration of naïve CDRH3 in synthetic HC	1×10^{10}	Hoet et al. 2005
pScUAGDcp3	semi-synthetic	scFv connected to Ck	3 step cloning with random CDR3 Primers	1.7×10^7	Akamatsu et al. 1993
pSEX81	naive	scFv (with N-terminus of CH1 and CL)	2 step cloning	4×10^7	Dörsam et al. 1997
pSEX81	naive	scFv (with N-terminus of CH1 and CL)	2 step cloning	4×10^9	Little et al. 1999
pSEX81	naive	scFv (with N-terminus of CH1 and CL)	4 step cloning	1.6×10^7/ 1.8×10^7/ 4×10^7	Schmiedel et al. 2000
pSEX81	naive	scFv (with N-terminus of CH1 and CL)	2 step cloning	6.4×10^9	Løset et al. 2005

with a fd fragment containing synthetic CDR1 and CDR2 in the human V_H3-23 framework and naive, origined from autoimmune patients, CDR3 regions. The fully synthetic libraries have a human framework with randomly integrated CDR cassettes (Hayashi et al. 1994). Antibody fragments selected from fully synthetic libraries exhibit affinities between 10^{-6} M and 10^{-11} M (Knappik et al. 2000). All library types – immune, naive, synthetic and their intermediates – are useful sources for the selection of antibodies for diagnostic and therapeutic purposes.

3.6 Generation of Phage Display Libraries

Various methods have been employed to clone the genetic diversity of antibody repertoires. After the isolation of mRNA from the desired cell type and the preparation of cDNA, the construction of immune libraries is usually done by a two step cloning or assembly PCR (see below). Naive libraries are constructed by two or three cloning steps. In the two step cloning strategy, the amplified repertoire of light chain genes is cloned into the phage display vector first, as the heavy chain contributes more to diversity, due to its highly variable CDRH3. In the second step the heavy chain gene repertoire is cloned into the phagemids containing the light chain gene repertoire (Johansen et al. 1995, Welschof et al. 1997, Little et al. 1999). In the three step cloning strategy, separate heavy and light chain libraries are engineered. The V_H gene repertoire has then to be excised and cloned into the phage display vector containing the repertoire of V_L genes (De Haardt et al. 1999). Another common method used for the cloning of naive (McCafferty et al. 1994, Vaughan et al. 1996), immune (Clackson et al. 1991) or hybridoma (Krebber et al. 1997) scFv phage display libraries is the assembly PCR. The V_H and V_L genes are amplified seperately and connected by a subsequent PCR, before the scFv encoding gene fragments are cloned into the vector. The assembly PCR is usually combined with a randomization of the CDR3 regions, leading to semi-synthetic libraries. To achieve this, oligonucleotide primers encoding various CDR3 and J gene segments were used for the amplication of the V gene segments of human germlines (Akamatsu et al. 1993). The CDRH3 is a major source of sequence variety (Shirai et al. 1999). Hoogenboom and Winter (1992) and Nissim et al. (1994) used degenerated CDRH3 oligonucleotide primers to produce a semi-synthetic heavy chain repertoire derived from human V gene germline segments and combined this repertoire with an anti-BSA light chain. In some cases a framework of a well known antibody was used as scaffold for the integration of randomly created CDRH3 and CDRL3 (Barbas et al. 1992, Desiderio et al. 2001). Jirholt et al. (1998) and Söderlind et al. (2000) amplified all CDR regions derived from B cells before shuffling them into one antibody framework in an assembly PCR reaction. An example for an entirely synthetic library, Knappik et al. (2000) utilized seven different V_H and V_L germline master frameworks combined with six synthetically created CDR cassettes. The construction of large naive and semi-synthetic libraries (Hoet et al. 2005, Løset et al. 2005,

Little et al. 1999, Sheets et al. 1998, Vaughan et al. 1996) requires significant effort to tunnel the genetic diversity through the bottleneck of *E. coli* transformation, e.g. 600 transformations were necessary for the generation of a 3.5×10^{10} phage library (Hoet et al. 2005).

To move the diversity potentiating step of random V_H/V_L combination behind the bottleneck of transformation, the Cre-lox or lamda phage recombination system has been employed (Waterhouse et al. 1993, Griffiths et al. 1994, Geoffrey et al. 1994). However, libraries with more than 10^{10} independent clones have now been accomplished by conventional transformation, rendering most of these complicated methods unnecessary in particular as they may result in decreased genetic stability. A remarkable exception is the use of a genomically integrated CRE recombinase gene (Sblattero and Bradbury 2000) which is expected to solve the instability issue and allows the generation of libraries with complexities above the limit achievable by conventional cloning.

In summary, antibodies with nanomolar affinities can be selected from either type of library, naive or synthetic. If the assembly by cloning or PCR and preservation of molecular complexity is carefully controlled at every step of its construction, libraries of more than 10^{10} independent clones can be generated.

Antibody phage display is delivering and will deliver high affinity human antibodies and antibody fragments for research as well as for diagnostic and therapeutic applications in the future.

References

Akamatsu, Y., Cole, M.S., Tso, J.Y., Tsurushita, N. (1993) Construction of a human Ig combinatorial library from genomic V segments and synthetic CDR3 fragments. *J Immunol* 151: 4651–4659.

Arndt, K.M., Müller, K.M., Plückthun, A. (1998) Factors influencing the dimer to monomer transition of an antibody single-chain Fv fragment. *Biochemistry* 37: 12918–12926.

Baek, H., Suk, K.H., Kim, Y.H., Cha, S. (2002) An improved helper phage system for efficient isolation of specific antibody molecules in phage display. *Nucleic Acids Res* 30: e18.

Barbas III, C.F., Kang, A.S., Lerner, R.A., Benkovic, S.J. (1991) Assembly of combinatorial antibody libraries on phages surfaces: the gene III site. *Proc Natl Acad Sci* USA 88: 7987–7982.

Barbas III, C.F., Bain, J.D., Hoekstra, M., Lerner, R.A. (1992) Semisynthetic combinatorial antibody libraries: a chemical solution to the diversity problems. *Proc Natl Acad Sci* USA 89: 4457–4461.

Barbas III, C.F., Amberg, W., Simoncsitis, A., Jones, T.M., Lerner, R.A. (1993) Selection of human anti-hapten antibodies from semisynthetic libraries. *Gene* 137: 57–62.

Barbas III, C.F., Burton, D.R., Scott, J.K., Silverman, G.J. (2001) Phage display: A laboratory manual. *Cold Spring Harbor Laboratory Press.*

Better, M., Chang, C.P., Robinson, R.R., Horwitz, A.H. (1988) Escherichia coli secretion of an active chimeric antibody fragment. *Science* 240: 1041–1043.

Bird, R.E., Hardman, K.D., Jacobsen, J.W., Johnson, S., Kaufman, B.M, Lee, S.M., Lee, T., Pope, S.H., Riordan, G.S., Whitlow, M. (1988) Single-chain antigen-binding proteins. *Science* 242: 423–426.

Bird, R.E., Walker, B.W. (1991) Single chain variable regions. *Trends Biotech* 9: 132–137.

Boder, E.T., Wittrip, K.D. (1997)Yeast surface display for screening combinatorial polypeptide libraries. *Nat Biotech* 15: 553–558.

Breitling, F., Dübel, S., Seehaus, T., Kleewinghaus, I., Little, M. (1991) A surface expression vector for antibody screening. *Gene* 104: 1047–1153.

Bujard, H., Gentz, R., Lanzer, M., Stueber, D., Mueller, M., Ibrahimi, I., Haeuptle, M. T., Dobberstein, B. (1987) A T5 promoter-based transcription-translation system for the analysis of proteins in vitro and in vivo. *Methods Enzymol* 155: 416–433.

Carter, P., Kelley, R.F., Rodrigues, M.L., Snedecor, B., Covarrubias, M., Velligan, M. D., Wong W.L., Rowland, A.M., Kotts, C.E., Carver, M.E., et al. (1992) High level Escherichia coli expression and production of a bivalent humanized antibody fragment. *Biotechnology* 10: 163–167.

Clackson, T., Hoogenboom, H.R., Griffiths, A.D., Winter, G. (1991) Making antibody fragments using phage display libraries. *Nature* 352: 624–628.

Crissman, J.W., Smith, G.P. (1984) Gene 3 protein of filamentous phages: evidences for a carboxyl-terminal domain with a role in morphogenesis. *Virology* 132: 445–455.

Courtenay-Luck, N.S., Epenetos, A.A., Moore, R., Larche, M., Pectasides, D., Dhokia, B., Ritter, M.A. (1986) Development of primary and secondary immune responses to mouse monoclonal antibodies used in the diagnosis and therapy of malignant neoplasms. *Cancer Res.* 46: 6489–6493.

Cwirla, S.E., Peters, E.A., Barrett, R.W., Dower, W.J. (1990) Peptides on phage: a vast library of peptides for identifying ligands. *Proc Natl Acad Sci USA* 87: 6378–6382.

Danner, S., Belasco, J.G. (2001) T7 phage display: A novel genetic selection system for cloning RNA-binding protein from cDNA libraries. *Proc Natl Acad Sci* USA 98: 12954–12959.

De Kruif, J. Boel, E., Logtenberg, T. (1995) Selection and application of human single chain Fv antibody fragments from a semi-synthetic phage antibody display library with designed CDR3 regions. *J Mol Biol* 248: 97–105.

De Haardt, H.J., van Neer, N., Reurst, A., Hufton, S.E., Roovers, R.C., Henderikx, P., de Bruine, A.P., Arends, J.-W., Hoogenboom, H.R. (1999) A large non-immunized human Fab fragment phage library that permits rapid isolation and kinetic analysis of high affinity antibodies. *J Biol Chem* 274: 18218–18230.

Den, W., Sompuram, S.R., Sarantopoulos, S., Sharon, J. (1999) A bidirectional phage display vector for the selection and mass transfer of polyclonal antibody libraries. *J Immunol Meth* 222: 45–57.

Desiderio, A., Franconi, R., Lopez, M., Villani, A.E., Viti, F., Chiaraluce, R., Consalvi, V., Neri, D., Benvenuto, E. (2001) A semi-synthetic repertoire of intrinsically stable antibody fragments derived from a single-framework scaffold. *J Mol Biol* 310: 603–615.

Dörsam, H., Rohrbach, P, Kürschner, T. Kipriyanov, S., Renner, S., Braunnagel, M., Welschof, M., Little, M. (1997) Antibodies to steroids from a small human naive IgM library. *FEBS Letters* 414: 7–13.

Dübel, S., Breitling, F., Fuchs, P., Braunagel, M., Klewinghaus, I., Little, M. (1993) A family of vectors for surface display and production of antibodies. *Gene* 128: 97–101.

Dziegiel, M., Nielsen, L.K., Andersen, P.S., Blancher, A., Dickmeiss, E., Engberg, J. (1995) Phage display used for gene cloning of human recombinant antibody against the erythrocyte surface antigen, rhesus D. *J Immunol Meth* 182: 7–19.

Engberg, J. Andersen, P.S., Nielsen, L.K., Dziegiel, M., Johansen, L.K., Albrechtsen, B. (1996) Phage-display libraries of murine and human Fab fragments. *Mol Biotechnol* 6: 287–310.

Felici, F., Castagnoli, L., Musacchio, A., Jappelli, R., Cesareni, G. (1991) Selection of antibody ligands from a large library of oligopeptides expressed on a multivalent exposition vector. *J Mol Biol* 222: 301–310.

Finnern, R., Pedrollo, E., Fisch, I., Wieslander, J., Marks, J.D., Lockwood, C. M., Ouwehand, W.H. (1997) Human autoimmune anti-proteinase 3 scFv from a phage display library. *Clin Exp Immunol* 107: 269–281.

Fishwild, D.M., O'Donnel, S.L., Bengoechea, T., Hudson, D.V., Harding, F., Bernhar, S. L., Jones, D., Kay, R.M., Higgins, K.M., Schramm, S.R., Lonberg, N. (1996) High-avidity human IgG kappa monoclonal antibodies from a novel strain of minilocus transgenic mice. *Nat Biotech* 14: 845–851.

Fuchs, P., Breitling, F., Dübel, S., Seehaus, T., Little, M. (1991) Targeting recombinant antibodies to the surface of E. coli: Fusion to a peptidoglycan associated lipoprotein. *Bio/Technology* 9: 1369–1372.

Gao, C., Mao, S., Lo, C.H., Wirsching, P., Lerner, R.A., Janda, K.D. (1999) Making artificial antibodies: a format for phage display of combinatorial heterodimeric arrays. *Proc Natl Acad Sci* USA 96: 6025–6030.

Garrard, L.J., Yang, M., O'Connel, M.P., Kelley, R., Henner, D.J. (1991) Fab assembly and enrichment in a monovalent phage display system. *Bio/Technology* 9: 1373–1377.

Garrard, L.J., Henner, D.J. (1993) Selection of an anti-IGF-1 Fab from a Fab phage library created by mutagenesis of multiple CDR loops. *Gene* 128: 103–109.

Ge, L., Knappik, A., Pack, P., Freund, C., Plückthun, A. (1995) Expressing antibodies in *Escherichia coli*, in: Antibody Engineering (Borrebaeck, C.A.K., ed.), Oxford University Press.

Geoffroy, F., Sodoyer, R., Aujame, L. (1994) A new phage display system to construct multicombinatorial libraries of very large antibody repertoires. *Gene* 151: 109–113.

Goletz, A., Cristensen, P.A., Kristensen, P., Blohm, D., Tomlinson, I., Winter, G., Karsten, U. (2002) Selection of large diversities of antiidiotypic antibody fragments by phage display. *J Mol Biol* 315: 1087–1097.

Griffiths, A.D., Williams, S.C., Hartley, O., Tomlinson, I.M., Waterhouse, P., Crosby, W., Kontermann, R.E., Jones, P.T., Low, N.M., Allison, T.J., Prospero, T.D., Hoogenboom, H.R., Nissim, A., Cox, J.P.L., Harrison, J.L., Zaccolo, M., Gherardi, E., Winter, G. (1994) Isolation of high affinity human antibodies directly from large synthetic repertoires. *EMBO J* 13: 3245–3260.

Haidaris, C.G., Malone, J., Sherrill, L.A., Bliss, J.M., Gaspari, A.A., Insel, R.A., Sullivan, M.A. (2001) Recombinant human antibody single chain variable fragments reactive with *Candida albicans* surface antigens. *J Immunol Meth* 257: 185–202.

Hanes, J., Plückthun, A. (1997) In vitro selection and evolution of functional proteins by using ribosome display. *Proc Natl Acad Sci* USA 94: 4937–4942.

Hawlisch, H., Müller, M., Frank, R., Bautsch, W., Klos, A., Köhl, J (2001) Sitespecific anti-C3a receptor single-chain antibodies selected by differential panning on cellulose sheets. *Analytical Biochemistry* 293: 142–145.

Hayashi, N., Welschoff, M., Zewe, M., Braunagel, M., Dübel, S., Breitling, F., Little, M. (1994) Simultaneous mutagenesis of antibody CDR regions by overlap extension and PCR. *Biotechniques* 17: 310–316.

Hoet, R.M., Cohen, E.H., Kent, R.B., Rookey, K., Schoonbroodt, S., Hogan, S., Rem, L., Frans, N., Daukandt, M., Pieters, H., van Hegelsom, R., Neer, N.C., Nastri, H.G., Rondon, I.J., Leeds, J.A., Hufton, S. E., Huang, L., Kashin, I., Devlin, M., Kuang, G., Steukers, M., Viswanathan, M., Nixon, A.E., Sexton, D.J., Hoogenboom, H. R., Ladner, R.C. (2005) Generation of high-affinity human antibodies by combining donor-derived and synthetic complementarity-determining-region diversity. *Nat Biotechnol* 23: 344–348.

Hoogenboom, H.R., Griffiths, A.D., Johnson, K.S., Chiswell, D.J., Hudson, P., Winter, G. (1991) Multi-subunit proteins on the surface of filamentous phage: methodologies for displaying antibody (Fab) heavy and light chains. *Nucl Acids Res* 19: 4133–4137.

Hoogenboom, H.R., Winter, G. (1992) By-passing immunisation: Human antibodies from synthetic repertoires of germline V_H gene segments rearranged *in vitro*. *J Mol Biol* 227: 381–388.

Hoogenboom, H.R. (1997) Designing and optimizing library selection strategies for generating high-affinity antibodies. *Trends Biotech* 15: 62–70.

Huse, W.D., Sastry, L., Iverson, S.A., Kang, A.S., Alting-Mees, M., Burton, D.R.,

Benkovic, S.J., Lerner, R. (1989) Generation of a large combinatorial library of the immunoglobulin repertoire in phage lambda. *Science* 246: 1275–1281.

Hust, M., Maiss, E., Jacobsen, H.-J., Reinard, T. (2002) The production of a genus specific recombinant antibody (scFv) using a recombinant Potyvirus protease. *J Virol Meth* 106: 225–233.

Hust, M., Dübel, S. (2004) Mating antibody phage display with proteomics. *Trends in Biotechnology* 22: 8–14.

Hust, M., Dübel, S. (2005) Phage display vectors for the in vitro generation of human antibody fragments in: Immunochemical Protocols, 3rd ed., Ed: Burns, R., *Meth Mol Biol* 295: 71–95.

Hust, M., Toleikis, L., Dübel, S. (2005) Antibody Phage Display. *Mod Asp Immunobiol* 15: 47–49.

Huston, J.S., Levinson, D., Mudgett, H.M., Tai, M.S., Novotny, J., Margolies, M.N., Ridge, R.J., Bruccoloreri, R.E., Haber, E., Crea, R., Oppermann, H. (1988) Protein engineering of antibody binding sites: recovery of specific activity in an anti-digosin single-chain Fv analogue produced in *Escherichia coli Proc Natl Acad Sci* USA 85: 5879–5883.

Iba, Y., Ito, W., Kurosawa, Y. (1997) Expression vectors for the introduction of higly diverged sequences into the six complementarity-determining regions of an antibody. *Gene* 194: 35–46.

Iliades, P., Dougan, D.A., Oddie, G.W., Metzger, D.W., Hudson, P.J., Kortt, A.A. (1998) Single-chain Fv of anti-idiotype 11-1G10 antibody interacts with antibody NC41 single-chain Fv with a higher affinity than the affinity for the interaction of the parent Fab fragments. *J Protein Chem* 17: 245–254.

Jacob, F., Monod, J. (1961) Genetic regulatory mechanism in the synthesis of proteins. *J Mol Biol* 3: 318–356.

Jakobovits, A. (1995) Production of fully human antibodies by transgenic mice. *Curr Opin Biotechnol* 6: 561–6.

Jespers, L.S., Messens, J.H., de Keyser, A., Eeckhout, D., van den Brande, I., Gansemans, Y.G., Lauwerey, M.J., Vlasuk, G.P., Stanssens, P.E. (1995) Surface expression and ligand based selection of cDNAs fused to filamentous phage gene VI. *Bio/Technology* 13: 378–381.

Jirholt, P., Ohlin, M., Borrebaeck, C.A.K., Söderlind, E. (1998) Exploiting sequences space: shuffling *in vivo* formed complementarity determining regions into a master framework. *Gene* 215: 471–476.

Johansen, L.K., Albrechtsen, B., Andersen, H.W., Engberg, J. (1995) pFab60: a new, efficient vector for expression of antibody Fab fragments displayed on phage. *Prot Engineering* 8: 1063–1067.

Jostock, T., Vanhove, M., Brepoels, E., Van Gool, R., Daukandt, M., Wehnert, A., Van Hegelsom, R., Dransfield, D., Sexton, D., Devlin, M., Ley, A., Hoogenboom, H., Mullberg, J. (2004) Rapid generation of functional human IgG antibodies derived from Fab-on-phage display libraries. *J Immunol Methods* 289: 65–80.

Jostock, T., Dübel, S. (2005) Screening of Molecular Repertoires by Microbial Surface Display. *Comb Chem High Throughput Screen* 8: 127–133.

Kang, A.S., Jones, T.M., Burton, D.R. (1991a) Antibody redesign by chain shuffling from random combinatorial immunoglobulin libraries. *Proc Natl Acad Sci* USA 88: 11120–11123.

Kang, A.S., Barbas, C.F., Janda, K.D., Bencovic, S.J., Lerner, R.A. (1991b) Linkage of recognition and replication functions by assembling combinatorial antibody Fab libraries along phage surfaces. *Proc Natl Acad Sci* USA 88: 4363–4366.

Kirsch, M., Zaman, M., Meier, D., Dübel, S., Hust, M. (2005) Parameters affecting the display of antibodies on phage. *J Immunol Meth* 301: 173–185.

Knappik, A., Ge, L., Honegger, A., Pack, P., Fischer, M., Wellnhofer, G., Hoess, A., Wölle, J., Plückthun, A., Virnekäs, B. (2000) Fully synthetic human combinatorial antibody libraries (HuCAL) based on modular consensus framework and CDRs randomized with trinucleotides. *J Mol Biol* 296: 57–86.

Köhler, G., Milstein, C. (1975) Continous cultures of fused cells secreting antibody of predefined specificity. *Nature* 256: 495–497.

Kobayashi, N. Söderlind, E., Borrebaeck, C. A.K. (1997) Analysis of assembly of synthetic antibody fragments: Expression of functional scFv with predifined specificity. *BioTechniques* 23: 500–503.

Kontermann, R., Dübel, S. (eds.) (2001) Antibody Engineering. Springer-Verlag, New York.

Konthur, Z., Hust, M., Dübel, S. (2005) Perspectives for systematic in vitro antibody generation. *Gene* 364: 19–29.

Kortt, A.A., Lah, M., Oddie, G.W., Gruen, L. C., Burns, J.E., Pearce, L.A., Atwell, J.L., McCoy, A.J., Howlett, G.J., Metzger, D.W., Webster, R.G., Hudson, P.J. (1997) Single chain Fv fragments of anti-neurominidase antibody NC10 containing five and ten residue linkers form dimers and with zero residue linker a trimer. *Prot Eng* 10: 423–428.

Kramer, K., Fiedler, M., Skerra, A., Hock, B. (2002) A generic strategy for subcloning antibody variable regions from the scFv phage display vector pCANTAB 5 E into pASK85 permits the economical production of Fab fragments and leads to improved recombinant immunoglobulin stability. *Biosensors & Bioelectronics* 17: 305–313.

Krebber, A., Bornhauser, S., Burmester, J., Honegger, A., Willuda, J., Bosshard, H.R., Plückthun, A. (1997) Reliable cloning of functional antibody variable domains from hybridomas and spleen cell repertoires employing a reengineered phage display system. *J Immunol Meth* 201: 35–55.

Lauer, B., Ottleben, I., Jacobsen, H.J., Reinard, T. (2005) Production of a single-chain variable fragment antibody against fumonisin B1. *J Agric Food Chem* 53: 899–904.

Le Gall, F., Kipriyanov, S.M., Moldenhauer, G., Little, M. (1999) Di-, tri- and tetrameric single chain Fv antibody fragments against human CD19: effect of valency on cell binding. *FEBS Letters* 453: 164–168.

Lei, S.-P., Lin, H.-C, Wang, S.-S., Callaway, J., Wilcox, G. (1987) Characterization of the *Erwinia caratovora pelB* gene and its product pectate lyase. *J Bacteriol* 169: 4379–4383.

Little, M., Welschof, M., Braunagel, M., Hermes, I., Christ, C., Keller, A., Rohrbach, P., Kürschner, T., Schmidt, S., Kleist, C., Terness, P. (1999) Generation of a large complex antibody library from multiple donors. *J Immunol Meth* 231: 3–9.

Lonberg, N., Huszar, D. (1995) Human antibodies from transgenic mice. *Int Rev Immunol* 13: 65–93.

Løset, G.Å., Løbersli, I., Kavlie, A., Stacy, J. E., Borgen, T., Kausmally, L., Hvattum, E., Simonsen, B., Hovda M.B., Brekke, O.H. (2005) Construction, evaluation and refinement of a large human antibody phage library based on the IgD and IgM variable gene repertoire. *J Immunol Meth* 299: 47–62.

Lowman, H.B., Bass, S.H., Simpson, N., Wells, J.A. (1991) Selecting High-Affinity binding proteins by monovalent phage display. *Biochemistry* 30: 10832–10838.

Marks, J.D., Hoogenboom, H.R., Bonnert, T. P., McCafferty, J., Griffiths, A.D., Winter, G. (1991) By-passing immunization: human antibodies from V-gene libraries diplayed on phage. *J Mol Biol* 222: 581–597.

Marks, J.D., Griffiths, A.D., Malmqvist, M., Clackson, T.P., Bye, J.M., Winter, G. (1992a) By-passing immunization: building high affinity human antibodies by chain shuffling. *Bio/Technology* 10: 779–783.

Marks, J.D., Hoogenboom, H.R., Griffiths, A.D., Winter, G. (1992b) Molecular evolution of proteins on filamentous phage. *J Biol Chem* 267: 16007–16010.

McCafferty, J., Griffiths, A.D., Winter, G., Chiswell, D.J. (1990) Phage antibodies: filamentous phage displaying antibody variable domain. *Nature* 348: 552–554.

McCafferty, J., Fitzgerald, K.J., Earnshaw, J., Chiswell, D.J., Link, J., Smith, R., Kenten, J. (1994) Selection and rapid purification of murine antibody fragments that bind a transition-state analog by phage-display. *Appl Biochem Biotech* 47: 157–173.

McCafferty, J., Hoogenboom, H.R., Chiswell, D.J. (eds.) (1996) Antibody Engineering. IRL Press, Oxford.

Moghaddam, A., Borgen, T., Stacy, J., Kausmally, L., Simonsen, B., Marvik, O.J., Brekke, O.H., Braunagel, M. (2003) Identification of scFv antibody fragments that specifically recognise the heroin metabolite 6-monoacetylmorphine but not morphine. *J Immunol Meth* 280: 139–155.

Mullinax, R.L., Gross, E.A., Amberg, J.R., Hay, B.N., Hogreffe, H.H., Kubitz, M.M., Greener, A., Alting-Mees, M., Ardourel, D., Short, J.M., Sorge, J.A., Shopes, B. (1990) Identification of human antibody fragment clones specific for tetanus toxoid in a bacteriophage λ immunoexpression library. *Proc Natl Acad Sci* USA 87: 8095–8099.

Nissim, A., Hoogenboom, H.R., Tomlinson, I.M., Flynn, G., Midgley, C., Lane, D., Winter, G. (1994) Antibody fragments from a "single pot" phage display library as immunochemical reagents. *EMBO J* 13: 692–698.

O'Connel, D., Becerril, B., Roy-Burman, A., Daws, M., Marks, J.D. (2002) Phage versus phagemid libraries for generation of human monoclonal antibodies. *J Mol Biol* 321: 49–56.

Ørum, H., Andersen, P.S., Øster, A., Johansen, L.K., Riise, E., Bjørnevad, M., Svendsen, I., Engberg, J. (1993) Efficient method for constructing comprehensive murine Fab antibody libraries displayed on phage. *Nucl Acids Res* 21: 4491–4498.

Parmley, S.F., Smith, G.P. (1988) Antibody selectable filamentous fd phage vectors: affinity purification of target genes. *Gene* 73: 305–318.

Paschke, M., Zahn, G., Warsinke, A., Höhne, W. (2001) New series of vectors for phage display and prokaryotic expression of proteins. *BioTechniques* 30: 720–726.

Persson, M.A.A., Caothien, R.H., Burton, D. R. (1991) Generation of diverse high-affinity human monoclonal antibodies by repertoire cloning. *Proc Natl Acad Sci* USA 88: 2432–2436.

Pini, A, Viti, F., Santucci, A., Carnemolla, B., Zardi, L., Neri, P., Neri, D. (1998) Design and use of a phage display library. *J Biol Chem* 273: 21769–21776.

Plückthun, A. (1990) Antibodies from *Escherichia Coli Nature* 347: 497–498.

Plückthun, A. (1991) Antibody Engineering: Advances from the use of *Escherichia coli* expression systems. *Bio/Technology* 9: 545–551.

Roberts, R.W., Szostak, J.W. (1997) RNA-peptide fusions for the in vitro selection of peptides and proteins. *Proc Natl Acad Sci* USA 94: 12297–12302.

Rojas, Gertrudis, Almagro, J.C., Acevedo, B., Gavilondo, J.V. (2002) Phage antibody fragments library combining a single human light chain variable region with immune mouse heavy chain variable regions. *J Biotech* 94: 287–298.

Rondot, S., Koch, J., Breitling, F., Dübel, S. (2001) A helperphage to improve single chain antibody presentation in phage display. *Nat Biotech* 19: 75–78.

Sblattero, D., Bradbury, A. (2000) Exploiting recombination in single bacteria to make large phage antibody libraries. *Nat Biotech* 18: 75–80.

Schirrmann, T., Pecher, G. (2002) Human natural killer cell line modified with a chimeric immunoglobulin T-cell receptor gene leads to tumor growth inhibition *in vivo. Cancer Gene Ther* 4: 390–398.

Schmiedl, A., Breitling, F., Dübel, S. (2000) Expression of a bispecific dsFv-dsFv′ antibody fragment in Escherichia coli. *Protein Engineering* 13: 725–734.

Sheets, M.D., Amersdorfer, P., Finnern, R., Sargent, P., Lindqvist, E., Schier, R., Hemingsen, G., Wong, C., Gerhart, J.C., Marks, J.D. (1998) Efficient construction of a large nonimmune phage antibody library: the production of high-affinity human single-chain antibodies to protein antigens. *Proc Natl Acad Sci* USA 95: 6157–6162.

Shirai, H., Kidera, A., Nakamura, H. (1999) H3-rules: identification of CDR3-H3 structures in antibodies. *FEBS Letters* 455: 188–197.

Simmons, L.C., Reilly, D., Klimowski, L., Raju, T.S., Meng, G., Sims, P., Hong, K., Shields, R.L., Damico, L.A., Rancatore, P., Yansura, D.G. (2002) Expression of full-length immunoglobulins in Escherichia coli: rapid and efficient production of aglycosylated antibodies. *J Immunol Meth* 263: 133–47.

Skerra, A., Plückthun, A. (1988) Assembly of a functional immunoglobulin Fv fragment in *Escherichia coli Science* 240: 1038–1041.

Skerra, A., Pfitzinger, I., Plückthun, A. (1993) The functional expression of antibody Fv fragments in *Escherichia coli*:

improved vectors and a generally applicable purification technique. *Bio/Technology* 9: 273–278.

Skerra, A., Schmidt, T.G.M. (1999) Applications of a peptide ligand for streptavidin: the Strep-tag. *Biomol Engineer* 16: 79–86.

Smith, G.P. (1985) Filamentous fusion phage: novel expression vectors that display cloned antigens on the virion surface. *Science* 228: 1315–1317.

Söderlind, E., Lagerkvist, A.C.S., Dueñas, M., Malmborg, A.-C., Ayala, M., Danielsson, L., Borrebaeck, C.A.K. (1993) Chaperonin assisted phage display of antibody fragments of filamentous bacteriophages. *Bio/Technology* 11: 503–507.

Söderlind, E., Strandberg, L., Jirholt, P., Kobayashi, N., Alexeiva, V., Aberg, A.-M., Nilsson, A., Jansson, B., Ohlin, M., Wingren, C., Danielsson, L., Carlsson, R., Borrebaeck, C.A.K. (2000) Recombining germline-derived CDR sequences for creating diverse single-framework antibody libraries. *Nat Biotech* 18: 852–856.

Spada, S., Plückthun, A. (1997) Selectivity infective phage (SIP) technology: A novel method for *in vivo* selection of interacting protein-ligand pairs. *Nature Medicine* 3: 694–696.

Studnicka, G.M., Soares, S., Better, M., Williams, R.E., Nadell, R., Horwitz, A.H. (1994) Human-engineered monoclonal antibodies retain full specific binding activity by preserving non-CDR complementarity-modulating residues. *Protein Eng* 6: 805–14.

Toleikis, L., Broders, O., Dübel, S. (2004) Cloning single-chain antibody fragments (scFv) from hybridoma cells. *Meth Mol Med* 94: 447–458.

Tsurushita, N., Fu, H., Warren, C. (1996) Phage display vectors for *in vivo* recombination of immunoglobulin heavy and light chain genes to make large combinatoríal libraries. *Gene* 172: 59–63.

Vaughan, T.J., Williams, A.J., Pritchard, K., Osbourn, J.K., Pope, A.R., Earnshaw, J.C., McCafferty, J., Hodits, R.A., Wilton, J., Johnson, K.S. (1996) Human antibodies with sub-nanomolar affinities isolated from a large non-immunized phage display library. *Nat Biotech* 14: 309–314.

Vieira, J., Messing, J. (1987) Production of single-stranded plasmid DNA. *Methods Enzymol* 153: 3–11.

Viti, F., Nilsson, V., Demartis, S., Huber, A., Neri, D. (2000) Design and use of phage display libraries for the selection of antibodies and enzymes. *Meth Enzymol* 326: 480–497.

Von Behring, E., Kitasato, S. (1890) Über das Zustandekommen der Diphtherie-Immunität und der Tetanus-Immunität bei Thieren. *Deutsche Medizinische Wochenzeitschrift* 16: 1113–1114.

Ward, E.S. (1993) Antibody engineering using *Escherichia coli* as host. *Adv Pharmacol* 24: 1–20.

Waterhouse, P., Griffiths, A.D., Johnson, K.S., Winter, G. (1993) Combinatorial infection and *in vivo* recombination: a stratagie for making large phage antibody repertoires. *Nucl Acids Res* 21: 2265–2266.

Welschof, M., Terness, P., Kipriyanov, S., Stanescu, D., Breitling, F., Dörsam, H., Dübel, S., Little, M., Opelz, G. (1997) The antigen binding domain of a human IgG-anti-F(ab')2 autoantibody. *Proc Natl Acad Sci* USA 94: 1902–1907.

Winter, G., Milstein, C. (1991) Man-made antibodies. *Nature* 349: 293–299.

Zahn, G., Skerra, A., Höhne, W. (1999) Investigation of a tetracycline-regulated phage display system. *Protein Eng* 12: 1031–1034.

4
Selection Strategies III: Transgenic Mice

Marianne Brüggemann, Jennifer A. Smith, Michael J. Osborn, and Xiangang Zou

4.1
Introduction

Antibodies of diverse specificity are produced in all mammals. Indeed, all jawed vertebrates contain the genes to allow combinatorial immune responses and can produce specific antibodies following immunization. The primary repertoire is generated after V (variable) D (diversity) J (joining) gene rearrangement of the Ig (immunoglobulin) heavy (H) chain locus and VJ joining of a κ or λ light (L) chain locus. This usually creates a limited diversity of low-affinity binders. In a second wave of diversification, initiated by repeated antigen encounter, low-affinity antibodies can be edited by undergoing somatic hypermutation. *In vivo* strategies using transgenic mice have been successful in obtaining high-affinity human antibodies while *in vitro* methods have allowed the selection of single-domain binders by library display technology (Neuberger and Brüggemann 1997).

Here we present a broad review of the methodologies used to express fully human antibody repertoires in mice. Transgenic animals carrying human Ig H and L chain loci on YACs or chromosome fragments have been derived from manipulated embryonic stem cells, oocytes, and fibroblasts (Fig. 4.1). This allows the expression of diverse repertoires, somatic mutation, and class switching in a background with silenced endogenous gene loci. In addition, alternative technologies are emerging that utilize the expression of specific single-chain antibodies and iterative affinity maturation of rearranged Ig genes transfected into cell lines.

4.2
Human Ig Genes and Loci

The human Ig loci (IgH, Igκ, and Igλ) have been cloned and all the genes have been sequenced (summarized in Lefranc and Lefranc 2001). The human IgH

Handbook of Therapeutic Antibodies. Edited by Stefan Dübel

ISBN 978-3-527-31453-9

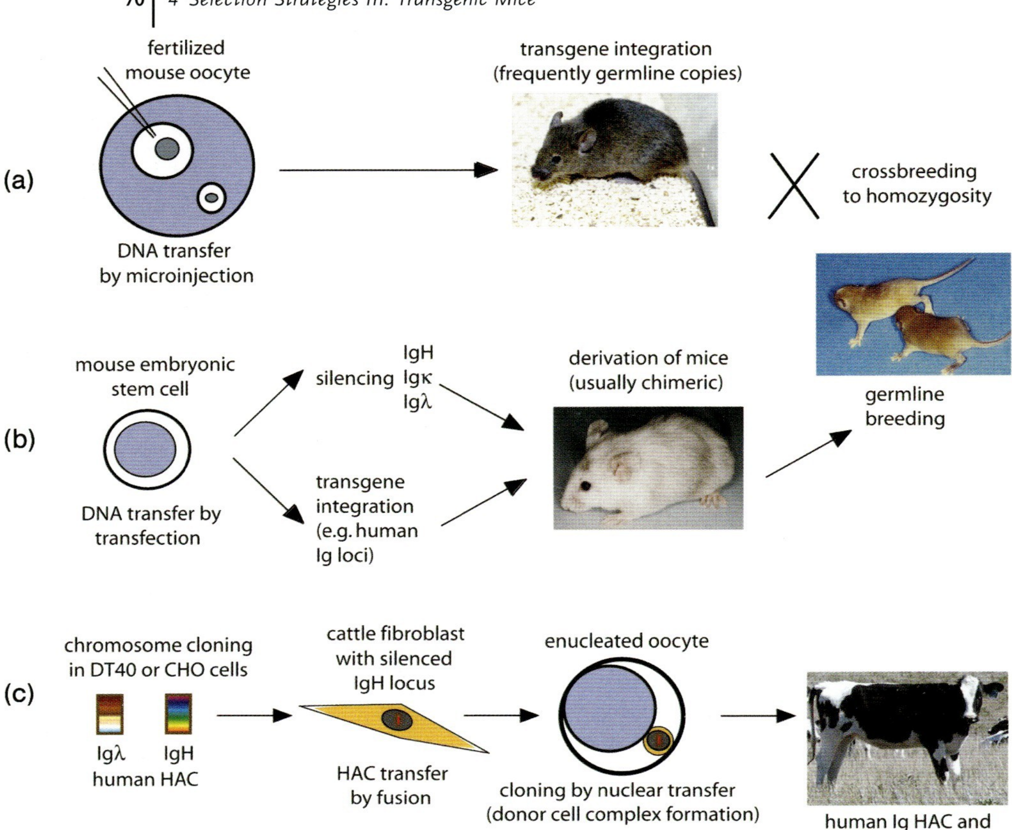

Fig. 4.1 Derivation of new mouse lines using (a) DNA microinjection into oocytes, (b) gene targeting in embryonic stem (ES) cells, and (c) nuclear transfer of manipulated fibroblasts. DNA microinjection into the male pronucleus of fertilized mouse oocytes is used for the production of transgenic animals. ES cells can be manipulated by transfection (e.g. DNA electroporation or protoplast fusion using YAC-containing yeast cells), which facilitates random integration and, using homology constructs, site-specific integration. From manipulated ES cells chimeric mice can be produced and further breeding may establish germline transmission and homozygous animals. A chromosome cloning system using DT40 or CHO host cells was used for the integration of human chromosome fragments or artificial chromosomes (HACs) containing the IgH and IgL loci. Human HACs were transferred into cattle fetal fibroblast cells by fusion. In addition gene targeting of the IgH locus has been achieved in fibroblasts. Animals expressing human Ig were produced from the manipulated fibroblasts by nuclear transfer. Extensive cross-breeding allowed the expression of, for example, human antibody H- and L-chains in a knockout background where equivalent endogenous genes are non-functional (Brüggemann et al. 1989; Taylor et al. 1994; Mendez et al. 1997; Nicholson et al. 1999; Kuroiwa et al. 2002, 2004).

locus is about 1.3 Mb in size with 38–46 functional Vs, 23 Ds, 6 Js, and 9 C region genes (Hofker et al. 1989; Cook et al. 1994; Corbett et al. 1997; Matsuda et al. 1998). The human κ L chain locus is accommodated on a ~1.8 Mb region with 17–19 functional Vs located at the proximal cluster, separated by about 800 kb from the distal V_κ cluster with 15–17 functional genes, followed by 5 Js and one C_κ (Roschenthaler et al. 2000; Zachau 2000). The human Igλ locus is just over 1 Mb in size with 29–33 functional Vs and 4–5 J-C genes (Frippiat et al. 1995; Kawasaki et al. 1995). The V_λ families form group clusters while members of the different V_H and V_κ gene families are interspersed. Members of the IgH as well as Igλ chain gene segments are assembled in the same transcriptional orientation, which permits conventional DNA rearrangement and deletion, whilst V_κ genes are organized in both transcriptional orientations, which allows deletional and inversional joining (Frippiat et al. 1995; Weichhold et al. 1990). In translocus mice (animals bearing an introduced [Ig] locus), rearrangement and expression of human V_κ genes assembled in either transcriptional orientation has been achieved (Xian et al. 1998). The layout of the human Ig loci and the established transloci with their gene content, using plasmids, YACs, and chromosome fragments, is illustrated in Fig. 4.2 and Table 4.1.

4.2.1
Minigene Constructs

Phenotypic modification of mice by gene transfer of microinjected DNA into fertilized oocytes was established in the early 1980s (Gordon and Ruddle 1983). The use of embryonic stem (ES) cells (Evans 1989) for gene targeting followed a few years later (Capecchi 1989) and more recent approaches allowed the removal or introduction of quite large gene loci in mice (Davies et al. 1993; Ren et al. 2004). The early experiments demonstrated that exogenous genes could be expressed but it had to be determined how efficiently they would interact with endogenous polypeptides. For example, could human Ig genes introduced into mice rearrange and interact with the cellular signaling machinery to permit B-cell receptor assembly and antigen-induced differentiation events?

The size limit for the different IgH and IgL minigene constructs (15–180 kb) restricted both the number of genes and the distances between gene segments or exons that could be included. For this reason human V, (D), J, and C gene segments in germline configuration were placed in artificially close proximity. The number of V_H and V_κ genes was between one and five and the number of D segments for the IgH chain was between 3 and 15 (Brüggemann et al. 1989, 1991; Taylor et al. 1992, 1994; Lonberg et al. 1994; Xian et al. 1998). In the initial IgH gene constructs a ~15 kb region including DQ52, J_H, Eμ, switch μ and C_μ was largely retained, while for an Igκ construct a V_κ was added to the J_κ–C_κ cluster (Fig. 4.2). The advantage of maintaining the region between J and C was that functional activity of the intronic enhancer, important for transcriptional activation after DNA rearrangement, could be preserved. Thus, with rather small constructs, DNA rearrangement and expression was obtained whilst the addition of

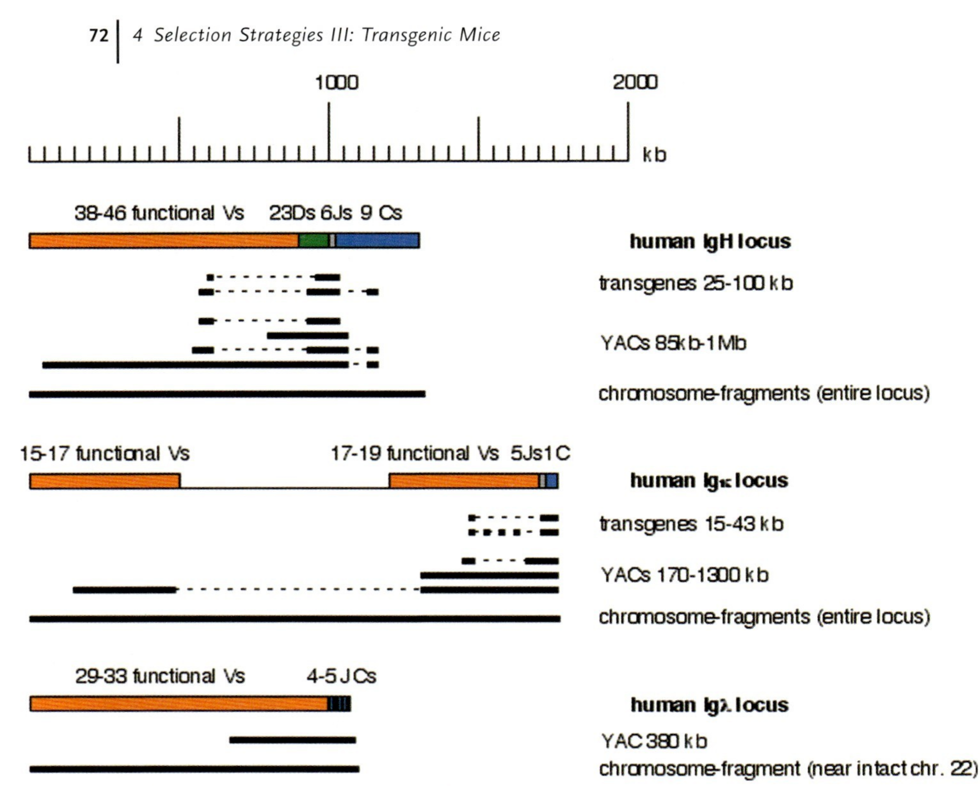

Fig. 4.2 Layout of the introduced human Ig loci expressed in mice. Linear constructs were obtained by gene assembly with size and gene content indicated. Human IgH, Igκ and Igλ loci in germline configuration are illustrated above (Ig loci summarized in Lefranc and Lefranc 2001; Ig transgenes summarized in Brüggemann 2004).

another C_H gene showed that switching from C_μ to $C_\gamma1$ was possible (Taylor et al. 1992).

Transgenic animals, produced by microinjection, sometimes contain many (at times several hundred) identical copies of a human Ig construct resulting in higher levels of expression. Also, tandem integration events can occur when two or more constructs are injected together. For example, this has resulted in head to tail integration of two cosmids creating a ~100 kb IgH locus which rearranged and expressed a combination of segments from both cosmids (Brüggemann et al. 1991; Wagner et al. 1994a,b). A further increase in size was achieved by the use of the P1 cloning system and microinjection of three overlapping regions of ~80 kb each (Wagner et al. 1996). Following homologous recombination with each other, ~180 kb of the human IgH locus was reconstituted. This comprised five V_H genes, all D and J segments and C_μ and C_δ, containing the core region of the

Table 4.1 V and C gene content of mice carrying human IgH, Igκ and Igλ transloci.

	V genes	*C genes*	*Reference*
H chain			
HuIgH	2	μ	Brüggemann et al. 1989; Wagner et al. 1994a,b
HuIgH^{cos}	2	μ	Brüggemann et al. 1991; Wagner et al. 1994a,b
HC1	1	μ, γ1	Taylor et al. 1992
HC2	4	μ, γ1	Lonberg et al. 1994; Taylor et al. 1994
J1-3	2	μ, δ	Choi et al. 1993
HuIgH^{pl-2}	5	μ, *d	Wagner et al. 1996
HuIgH	5	μ, δ	Nicholson et al. 1999; Mundt et al. 2001
yH1	5	μ, δ	Green et al. 1994
yH2	~40	μ, δ, γ2 or γ1 or γ4	Mendez et al. 1997; Davis et al. 1999; Green 1999
hCF(SC20)	~40 (whole locus)	μ, δ, γs, ε, αs	Tomizuka et al. 2000; Tomizuka et al. 1997
κ L chain			
KC1	1	1 κ	Taylor et al. 1992
KCo4	4	1 κ	Lonberg et al. 1994
HuIgHκML	5	1 κ	Xian et al. 1998
yK1	2	1 κ	Green et al. 1994
HuIgκYAC	2	1 κ	Davies et al. 1993; Xian et al. 1998
HucosIgκYAC	~80	1 κ	Xian et al. 1998; Zou et al. 1996
KCo5	~26	1 κ	Fishwild et al. 1996
yK2	~25	1 κ	Mendez et al. 1997
hCF(2-W23)	~50 (whole locus)	1 κ	Tomizuka et al. 2000; Tomizuka et al. 1997
λ L chain			
Igl	15	7 Jλ (incl. 3 ψλs)	Popov et al. 1999
hCF[MH(ES)22-1]	~30 (whole locus)	4–5 Jλ	Tomizuka et al. 1997

locus required for DNA rearrangement and expression. For the Igκ locus, co-injection of two minigene constructs and homologous recombination between V genes achieved integration of a contiguous 43 kb translocus (Lonberg et al. 1994).

These early human Ig constructs, some with tightly assembled exons and control regions, established that human Ig gene segments in germline configuration could be rearranged and expressed in mouse lymphocytes. Additional mouse sequences, initially assumed to be beneficial to drive transgene expression, appeared to be unnecessary if equivalent human sequences, such as enhancer regions, were included. However, small constructs carrying few gene segments express relatively poorly and fully human antibody repertoires have not been obtained in mice that still rearrange and express their endogenous Ig genes.

4.2.2
Yeast Artificial Chromosomes (YACs)

In yeast, YACs can be used as cloning vehicles to accommodate large loci. In addition, YACs can be easily modified by site-specific recombination (Anand 1992; Davies et al. 1992). To create antibody repertoires comparable to the diversity obtained in humans, a large number of genes and perhaps the transfer of whole Ig loci may be necessary. For this reason much attention has been focused on using YACs to accommodate large genomic regions to which individual V genes of the different families, either on minigene constructs (for example, obtained by PCR) or in authentic configuration, could be added (Lonberg et al. 1994; Fishwild et al. 1996; Popov et al. 1996, 1999; Mendez et al. 1997). Defined genomic regions, ranging from a few hundred kilobases to well over 1 Mb, can be assembled from overlapping BACs (bacterial artificial chromosomes) or cosmids, or from direct cloning of human genomic DNA into YAC libraries. The cloned DNA can be modified by targeted retrofitting or sequence removal, and also can be easily extended by mating of yeast clones carrying overlapping YACs (Markie 1996). The core regions of the human IgH locus and the Igκ locus have been cloned separately on YACs of up to ~300 kb (reviewed in Brüggemann and Neuberger 1996). Impressive V gene additions by stepwise recombination have been made to these YACs (Zou et al. 1996; Mendez et al. 1997). This has resulted in a ~1 Mb human IgH YAC with ~66 V_H genes, Ds, J_Hs, C_μ, and C_δ, and a ~800 kb Igκ YAC with ~32 V_κ genes, J_κs, and C_κ all in authentic configuration (Mendez et al. 1997). The efficiency of the yeast host in homologous integration has also been exploited by extending a 300 kb Igκ YAC (Davies and Brüggemann 1993) to 1.3 Mb by multiple integration of a 50 kb cosmid with five V_κ genes (Zou et al. 1996). As no YAC containing the core region of the human Igλ locus (V_λ–JC_λs) was available, this had to be constructed from overlapping cosmids. Three cosmids with their 5′ and 3′ regions ligated to YAC vector arm sequences were co-transfected into yeast and YACs with correctly reconstituted Igλ core region were identified by Southern hybridization (Popov et al. 1996). Further extension was achieved by yeast mating, which allowed homologous recombination of the overlapping region and resulted in Igλ YACs with an authentic 380 kb region accommodating 15 functional V_λ genes (Popov et al. 1999). Therefore manipulation of YACs using stepwise recombination has been a valuable technique in creating human Ig loci with large or near authentic regions over 1 Mb in size.

Several strategies have been used successfully for YAC introduction into the mouse germline: DNA purification and microinjection into fertilized oocytes (Fishwild et al. 1996); co-lipofection of ES cells with a mixture of size-fractionated YAC DNA in agarose and a selectable marker gene (Choi et al. 1993); and fusion of yeast protoplasts with ES cells (Davies et al. 1993, 1996). Despite establishing very efficient methods for YAC purification and microinjection (Schedl et al. 1993) large DNA molecules are difficult to integrate as one complete copy of a human Ig translocus (Taylor et al. 1992). Improvements allowing complex and large (human Ig) loci in their intact form to be transferred into the mouse genome came with the use and manipulation of ES cells (Hogan et al. 1994). Lipid-mediated DNA transfer (lipofection) of ES cells avoids the handling of naked DNA and if co-transfection is applied, removes the requirement to retrofit a selectable marker into the YAC or BAC (Choi et al. 1993). However, a recurring problem is that frequently only a portion of the introduced region is integrated into the host genome. To overcome this difficulty, protoplast fusion previously used for YAC integration into differentiated mammalian cells (Pachnis et al. 1990; Pavan et al. 1990) has been adapted for YAC transfer into ES cells (Davies et al. 1993, 1996; Zou et al. 1996). The preparation of YAC-containing yeast spheroplasts does not involve DNA handling or gel separation, but requires the YAC to be retrofitted with a selectable marker gene. Fusion of yeast protoplasts with ES cells is similar to the method employed for the generation of hybridomas, but usually only a few clones are obtained. Nevertheless, the approach achieved a reliable integration of complete single-copy Ig YACs into the mouse genome (Mendez et al. 1997; Nicholson et al. 1999).

The expression of multiple V_H and V_L transgenes in mice allowed the production of different combinations of V region pairs, generating a diverse antibody spectrum. Specific antibodies were obtained by hybridoma and/or PCR technologies which, upon re-expression, allowed bulk production of monoclonal human antibodies (reviewed in Maynard and Georgiou 2000). Initially, less emphasis was put on the addition of C_H region genes to permit class switching. The reason was that even small constructs, with a limited number of Vs, Ds, and Js, rearranged and expressed as μ H-chain which permitted surface IgM or B-cell receptor expression. This was followed by normal differentiation events to produce secreted Ig and diversification of V_H gene sequences by hypermutation (Wagner et al. 1996). Some larger IgH loci, identified from YAC libraries, included C_δ, downstream of C_μ. However, these rarely produced specific IgD antibodies (Choi et al. 1993; Green et al. 1994; Wagner et al. 1996; Brüggemann and Neuberger 1996). Further efforts were made to include one C_γ gene (Mendez et al. 1997; Davis et al. 1999; Green 1999) to allow isotype switching. Differentiation events associated with IgG expression allow a cell to enter the recirculating B-cell pool and become established as a long-lived memory B cell accumulating high levels of somatic mutation. Engineering mouse lines that express particular C_γ genes provides antibodies with tailor-made effector functions. For example, strong (IgG1 and IgG3) or weak (IgG2 and IgG4) binding to the Fc receptors ($Fc_\gamma R$ I, II, and III) could be useful to permit or avoid, respectively, interaction with macrophages or natural killer (NK) cells. Similarly, IgG1 and IgG3 antibodies have the ability to fix complement and can initiate extensive hemolytic activity which, in context, may or may not be a useful attribute. To produce defined isotype repertoires, such as IgG1 for the destruction of target cells and IgG2 or IgG4 to block or compete for binding without initiating cell lysis, the respective C_γ gene was placed immediately 3′ of C_δ (Mendez et al. 1997; Davis et al. 1999). Although switching from μ to γ was achieved, such close proximity of two C genes disregards the importance of many intervening regions exerting important control functions critical in securing high expression levels. In the human IgH locus the ~50 kb C_δ–$C_\gamma 3$

interval is such a region, rich in transcription factor-binding motifs and with lymphocyte-specific enhancer activity (Mundt et al. 2001). It would be interesting to compare antibody titers and gene usage in mice carrying either closely assembled C genes or C genes in authentic configuration. As yet, no parallel immunizations with the same antigen have been carried out using the different human Ig mouse lines.

To secure good expression levels the IgH and IgL chain YACs contain at least one but usually two transcriptional enhancers. For the (human) IgH locus the activity of four enhancers (E) has been described: EDQ52, Eμ, Eδ–γ3, and E3′α in a multiple E site region (reviewed in Magor et al. 1999; Mundt et al. 2001; Arulampalam et al. 1997 and references therein). Human Eμ is present on all IgH transloci and as a second control region the rat or mouse enhancer downstream of the last C gene, E3′α (also termed HS1,2 to identify the precise location), has been added 3′ of C_{γ} (Taylor et al. 1994; Mendez et al. 1997). In the IgL loci there are transcriptional enhancers proximal to both C_{κ} and C_{λ}, making it unnecessary to add further enhancers to these YACs. Human, mouse, and rat enhancer sequences have all been used to drive human IgH and IgL chain transcription but there is little information about their requirement. For example, with the H chain the inclusion of additional enhancers in the presence of Eμ may be unnecessary because similar expression levels have been obtained in mice with and without a second enhancer (Brüggemann and Neuberger 1996).

4.2.3 Chromosome Fragments

Human chromosome fragments or human artificial chromosomes (HACs) can be maintained in the cell as distinct minichromosomes, which have the advantage that they provide whole loci when expressed in the mouse. In contrast, minigene constructs or YACs are integrated into a host chromosome and do not usually provide the full gene content. Individual chromosomes or their fragments, tagged with a selectable marker gene (neomycin, hygromycin, or puromycin) introduced by transfection, have been transferred by microcell-mediated fusion of human fibroblasts with somatic cell lines such as the mouse fibroblast line A9, the Chinese hamster ovary cell line CHO and the avian leukosis virus (ALV)-induced chicken tumor line DT40 (Fournier and Ruddle 1977; Koi et al. 1989; Shinohara et al. 2000). Hybrid libraries with different chromosomal regions, maintained under selection, include the human IgH, Igκ, and Igλ locus and adjacent loci identified by chromosomal marker analysis. Microcell-mediated fusion into ES cells allowed the derivation of chimeric mice (Tomizuka et al. 1997). Due to stability problems, maintenance of the complete transferred chromosome varies extensively, ranging from only a few per cent of the original sequence to nearly an intact transchromosomal region (Tomizuka et al. 1997).

Transchromosome stability and germline transmission were first obtained with a human chromosome 2-derived fragment containing the Igκ locus (Tomizuka et al. 1997). Germline transmission was essential to achieve homozygous breeding into Ig knockout lines. This was accomplished for the IgH locus with a fragment of human chromosome 14 and for Igκ using a minichromosome (Tomizuka et al. 2000).

Further efforts have concentrated on site-directed chromosome truncation to overcome instability (somatic mosaicism) of HACs, which would prevent germline transmission (Kuroiwa et al. 1998, 2000; Kakeda et al. 2005). In addition, attempts have been made to perform recombination of two nonhomologous chromosomal fragments to increase the size of the HACs. Kuroiwa and coworkers achieved both these aims by modifying the unstable human Igλ locus from chromosome 22 by targeted insertion of a new telomeric region and *loxP* sequences either side of the locus. Recombination-proficient cells (DT40) containing this altered chromosome fragment were fused with cells carrying a stable HAC accommodating the human IgH locus and a *loxP*-modified RNR2 locus. After transfection with Cre, an HAC of ~10 Mb containing both the Igλ and IgH loci was produced (Kuroiwa et al. 2000). Integration of the HACs into ES cells established mouse lines expressing human IgH and IgL chain loci (Kuroiwa et al. 2000). Transchromosomic calves expressing Ig from human IgH and Igλ loci have been produced using a similarly generated HAC transfected into bovine fetal fibroblasts followed by transfer into bovine oocytes (Kuroiwa et al. 2002).

4.3 Transgenic Ig Strains

Experiments to derive human antibody repertoires in transgenic mice started almost 20 years ago (Brüggemann et al. 1989). From the early approaches it became clear that introduced human Ig genes in germline configuration could undergo rearrangement and be expressed in the mouse. Although human antibodies were clearly detectable in mouse serum their concentration was quite low compared with the level of endogenous Ig (summarized in Brüggemann and Neuberger 1996). The reason for this was that only a few per cent of transgenic mouse B cells were positive for human IgH, the majority of lymphocytes expressing mouse Ig. Despite this drawback human IgM titers of up to 100 $\mu g\, mL^{-1}$ have been achieved in some transgenic lines (Brüggemann et al. 1989; Brüggemann and Neuberger 1991). Comparing expression levels achieved using Ig (mini)loci suggested that larger regions may favor better expression. Furthermore, L chain loci may be more efficiently expressed than H chain loci, perhaps because of their more compact layout with the presence of enhancers in the natural configuration.

Significant improvements in the levels of human Ig expression in the mouse were achieved by the use of gene targeting technology and the derivation of knockout animals (Capecchi 1989). Silencing of the mouse Ig loci, first accomplished by Rajewsky and coworkers for the H chain locus and later for the κ locus (Kitamura et al. 1991; Zou et al. 1993b), proved invaluable to secure human antibody expression without mouse H and κ L chain interference. The currently used mouse strains that express fully human antibody repertoires have been produced by the integration of human H and L chain (κ and/or λ) YACs and/or HACs, and crossing with animals in which the endogenous H and κ L chain loci have been silenced by gene targeting (Lonberg et al. 1994; Mendez et al. 1997; Nicholson et al. 1999; Tomizuka et al. 2000). Recently the mouse Igλ locus has been silenced (Zou et al. 2003) and suitable breeding could secure a mouse strain entirely free of any endogenous Ig production. Figure 4.2 summarizes the layout of the various human IgH, Igκ, and Igλ transloci. Websites illustrating the generation and use of these mice include http://www.babraham.ac.uk, http://www.abgenix.com, http://www.medarex.com, and http://www.tcmouse.com.

4.3.1 Stability of the Transloci

Miniloci and YACs have been transferred by both microinjection into oocytes, and ES cell technology. The former produces germline mice, but the latter generates chimeric animals, which require further breeding to establish heterozygous and homozygous mouse strains. Homozygosity for a combination of five features (human IgH, human Igκ, human Igλ, mouse IgH knockout, mouse Igκ knockout) has been readily obtained, indicating that these loci largely integrate in random fashion and not at preferred sites (Nicholson et al. 1999). No reports show the actual chromosomal integration sites of minigene constructs or YACs, while the maintenance of introduced HACs, as separate single units, has been well documented (Tomizuka et al. 2000).

Both YACs and HACs have advantages and disadvantages when used for the integration and expression of large human loci in the mouse. The advantage of transferring YACs is that integration into a mouse chromosome secures perfect stability and transmission, and in addition the gene content of the YAC is essentially known from sequence analysis. A disadvantage is that current YACs are hardly larger than 1 Mb and may not accommodate complete Ig loci. The generation of HACs offers the advantage of allowing the transfer of defined large regions. However, a drawback of using HACs is their somatic mosaicism, resulting in unpredictable transmission rates: HACs can be easily maintained in ES cells under selection but their germline transmission and maintenance in somatic cells appears to be significantly reduced (Shinohara et al. 2000). This means that every resulting mouse has to be analyzed for the level of HAC-positive B cells. Breeding analyses revealed a variable transmission efficiency that reached an impressive 38% for one particular HAC compared with the ideal 50% transmission rate of conventional genes in heterozygous configuration. One reason for HAC instability appears to be imprecise separation at mitosis due to poor centromere function (Shen et al. 1997).

4.3.2
Silenced Endogenous Loci

Establishing extensive fully human antibody repertoires has not been possible in the normal mouse background. The two major reasons are a generally low-level expression of human Ig and the presence of mixed molecules, such as human H chain associated with mouse L chain. Both problems have been overcome by breeding human IgH and IgL mice with animals in which the endogenous Ig genes had been silenced by gene targeting. In μMT mice (Kitamura et al. 1991) the μ transmembrane exons were disabled by the introduction of a selectable marker gene, which impeded B-cell development in the C57BL/6 background. The μMT knockout strain has been used to allow the production of authentic human Igs (Nicholson et al. 1999) although when breeding this particular knockout feature into other mouse backgrounds (e.g. BALB/c), endogenous IgG is still produced (Hasan et al. 2002; Orinska et al. 2002). The endogenous mouse IgH locus can interfere with human antibody production by trans-switching or trans-splicing events, allowing the expression of human–mouse chimeric heavy chains (Wagner et al. 1994b; Knight et al. 1995; Brüggemann and Taussig 1997). In the other extensively used IgH knockout strain, the J_H segments have been removed by gene targeting (Chen et al. 1993b; Jakobovits et al. 1993; Lonberg et al. 1994). The advantage of the J_H deficiency is that DNA rearrangement of the endogenous H chain locus is prohibited. An alternative approach, to prevent any undesired usage, has been the deletion of all constant region genes from the IgH locus (200 kb in total) (Ren et al. 2004).

Most attempts at silencing mouse L chain production have focused on the Igκ locus. These have resulted in knockout strains without J_κs and C_κ (Chen et al. 1993a) or a removed or disrupted Cκ (Zou et al. 1993b, 1995; Sanchez et al. 1994). Although κ L chain Ig is predominantly expressed in the mouse, silencing of the κ locus leads to a much-increased production of Igλ. Silencing of the λ L chain locus has been achieved by Cre-*loxP*-mediated removal of all J_λs and C_λs in a ~120 kb region (Zou et al. 2003). Complete ablation of IgL by breeding of κ and λ knockout mice resulted in a block in B cell development which can be overcome by expression of human L chain loci.

4.3.3
Immune Responses and Affinity of Human Ig

The introduced human IgH and IgL (κ and λ) loci are well expressed in a background where the endogenous IgH and Igκ loci have been silenced by targeted gene removal or disruption. Whilst human IgH and IgL chains can be expressed from minigene constructs containing one or a few of each of the V, (D), J, and C genes and one C proximal enhancer, better expression and repertoire formation is achieved from integration of larger Ig regions. Human antibody expression in the normal mouse background can reach up to $50\,\mu g\,mL^{-1}$ for Ig with human H chain and a somewhat higher level, $15–400\,\mu g\,mL^{-1}$ for Ig with human L chains.

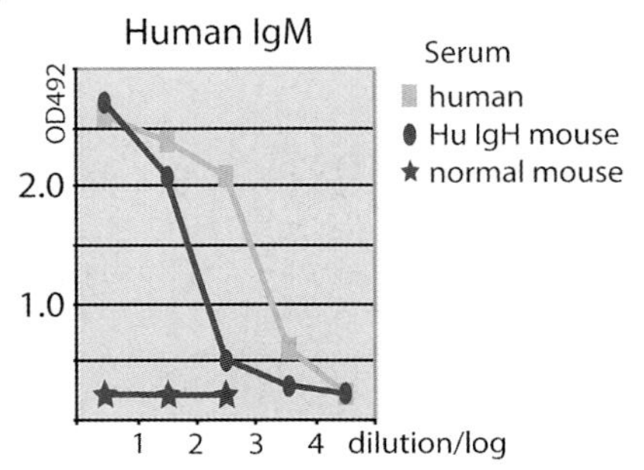

Fig. 4.3 Serum IgM in mice carrying a human H-chain locus in a background where the endogenous mouse IgH locus had been silenced by gene targeting. The titration was carried out in an ELISA assay with the level of human IgM produced in translocus mice >100 μg mL^{-1}.

However, as mouse Ig in the mg mL^{-1} range dominates expression, transgenic human Ig H and L chains are rarely associated with each other when forming antibodies. In the knockout background the level of human Ig reaches in many cases a few hundred μg mL^{-1} (Fig. 4.3), which suggests that expression levels are determined by the transferred Ig core region. Further improvements in Ig expression can be seen in YAC-based mouse strains where human antibody levels can exceed 1000 μg mL^{-1}, suggesting that larger Ig transloci with more gene segments are potentially better expressed. Disappointingly, Ig expression in the transchromosomal mice carrying complete H and L chain loci (Tomizuka et al. 1997 and 2000) is not increased further but is only about half the level found in IgH and Igλ YAC mice (Mendez et al. 1997; Popov et al. 1999). A plausible reason is that maintenance of the transferred HAC as a separate human chromosome in the mouse cells may interfere with recognition by the cellular machinery affecting chromatin structure and locus accessibility (Jenuwein and Allis, 2001). In all the transgenic Ig strains a large proportion of the "human" antibodies contain mouse λ L chain and thus are in chimeric configuration (Fishwild et al. 1996; Mendez et al. 1997; Nicholson et al. 1999; Magadán et al. 2002). Although the mouse Igλ locus has been silenced by gene targeting (Zou et al. 2003), human Ig mice with a background in which all three endogenous mouse Ig loci (IgH, Igκ and Igλ) have been rendered non-functional, and are thus unable to express any endogenous mouse Ig, have not yet been established by cross-breeding.

The high avidity but generally low affinity of IgM can be advantageous for some applications (Okada and Okada 1999), but other isotypes are also desirable. Early experiments adding a γ1 C gene to allow switching from IgM proved successful and established genomic recombination between the transgene μ and γ1 switch regions (Taylor et al. 1992, 1994). The human IgM and IgG1 concentration in serum improved significantly when the animals were crossed with endogenous H and κ L chain knockout mice. Also, the addition of more V_H and V_κ genes increased human antibody levels (Lonberg et al. 1994; Taylor et al. 1994). The most notable advantages of an introduced human IgH locus on a 1 Mb YAC have been described by Mendez *et al.* (1997): good expression, extensive V_H gene diversity (including hypermutation) and switching to the desired isotype. However, although switching from C_μ to the added C_γ is achieved, the serum levels of human IgG are considerably lower than those of human IgM and do not represent the ratio or levels (1–2 mg mL^{-1} IgM and ~10 mg mL^{-1} IgG) found in human serum (Frazer and Capra 1999).

B-cell development in mice, carrying a human IgH, human Igκ, and human Igλ translocus in a background where the function of the endogenous IgH and Igκ loci have been disabled, is illustrated in Fig. 4.4. Flow cytometry analysis of these five-feature mice shows an up to two-thirds B cell recovery rate with, for example, the number of Ig^+ splenic lymphocytes being ~20% in these human Ig mice compared with ~30% found in normal mice kept under the same conditions (Nicholson et al. 1999; Brüggemann, 2004). With the availability of an introduced Igκ and Igλ locus Ig^+ lymphocytes predominantly express human λ L chain. This is linked to a more efficient B cell recovery and much reduced expression of endogenous λ L chain. Although human Igκ antibodies can be easily obtained from five-feature mice, it may be an advantage to use individual four-feature strains, which either express human IgH,λ or human IgH,κ antibodies, to select the type of response (Lonberg et al. 1994; Mendez et al. 1997; Nicholson et al. 1999; Magadán et al. 2002; Protopapadakis et al. 2005).

The therapeutic applications of antibodies necessitate that a diverse range of specificities can be easily obtained (e.g. high-affinity monoclonal antibodies after immunization) and that the immunogenicity of such antibodies is abolished or at least reduced (Waldmann and Cobbold 1993; Klingbeil and Hsu 1999). Expression of human H and L chain loci in an animal not expressing its own Ig genes and which can be immunized meets these demands. Immunizations of mice expressing human antibodies have been carried out in the same way as described for conventional mouse strains. However, evaluation of immune responses

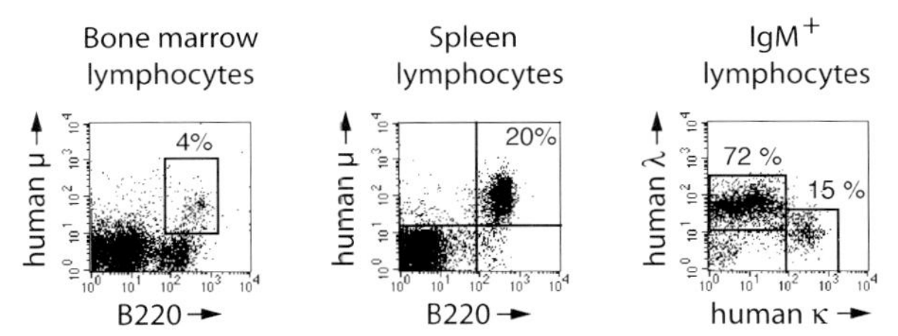

Fig. 4.4 Flow cytometry analysis of five-feature mice (carrying three human transloci, IgH, Igκ, and Igλ and two disabled endogenous loci mouse IgH and Igκ). The analysis shows a 1/3 to 2/3 B-cell recovery rate with, for example, the number of Ig^+ splenic lymphocytes being ~20% in the human Ig mice compared with ~30% found in normal mice kept under the same conditions (Nicholson et al. 1999; Brüggemann 2004).

showed that human antibody titers are reduced compared with those of normal nonmanipulated animals (Lonberg et al. 1994; Wagner et al. 1994a,b; Jakobovitz et al. 1995; Magadán et al. 2002). Despite low-serum titers, human Ig loci mice are capable of mounting antibody responses to a wide range of antigens and, similar to normal mice, increased levels of specific antibodies are visible 2–3 weeks after primary immunization and can be further increased by secondary immunization. Some antigen-specific antibodies were fully human (human H and L chain), while others were chimeric (consisting of human/mouse H/L chain combinations or a mixed H chain obtained by trans-switching).

The formation of mixed molecules or polypeptides led to disappointment particularly when some immunizations produced a substantial number of chimeric human antibodies with mouse λ L chain (Russell et al. 2000; Magadán et al. 2002). Nevertheless antigen-specific fully human antibodies were produced and the use of hybridoma technology (for background and methods see King 1998) established a large number of human monoclonal antibodies (mAbs) of good affinity (Table 4.2). Expression levels varied from a few $\mu g\,mL^{-1}$ in conventional tissue culture plates up to $400\,mg\,L^{-1}$ in serum-free, fed-spinner cultures (Ball et al. 1999; Nicholson et al. 1999; Davis et al. 1999; Green 1999). The mAbs exhibited a good usage of the different V, (D), J, and C genes. Hypermutation and the addition of N-sequences, to establish junctional diversity, allowed the creation of extensive repertoires. The length of the complementarity determining region (CDR) 3 regions established by nucleotide additions at the V to D and D to J joins (7–19 amino acids) is comparable to those identified in humans and considerably longer than those found in the mouse (Mendez et al. 1997; Nicholson et al. 1999).

Antibodies specific for the epidermal growth factor receptor (EGFR), overexpressed on many types of tumors, showed the preferential use of the closely related V_H and V_κ genes in combination with different D and J segments (Davis et al. 1999; Yang et al. 2001). The V genes were diversified by somatic mutation but most strikingly eight V_H genes coded for an aspartate residue in CDR1 at position 33. Analysis of the five-feature mice described by Nicholson et al. (1999)

Table 4.2 Antigen spectrum and affinity of fully human IgG monoclonal antibodies.

Antigen	*H chain*	*L chain*	*affinity*	*Reference*
Digoxin	$\gamma1$	κ	$2.5–22\,nmol\,L^{-1}$	Ball et al. 1999
Human CD4	$\gamma1$	κ	$11\,nmol\,L^{-1}–27\,pmol\,L^{-1}$	Lonberg et al. 1994; Fishwild et al. 1996
Human IL-8	$\gamma2$	κ	$0.2–0.9\,nmol\,L^{-1}$	Mendez et al. 1997; Green 1999
Human EGFR	$\gamma2$	κ	$0.8\,nmol\,L^{-1}–30\,pmol\,L^{-1}$	Mendez et al. 1997; Green 1999
Human TNFα	$\gamma2$	κ	$0.2–0.8\,nmol\,L^{-1}$	Mendez et at. 1997; Green 1999
CD4	γ[a]	κ	$32–77\,pmol\,L^{-1}$	Ishida et al. 2002
GCSF	γ[a]	κ	$0.2–0.3\,pmol\,L^{-1}$	Ishida et al. 2002

a isotype not defined.

revealed that the emergence of human IgH,κ or human IgH,λ Ig appears to be antigen-driven and that these mice do not produce significant levels of chimeric human antibodies with mouse λ L chain (Magadán et al. 2002; Brüggemann 2004). Immunization of transchromosomal mice resulted in human antibody responses including all Ig subclasses (Yoshida et al. 1999; Ishida et al. 2002). Problems with obtaining hybridomas, due to the instability of the Igκ locus-bearing chromosome fragment, were overcome by crossbreeding the H chain transchromosomal mice with mice carrying an Igκ YAC (Fishwild et al. 1996; Ishida et al. 2002).

Diverse human antibody repertoires with mAbs of desired specificity and high affinity in the picomolar range have been obtained from the translocus mouse strains. As an alternative to hybridoma production, immunization of translocus mice was combined with ribosome display technology to select high-affinity human V_H–V_κ fragments binding to progesterone (He et al. 1999). In this rapid approach antigen-specific V(D)J segments are selected which can be used for further manipulation, for example, the addition of a particular C_γ gene. But despite this success, immune responses of human translocus mice, their Ig levels and antibody diversity are not as refined as those of a normal animal. Extensive mutation in V_H and V_L have been described but it seems that the mouse strains with larger and indeed complete Ig loci (Ishida et al. 2002) may produce a more diverse repertoire with choice and selection optimized to produce superior antigen-binders.

Currently four promising human IgG,κ mAbs have reached phase III trials (Reichert et al. 2005; Lonberg 2005). Treatment with anti-EGF receptor antibodies is aimed at eradicating colorectal cancer, nonsmall cell lung cancer, and renal cell carcinoma; anti-CTLA-4 targets melanoma and possibly other cancers; anti-RANKL treatment is beneficial in osteoporosis; and anti-CD4 is used for the removal of lymphoma cells.

4.3.4 Ig Replacement

Attempts to retain the high levels of antibody expression and hypermutation of the mouse immune system for human Ig production concentrated on targeted substitution of mouse genes with human genes. This allowed site-specific integration of human C_κ and $C_\gamma1$ replacing mouse C_κ and $C_\gamma1$ or $C_\gamma2a$ (Zou et al. 1993a, 1994; Pluschke et al. 1998). In these two targeting constructs human C_κ and human γ1, adjacent to selectable marker genes, were flanked by the appropriate mouse homology sequences. This produced animals that rearranged and expressed chimeric Igκ and IgG1 antibodies with human C regions (Zou et al. 1993a; Pluschke et al. 1998). In another approach mouse $C_\gamma1$, excluding the transmembrane exons, was replaced by human $C_\gamma1$ (Zou et al. 1993a). Here the homology region and selectable marker gene were flanked by *loxP* sequences, which allowed their removal by Cre-mediated deletion. This permitted seamless insertion of a human γ1 C gene into the previous location of mouse γ1. The resulting mouse produced chimeric human IgG1 in serum at levels similar to those of mouse IgG1 in normal animals. Immunizations induced a normal immune response and produced a diverse repertoire of chimeric human IgH and IgL chains (Zou et al. 1994). Replacement of mouse with human genes appears to be successful when small DNA regions, up to a few kilobases in size, are exchanged. Future improvements may permit targeted replacements using larger human regions, perhaps whole loci, appropriately regulated by integration adjacent to endogenous control regions.

4.4 Complementary Strategies

Major efforts have focused on the expression of modified antibodies with the aim of altering the specificity of particular V genes by hypermutation. *In vivo* selection regimes can be used to increase the affinity of cloned mAbs either re-expressed in the mouse or transfected into cell lines. In another complementary attempt H chain antibodies, normally only produced in camelids, have been expressed in the mouse. This approach aims at generating a new type of antibody without L chain. H-chain-only antibodies have a propensity to recognize grooved surfaces found on many viruses rather than flat or less contoured areas predominantly recognized by conventional antibodies (van der Linden 2000).

4.4.1 H-chain-only Ig

Conventional antibodies consist of multiple units of paired H and L chains (Padlan 1994), which are present in all jawed vertebrates studied to date (Litman et al. 1999). In addition to these conventional heteromeric antibodies, sera of camelids (suborder Tylopoda, which includes camels, dromedaries, and llamas) contain a major type of IgG composed solely of paired H chains (Hamers-Casterman et al. 1993). Homodimeric H chain antibodies in camelids, illustrated in Fig. 4.5, lack the first C domain (C_H1), which is spliced out during mRNA maturation, and use distinctive yet diverse V region genes termed V_HH (Muyldermans et al. 1994; Nguyen et al. 2000). The lack of C_H1 in H chain antibodies is most likely to be the crucial factor in allowing their release from cells in the absence of L chains (Haas and Wabl 1983; Munro and Pelham, 1987; Hendershot, 1990). Using structural analysis it has been concluded that the hydrophilic nature of particular amino acids in V_HH–D–J_H prohibits association with L chain.

In conventional antibodies association between L and H chains is regarded as important in securing a diverse repertoire for antigen-binding. Nevertheless, antibodies composed solely of paired H chains are efficient antigen binders. In particular they recognize clefts on the antigen surface that are normally less immunogenic for conventional antibodies (Lauwereys et al. 1998). This may allow H chain antibodies to recognize viral structures much more readily, and perhaps

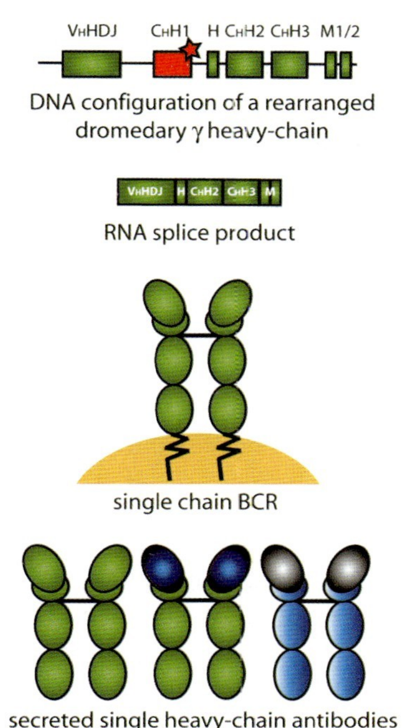

Fig. 4.5 Dromedary H-chain-only antibodies. The configuration of a rearranged dromedary γ H-chain gene is illustrated at the top. The C_HH1 exon, with the splice sequence mutation shown as a star, is indicated in red. The transcription product with removed C_HH1 is shown below. A repertoire of homodimeric H-chain antibodies with different V and C regions is illustrated at the bottom, with a hypothetical single-chain B-cell receptor (BCR) illustrated above.

exclusively, compared to large heterodimeric antibodies. In addition, H chain antibodies retain a dimeric configuration that is essential for antigen crosslinking, which can be followed by phagocytosis to remove an invader. H chain antibodies are absent in other mammals except in pathological situations, such as heavy chain disease, where they occur in mutated form (Alexander et al. 1982).

Recently it has been shown that antigen-specific dromedary H chain antibodies can be expressed in cultured cells and in mice (Nguyen et al. 2003; Zou et al. 2005). In the past, the aim of exploring how H-chain-only Ig could be expressed has been considered to present insurmountable obstacles and, indeed, without L chains, B-cell differentiation events are blocked (Zou et al. 2003). However, in transgenic mice carrying a dromedary H chain gene B cell development was found to be near normal and H chain antibodies were present on the cell surface and secreted in serum. This suggested that mice, just like camels, can remove the C_H1 exon by splicing. The finding that dromedary H chain antibodies can be readily produced is a crucial step towards the generation of a diverse human H-chain-only antibody repertoire in mice. Production of H chain antibodies by a mouse, which already has a human H chain locus integrated into the genome (Nicholson et al. 1999), would offer great advantages.

Already major advances towards expressing H-chain-only antibodies from a "camelized" human H chain locus have been made in the Drabek laboratory, Erasmus MC, Rotterdam, Netherlands (http://www2.eur.nl/fgg/ch1/cellbiology.

html). In the near future a spectrum of human antibodies with novel specificities may be available. For example, after immunization of the mice with human virus or viral proteins, not normally recognized by conventional antibodies, important therapeutic agents could be produced.

4.4.2
In vivo Mutation

H and L chain genes cloned from antibodies isolated by hybridoma or library display technology can be further modified by *in vivo* mutation and selection. Antibody expression in transgenic mice from rearranged Ig genes has been performed by random integration or by site-specific knockin strategies (Sohn et al. 1993; Li et al. 2005). In both approaches mature B cells expressed the transgene and underwent multiple cycles of hypermutation and/or receptor editing, which created extensive diversity (Bertrand et al. 1998; Jena et al. 2003). After immunization new specificities emerged from introduced V_HDJ_H segments (Li et al. 2005). The use of highly recombinogenic cells like the chicken DT40 line has allowed the selection of antigen-specific antibodies with nanomolar affinities generated by iterative affinity maturation in tissue culture (Cumbers et al. 2002; Harris et al. 2002). The mechanisms and key enzymes, such as activation-induced cytidine deaminase (AID), that operate on Ig diversification have been described in detail (Petersen-Mahrt 2005). Introduction of a *loxP*-flanked AID construct into $AID^{-/-}$ DT40 cells allowed induction of a reversible hypermutation regime (Kanayama et al. 2005). Similarly, AID expression in hybridoma cells induced a high rate of V gene mutation (Martin et al. 2002; Ronai et al. 2005). It appears that specificity and affinity of (human) antibodies can be proficiently improved by the above methods. In addition, cells expressing novel or high-affinity antigen binders can be easily identified by microarray technologies and flow cytometry.

4.5
Outlook

The success in expressing fully human antibody repertoires in mice and, more recently, in cattle is now extensively exploited by many different pharmaceutical and biotechnology companies. Immunized animals provide high-affinity antibodies and a diverse antibody repertoire, generating monoclonal human anti-human Ig from mice and highly specific polyclonal human Ig from large animals. The therapeutic human anti-human mAbs studied to date fall into three major categories: oncological, immunological, and anti-infective (Reichert et al. 2005). Emerging alternatives have focused on the production of single-chain Ig in the mouse. This has resulted in the expression of homodimeric H-chain-only antibodies retaining isotype effector functions. The small H chain Ig appears to recognize additional specificities not seen by conventional antibodies, which may provide novel therapeutic agents in combating viral infections. Approaches to tailor the affinity of individual antibodies rely on diversification by the immune system of the mouse and on the selection of new specificities in hypermutating cell lines.

Acknowledgments

The work in the laboratory is supported by the BBSRC and the Babraham Institute. We thank Drs Peter Kilshaw and Mike Taussig for critical reading of the manuscript.

The anti-EGF receptor antibody (page 83) has been approved by the U.S. Food and Drug Administration and is under the name of Vectibix for use in colon cancer patients. New results on the Generation of heavy-chain-only antibodies in mice[1] are described by Janssens et al., (2006) *Proc Natl Acad Sci USA* 103: 15130–15135.

References

Alexander, A., Steinmetz, M., Barritault, D., Frangione, B., Franklin, E.C., Hood, L., Buxbaum, J.N. (1982) γ Heavy chain disease in man: cDNA sequence supports partial gene deletion model. *Proc Natl Acad Sci USA* 79: 3260–3264.

Anand, R. (1992) Yeast artificial chromosomes (YACs) and the analysis of complex genomes. *Trends Biotechnol* 10: 35–40.

Arulampalam, V., Eckhardt, L., Pettersson, S. (1997) The enhancer shift: a model to explain the developmental control of IgH gene expression in B-lineage cells. *Immunol Today* 18: 549–554.

Ball, W.J. Jr., Kasturi, R., Dey, P., Tabet, M., O'Donnell, S., Hudson, D., Fishwild, D. (1999) Isolation and characterization of human monoclonal antibodies to digoxin. *J Immunol* 163: 2291–2298.

Bertrand, F.E., Golub, R., Wu, G.E. (1998) V_H gene replacement occurs in the spleen and bone marrow of non-autoimmune quasi-monoclonal mice. *Eur J Immunol* 28: 3362–3370.

Brüggemann, M. (2004) Human monoclonal antibodies from translocus mice. In: Honjo, T., Alt, F.W., Neuberger, M.S. (eds) *Molecular Biology of B-Cells*. Amsterdam: Elsevier, pp. 547–561.

Brüggemann, M., Caskey, H.M., Teale, C., Waldmann, H., Williams, G.T., Surani, M. A., Neuberger, M.S. (1989) A repertoire of monoclonal antibodies with human heavy-chains from transgenic mice. *Proc Natl Acad Sci USA* 86: 6709–6713.

Brüggemann, M., Neuberger, M.S. (1991) Generation of antibody repertoires in transgenic mice. *Methods: A Companion to Methods Enzymol* 2: 159–165.

Brüggemann, M., Neuberger, M.S. (1996) Strategies for expressing human antibody repertoires in transgenic mice. *Immunol Today* 17: 391–397.

Brüggemann, M., Spicer, C., Buluwela, L., Rosewell, I., Barton, S., Surani, M.A., Rabbitts, T.H. (1991) Human antibody production in transgenic mice: Expression from 100 kb of the human IgH locus. *Eur J Immunol* 21: 1323–1326.

Brüggemann, M., Taussig, M.J. (1997) Production of human antibody repertoires in transgenic mice. *Curr Opin Biotechnol* 8: 455–458.

Capecchi, M.R. (1989) The new mouse genetics: altering the genome by gene targeting. *Trends Genet* 5: 70–76.

Chen, J., Trounstine, M., Kurahara, C., Young, F., Kuo, C.-C., Xu, Y., Loring, J.F., Alt, F.W., Huszar, D. (1993a) B cell development in mice that lack one or both immunoglobulin κ light chain genes. *EMBO J* 12: 821–830.

Chen, J., Trounstine, M., Alt, F.W., Young, F., Kurahara, C., Loring, J.F., Huszar, D. (1993b) Immunoglobulin gene rearrangement in B cell deficient mice generated by targeted deletion of the J_H locus. *Int Immunol* 5: 647–656.

Choi, T.K., Hollenbach, P.W., Pearson, B.E., Ueda, R.M., Weddell, G.N., Kurahara, C. G., Woodhouse, C.S., Kay, R.M., Loring, J.F. (1993) Transgenic mice containing a human heavy chain immunoglobulin gene fragment cloned in a yeast artificial chromosome. *Nat Genet* 4: 117–123.

Cook, G.P., Tomlinson, I.M., Walter, G., Riethman, H., Carter, N.P., Buluwela, L., Winter, G., Rabbitts, T.H. (1994) A map of the human immunoglobulin VH locus completed by analysis of the telomeric region of chromosome 14q. *Nat Genet* 7: 162–168.

Corbett, S.J., Tomlinson, I.M., Sonnhammer, E.L., Buck, D., Winter, G. (1997) Sequence of the human immunoglobulin diversity (D) segment locus: a systematic analysis provides no evidence for the use of DIR segments, inverted D segments, "minor" D segments or D-D recombination. *J Mol Biol* 270: 587–597.

Cumbers, S.J., Williams, G.T., Davies, S.L., Grenfell, R.L., Takeda, S., Batista, F.D., Sale, J.E., Neuberger, M.S. (2002) Generation and iterative affinity maturation of antibodies in vitro using hypermutating B-cell lines. *Nat Biotechnol* 20: 1129–1134.

Davies, N.P., Brüggemann, M. (1993) Extension of yeast artificial chromosomes by cosmid multimers. *Nucleic Acids Res* 21: 767–768.

Davies, N.P., Rosewell, I.R., Brüggemann, M. (1992) Targeted alterations in yeast artificial chromosomes for inter-species gene transfer. *Nucleic Acids Res* 20: 2693–2698.

Davies, N.P., Rosewell, I.R., Richardson, J. C., Cook, G.P., Neuberger, M.S., Brownstein, B.H., Norris, M.L., Brüggemann, M. (1993) Creation of mice expressing human antibody light chains by introduction of a yeast artificial chromosome containing the core region of the human immunoglobulin κ locus. *Bio/technology* 11: 911–914.

Davies, N.P., Popov, A.V., Zou, X., Brüggemann, M. (1996) Human antibody repertoires in transgenic mice: Manipulation and transfer of YACs. In: McCafferty, J., Hoogenboom, H.R., Chiswell, D.J. (eds) *Antibody Engineering: A Practical Approach*. Oxford: IRL Press, pp. 59–76.

Davis, C.G., Gallo, M.L., Corvalan, R.F. (1999) Transgenic mice as a source of fully human antibodies for the treatment of cancer. *Cancer Metastasis Rev* 18: 421–425.

Evans, M.J. (1989) Potential for genetic manipulation of Mammals. *Mol Biol Med* 6: 557–565.

Fishwild, D.M., O'Donnell, S.L., Bengoechea, T., Hudson, D.V., Harding, F., Bernhard, S.L., Jones, D., Kay, R.M., Higgins, K.M., Schramm, S.R., Lonberg, N. (1996) High-avidity human Igκ monoclonal antibodies from a novel strain of minilocus transgenic mice. *Nat Biotechnol* 14: 845–851.

Fournier, R.E., Ruddle, F.H. (1977) Microcell-mediated transfer of murine chromosomes into mouse, Chinese hamster, and human somatic cells. *Proc Natl Acad Sci USA* 74: 319–323.

Frazer, J.K., Capra, J.D. (1999) Immunoglobulins: structure and function. In: Paul, W.E. (ed.) *Fundamental Immunology*, 4th edn. Philadelphia: Lippincott-Raven, pp. 37–74.

Frippiat, J.P., Williams, S.C., Tomlinson, I. M., Cook, G.P., Cherif, D., Le Paslier, D., Collins, J.E., Dunham, I., Winter, G., Lefranc, M.-P. (1995) Organization of the human immunoglobulin lambda light-chain locus on chromosome 22q11.2. *Hum Mol Genet* 4: 983–91.

Gordon, J.W., Ruddle, F.H. (1983) Gene transfer into mouse embryos: production of transgenic mice by pronuclear injection. *Methods Enzymol* 101: 411–433.

Green, L.L., Hardy, M.C., Maynard-Currie, C.E., Tsuda, H., Louie, D.M., Mendez, M.J., Abderrahim, H., Noguchi, M., Smith, D.H., Zeng, Y., David, N.E., Sasai, H., Garza, D., Brenner, D.G., Hales,

J.F., McGuinness, R.P., Capon, D.J., Klapholz, S., Jakobovits, A. (1994) Antigen-specific human monoclonal antibodies from mice engineered with human Ig heavy and light chain YACs. *Nat Genet* 7: 13–21.

Green, L.L. (1999) Antibody engineering via genetic engineering of the mouse: XenoMouse strains are a vehicle for the facile generation of therapeutic human monoclonal antibodies. *J Immunol Methods* 231: 11–23.

Haas, I.G., Wabl, M. (1983) Immunoglobulin heavy chain binding protein. *Nature* 306: 387–389.

Hamers-Casterman, C., Atarhouch, T., Muyldermans, S., Robinson, G., Hamers, C., Songa, E.B., Bendahman, N., Hamers, R. (1993) Naturally occurring antibodies devoid of light chains. *Nature* 363: 446–448.

Harris, R.S., Sale, J.E., Petersen-Mahrt, S.K., Neuberger, M.S. (2002) AID is essential for immunoglobulin V gene conversion in a cultured B cell line. *Curr Biol* 12: 435–438.

Hasan, M., Polic, B., Bralic, M., Jonjic, S., Rajewsky, K. (2002) Incomplete block of B cell development and immunoglobulin production in mice carrying the μMT mutation on the BALB/c background. *Eur J Immunol* 32: 3463–3471.

He, M., Menges, M., Groves, M.A., Corps, E., Liu, H., Brüggemann, M., Taussig, M.J. (1999) Selection of a human anti-progesterone antibody fragment from a transgenic mouse library by ARM ribosome display. *J Immunol Methods* 231: 105–117.

Hendershot, L.M. (1990) Immunoglobulin heavy chain and binding protein complexes are dissociated in vivo by light chain addition. *J Cell Biol* 111: 829–837.

Hofker, M.H., Walter, M.A., Cox, D.W. (1989) Complete physical map of the human immunoglobulin heavy chain constant region gene complex. *Proc Natl Acad Sci USA* 86: 5567–5571.

Hogan, B., Beddington, R., Costantini, F., Lacy, E. (1994) *Manipulating the Mouse Embryo: A Laboratory Manual*. Cold Spring Harbor, NY: Cold Spring Harbor Press.

Ishida, I., Tomizuka, K., Yoshida, H., Tahara, T., Takahashi, N., Ohguma, A., Tanaka, S., Umehashi, M., Maeda, H., Nozaki, C., Halk, E., Lonberg, N. (2002) Production of human monoclonal and polyclonal antibodies in transchromo animals. *Cloning Stem Cells* 4: 91–102.

Jakobovits, A., Vergara, G.J, Kennedy, J.L., Hales, J.F., McGuinness, R.P., Casentini-Borocz, D.E., Brenner, D.G., Otten, G.R. (1993) Analysis of homozygous mutant chimeric mice: deletion of the immunoglobulin heavy-chain joining region blocks B-cell development and antibody production. *Proc Natl Acad Sci USA* 90: 2551–2555.

Jakobovits, A., Green, L.L., Hardy, M.C., Maynard-Currie, C.E., Tsuda, H., Louie, D.M., Mendez, M.J., Abderrahim, H., Noguchi, M., Smith, D.H., Zeng, Y., David, N.E., Sasai, H., Garza, D., Brenner, D.G., Hales, J.F., McGuinnes, R.P., Capaon, D.J., Klapholz, S. (1995) Production of antigen-specific human antibodies from mice engineered with human heavy and light chain YACs. *Ann N Y Acad Sci* 764: 525–535.

Jena, P.K., Smith, D.S., Zhang, X., Aviszus, K., Durdik, J.M., Wysocki, L.J. (2003) Somatic translocation and differential expression of Ig μ transgene copies implicate a role for the *IgH* locus in memory B cell development. *Mol Immunol* 39: 885–897.

Jenuwein, T., Allis, C.D. (2001) Translating the histone code. *Science* 293: 1074–1080.

Kakeda, M., Hiratsuka, M., Nagata, K., Kuroiwa, Y., Kakitani, M., Katoh, M., Oshimura., M., Tomizuka, K. (2005) Human artificial chromosome (HAC) vector provides long-term therapeutic transgene expression in normal human primary fibroblasts. *Gene Ther* 12: 852–856.

Kanayama, N., Todo, K., Reth, M., Ohmori, H. (2005) Reversible switching of immunoglobulin hypermutation machinery in a chicken B cell line. *Biochem Biophys Res Commun* 327: 70–75.

Kawasaki, K., Minoshima, S., Schooler, K., Kudoh, J., Asakawa, S., de Jong, P.J., Shimizu, N. (1995) The organization of the human immunoglobulin λ gene locus. *Genome Res* 5: 125–135.

King, D.J. (1998) *Applications and Engineering of Monoclonal Antibodies*. London: Taylor and Francis.

Kitamura, D., Roes, J., Kühn, R., Rajewsky, K. (1991) A B cell-deficient mouse by targeted disruption of the membrane exon of the immunoglobulin μ chain gene. *Nature* 350: 423–426.

Klingbeil, C., Hsu, D.H. (1999) Pharmacology and safety assessment of humanized monoclonal antibodies for therapeutic use. *Toxicol Pathol* 27: 1–3.

Knight, K.L., Kingzette, M., Crane, M.A., Zhai, S.K. (1995) Transchromosomally derived Ig heavy chains. *J Immunol* 155: 684–691.

Koi, M., Shimizu, M., Morita, H., Yamada, H., Oshimura, M. (1989) Construction of mouse A9 clones containing a single human chromosome tagged with neomycin-resistance gene via microcell fusion. *Jpn J Cancer Res* 80: 413–418.

Kuroiwa, Y., Shinohara, T., Notsu, T., Tomizuka, K., Yoshida, H., Takeda, S., Oshimura, M., Ishida, I. (1998) Efficient modification of a human chromosome by telomere-directed truncation in high homologous recombination-proficient chicken DT40 cells. *Nucleic Acids Res* 26: 3447–3448.

Kuroiwa, Y., Tomizuka, K., Shinohara, T., Kazuki, Y., Yoshida, H., Ohguma, A., Yamamoto, T., Tanaka, S., Oshimura, M., Ishida, I. (2000) Manipulation of human minichromosomes to carry greater than megabase-sized chromosome inserts. *Nat Biotechnol* 18: 1086–1090.

Kuroiwa, Y., Kasinathan, P., Choi, Y.J., Naeem, R., Tomizuka, K., Sullivan, E.J., Knott, J.G., Duteau, A., Goldsby, R.A., Osborne, B.A., Ishida, I., Robl, J.M. (2002) Cloned transchromosomic calves producing human immunoglobulin. *Nat Biotechnol* 20: 889–894.

Kuroiwa, Y., Kasinathan, P., Matsushita, H., Sathiyaselan, J., Sullivan, E.J., Kakitani, M., Tomizuka, K., Ishida, I., Robl, J.M. (2004) Sequential targeting of the genes encoding immunoglobulin-μ and prion protein in cattle. *Nat Genet* 36: 775–780.

Lauwereys, M., Ghahroudi, M.A., Desmyter, A., Kinne, J., Holzer, W., De Genst, E., Wyns, L., Muyldermans, S. (1998) Potent enzyme inhibitors derived from dromedary heavy-chain antibodies. *EMBO J* 17: 3512–3520.

Lefranc, M.-P., Lefranc, G. (2001) *The Immunoglobulin Factsbook*. London: Academic Press, pp. 45–68.

Li, J., Geissal, E.D., Li, W., Stollar, D. (2005) Repertoire diversification in mice with an IgH-locus-targeted transgene for the rearranged VH domain of a physiologically selected anti-ssDNA antibody. *Mol Immunol* 42: 1475–1484.

Linden, van der R., Geus, de B., Stok, W., Bos, W., Wassenaar, van D., Verrips, T., Frenken, L. (2000) Induction of immune responses and molecular cloning of the heavy chain antibody repertoire of *Lama glama*. *J Immunol Methods* 240: 185–195.

Litman, G.W., Anderson, M.K., Rast, J.P. (1999) Evolution of antigen binding receptors. *Annu Rev Immunol* 17: 109–147.

Lonberg, N. (2005) Human antibodies from transgenic animals. *Nature Biotechnol* 23: 1117–1125.

Lonberg, N., Taylor, L.D., Harding, F.A., Troustine, M., Higgins, K.M., Schramm, S.R., Kuo, C.-C., Mashayekh, R., Wymore, K., McCabe, J.G., Munoz-O'Regan, D., O'Donnell, S.L., Lapachet, E.S.G., Bengoechea, T., Fishwild, D.M., Carmack, C.E., Kay, R.M., Huszar, D. (1994) Antigen-specific human antibodies from mice comprising four distinct genetic modifications. *Nature* 368: 856–859.

Magadán, S., Valladares, M., Suarez, E., Sanjuán, I., Molina, A., Ayling, C., Davies, S.L., Zou, X., Williams, G.T., Neuberger, M.S., Brüggemann, M., Gambón, F., Díaz-Espada, F., González-Fernández, A. (2002) Production of antigen-specific human monoclonal antibodies: comparison of mice carrying IgH/Iκ or IgH/Iκ/Iλ transloci. *BioTechniques* 33: 680–690.

Magor, B.G., Ross, D.A., Pilstrom, L., Warr, G.W. (1999) Transcriptional enhancers and the evolution of the IgH locus. *Immunol Today* 20: 13–17.

Markie, D. (1996) YAC protocols. In: *Methods in Molecular Biology*, Vol. 54. Totowa, NJ: Humana Press.

Martin, A., Bardwell, P.D., Woo, C.J., Fan, M., Shulman, M.J., Scharff, M.D. (2002) Activation-induced cytidine deaminase turns on somatic hypermutation in hybridomas. *Nature* 415: 802–806.

Matsuda, F., Ishii, K., Bourvagnet, P., Kuma, K., Hayashida, H., Miyata, T., Honjo, T. (1998) The complete nucleotide sequence of the human immunoglobulin heavy chain variable region locus. *J Exp Med* 188: 2151–2162.

Maynard, J., Georgiou, G. (2000) Antibody engineering. *Annu Rev Biomed Eng* 2: 339–376.

Mendez, M.J., Green, L.L, Corvalan, J.R., Jia, X.-C., Maynard-Currie, C.E., Yang, X.-D., Gallo, M.L., Louie, D.M., Lee, D.V., Erickson, K.L., Luna, J., Roy, C.M., Abderrahim, H., Kirschenbaum, F., Noguchi, M., Smith, D.H., Fukushima, A., Hales, J.F., Klapholz, S., Finer, M.H., Davis, C.G., Zsebo, K.M., Jakobovits, A. (1997) Functional transplant of megabase human immunoglobulin loci recapitulates human antibody response in mice. *Nat Genet* 15: 146–156.

Mundt, C.A., Nicholson, I., Zou, X., Popov, A.V., Ayling C., Brüggemann, M. (2001) Novel control motif cluster in the IgH δ-γ3 interval exhibits B-cell specific enhancer function in early development. *J Immunol* 166: 3315–3323.

Muyldermans, S., Atarhouch, T., Saldanha, J., Barbosa, J.A., Hamers, R. (1994) Sequence and structure of VH domain from naturally occurring camel heavy chain immunoglobulins lacking light chains. *Protein Eng* 7: 1129–1135.

Munro, S., Pelham, H.R. (1987) A C-terminal signal prevents secretion of luminal ER proteins. *Cell* 48: 899–907.

Neuberger, M., Brüggemann, M. (1997) Mice perform a human repertoire. *Nature* 386: 25–26.

Nguyen, V.K., Hamers, R., Wyns, L., Muyldermans, S. (2000) Camel heavy-chain antibodies: diverse germline V_HH and specific mechanisms enlarge the antigen-binding repertoire. *EMBO J* 19: 921–930.

Nguyen, V.K., Zou, X., Lauwereyes, M., Brys, L., Brüggemann, M., Muyldermans, S. (2003) Heavy-chain only antibodies derived from dromedary are secreted and displayed by mouse B-cells. *Immunology* 109: 93–101.

Nicholson, I.C., Zou, X., Popov, A.V., Cook, G.P., Corps, E.M., Humphries, S., Ayling, C., Xian, J., Taussig, M.J., Neuberger, M.S., Brüggemann, M. (1999) Antibody repertoires of four and five feature translocus mice carrying human immunuglobulin heavy chain and kappa and lambda light chain YACs. *J Immunol* 163: 6898–6906.

Okada, N., Okada, H. (1999) Human IgM antibody therapy for HIV-1 infection. *Microbiol Immunol* 43: 729–736.

Orinska, Z., Osiak, A., Lohler, J., Bulanova, E., Budagian, V., Horak, I., Bulfone-Paus, S. (2002) Novel B cell population producing functional IgG in the absence of membrane IgM expression. *Eur J Immunol* 32: 3472–3480.

Pachnis, V., Pevny, L., Rothstein, R., Costantini, F. (1990) Transfer of a yeast artificial chromosome carrying human DNA from *Saccharomyces cerevisiae* into mammalian cells. *Proc Natl Acad Sci USA* 87: 5109–5113.

Padlan, E.A. (1994) Anatomy of the antibody molecule. *Mol Immunol* 31: 169–217.

Pavan, W.J., Hieter, P., Reeves, R.H. (1990) Modification and transfer into an embryonal carcinoma cell line of a 360-kilobase human-derived yeast artificial chromosome. *Mol Cell Biol* 10: 4163–4169.

Petersen-Mahrt, S. (2005) DNA deamination in immunity. *Immunol Rev* 203: 80–97.

Pluschke, G., Joss, A., Marfurt, J., Daubenberger, C., Kashala, O., Zwickl, M., Stief, A., Sansig, G., Schläpfer, B., Linkert, S., van der Putten, H., Hardman, N., Schröder, M. (1998) Generation of chimeric monoclonal antibodies from mice that carry human immunoglobulin Cγ1 heavy or Cκ light chain gene segments. *J Immunol Methods* 215: 27–37.

Popov, A.V., Bützler, C., Frippiat, J.-P., Lefranc, M.-P., Brüggemann, M. (1996) Assembly and extension of yeast artificial chromosomes to build up a large locus. *Gene* 177: 195–205.

Popov, A.V., Zou, X., Xian, J., Nicholson, I. C., Brüggemann, M. (1999) A human immunoglobulin λ locus is similarly well expressed in mice and humans. *J Exp Med* 189: 1611–1620.

Protopapadakis, E., Kokla, A., Tzartos, S.J., Mamalaki, A. (2005) Isolation and characterization of human anti-acetylcholine receptor monoclonal antibodies from transgenic mice expressing human immunoglobulin loci. *Eur J Immunol* 35: 1960–1968.

Reichert, J.M., Rosensweig, C.J., Faden, L. B., Dewitz, M.C. (2005) Monoclonal antibody success in the clinic. *Nature Biotechnol* 23: 1073–1078.

Ren, L., Zou, X., Smith, J.A., Brüggemann, M. (2004) Silencing of the immunoglobulin heavy chain locus by removal of all 8 constant region genes on a 200 kb region. *Genomics* 84: 686–695.

Ronai, D., Iglesias-Ussel, M.D., Fan, M., Shulman, M.J., Scharff, M.D. (2005) Complex regulation of somatic hypermutation by cis-acting sequences in the endogenous IgH gene in hybridoma cells. *Proc Natl Acad Sci USA* 102: 11829–11834.

Roschenthaler, F., Hameister, H., Zachau, H.G. (2000) The 5′ part of the mouse immunoglobulin kappa locus as a continuously cloned structure. *Eur J Immunol* 30: 3349–3354.

Russell, N.D., Corvalan, J.R., Gallo, M.L., Davis, C.G., Pirofski, L. (2000) Production of protective human antipneumococcal antibodies by transgenic mice with human immunoglobulin loci. *Infect Immun* 68: 1820–1826.

Sanchez, P., Drapier, A.M., Cohen-Tannoudji, M., Colucci, E., Babinet, C., Cazenave, P.A. (1994) Compartmentalization of lambda subtype expression in the B cell repertoire of mice with a disrupted or normal C kappa gene segment. *Int Immunol* 6: 711–719.

Schedl, A., Larin, Z., Montoliu, L., Thies, E., Kelsey, G., Lehrach, H., Schütz, G. (1993) A method for the generation of YAC transgenic mice by pronuclear microinjection. *Nucleic Acids Res* 21: 4783–4787.

Shen, M.H., Yang, J., Loupart, M.L., Smith, A., Brown, W. (1997) Human mini-chromosomes in mouse embryonal stem cells. *Hum Mol Genet* 6: 1375–1382.

Shinohara, T., Tomizuka, K., Takehara, S., Yamauchi, K., Katoh, M., Ohguma, A., Ishida, I., Oshimura, M. (2000) Stability of transferred human chromosome fragments in cultured cells and in mice. *Chromosome Res* 8: 713–725.

Sohn, J., Gerstein, R.M., Hsieh, C.L., Lemer, M., Selsing, E. (1993) Somatic hypermutation of an immunoglobulin μ heavy chain transgene. *J Exp Med* 177: 493–504.

Taylor, L.D., Carmack, C.E., Schramm, S.R., Mashayekh, R., Higgins, K.M., Kuo, C.-C., Woodhouse, C., Kay, R.M., Lonberg, N. (1992) A transgenic mouse that expresses a diversity of human sequence heavy and light chain immunoglobulins. *Nucleic Acids Res* 20: 6287–6295.

Taylor, L.D., Carmack, C.E., Huszar, D., Higgins, K.M., Mashayekh, R., Sequar, G., Schramm, S.R., Kuo, C.-C., O'Donnell, S.L., Kay, R.M., Woodhouse, C.S., Lonberg, N. (1994) Human immunoglobulin transgenes undergo rearrangement, somatic mutation and class switching in mice that lack endogenous IgM. *Int Immunol* 6: 579–591.

Tomizuka, K., Yoshida, H., Uejima, H., Kugoh, H., Sato, K., Ohguma, A., Hayasaka, M., Hanaoka, K., Oshimura, M., Ishida, I. (1997) Functional expression and germline transmission of a human chromosome fragment in chimaeric mice. *Nat Genet* 16: 133–143.

Tomizuka, K., Shinohara, T., Yoshida, H., Uejima, H., Ohguma, A., Tanaka S., Sato, K., Oshimura, M., Ishida, I. (2000) Double trans-chromosomic mice: Maintenance of two individual human chromosome fragments containing immunoglobulin heavy and kappa loci and expression of fully human antibodies. *Proc Natl Acad Sci USA* 97: 722–727.

Wagner, S.D., Williams, G.T., Larson, T., Neuberger, M.S., Kitamura, D., Rajewsky, K., Xian, J., Brüggemann, M. (1994a) Antibodies generated from human immunoglobulin miniloci in transgenic mice. *Nucleic Acids Res* 22: 1389–1393.

Wagner, S.D., Popov, A.V., Davies, S.L., Xian, J., Neuberger, M.S., Brüggemann, M. (1994b) The diversity of antigen-specific monoclonal antibodies from transgenic mice bearing human immunoglobulin gene miniloci. *Eur J Immunol* 24: 2672–2681.

Wagner, S.D., Gross, G., Cook, G.P., Davies, S.L., Neuberger, M.S. (1996) Antibody expression from the core region of the human IgH locus reconstructed in transgenic mice using bacteriophage P1 clones. *Genomics* 35: 405–414.

Waldmann, H., Cobbold, S. (1993) The use of monoclonal antibodies to achieve immunological tolerance. *Immunol Today* 14: 247–251.

Weichhold, G.M., Klobeck, H.G., Ohnheiser, R., Combriato, G., Zachau, H.G. (1990) Megabase inversions in the human genome as physiological events. *Nature* 347: 90–92.

Xian, J., Zou, X., Popov, A.V., Mundt, C.A., Miller, N., Williams, G.T., Davies, S.L., Neuberger, M.S., Brüggemann, M. (1998) Comparison of the performance of a plasmid-based human Igκ minilocus and YAC-based human Igκ transloci for the production of a human antibody repertoire in transgenic mice. *Transgenics* 2: 333–343.

Yang, X.D., Jia, X.C., Corvalan, J.R., Wang, P., Davis, C.G. (2001) Development of ABX-EGF, a fully human anti-EGF receptor monoclonal antibody, for cancer therapy. *Crit Rev Oncol Hematol* 38: 17–23.

Yoshida, H., Tomizuka, K., Uejima, H., Satoh, K., Ohguma, A., Oshimura, M., Ishida, I. (1999) Creation of hybridomas from mice expressing human antibody by introduction of a human chromosome. In: Kitagawa, Y., Matuda, T., Iijima, S. (eds) *Animal Cell Technology: Basic and Applied Aspects*, Vol. 10. Dordrecht: Kluwer Academic Publishers, pp. 69–73.

Zachau, H.G. (2000). The immunoglobulin kappa gene families of human and mouse: a cottage industry approach. *Biol Chem* 381: 951–954.

Zou, Y.R., Gu, H., Rajewsky, K. (1993a) Generation of a mouse strain that produces immunoglobulin kappa chains with human constant region. *Science* 262: 1271–1274.

Zou, Y.R., Takeda, S., Rajewsky, K. (1993b) Gene targeting in the Ig kappa locus: efficient generation of lambda chain-expressing B cells, independent of gene rearrangements in Ig kappa. *EMBO J* 12: 811–820.

Zou, Y.R., Müller, W., Gu, H., Rajewsky, K. (1994) Cre-loxP-mediated gene replacement: a mouse strain producing humanized antibodies. *Curr Biol* 4: 1099–1103.

Zou, X., Xian, J., Popov, A.V., Rosewell, I.R., Müller, M., Brüggemann, M. (1995) Subtle differences in antibody responses and hypermutation of λ light chains in mice with a disrupted κ constant region. *Eur J Immunol* 25: 2154–2162.

Zou, X., Xian, J., Davies, N.P., Popov, A.V., Brüggemann, M. (1996) Dominant expression of a 1.3 Mb human Igκ locus replacing mouse light chain production. *FASEB J* 10: 1227–1232.

Zou, X., Piper, T.A., Smith, J.A., Allen, N. D., Xian, J., Brüggemann, M. (2003) Block in development at the pre B-II to immature B-cell stage in mice without Igκ and Igλ light chain. *J Immunol* 170: 1354–1361.

Zou, X., Smith, J.A., Nguyen, V.K., Ren, L., Muyldermans, S., Brüggemann, M. (2005) Expression of a dromedary heavy-chain-only antibody and B-cell development in the mouse. *J Immunol* 175: 3769–3779.

5
Bioinformatics Tools for Antibody Engineering

Andrew C.R. Martin and James Allen

5.1
Introduction

From the bioinformatics perspective, antibodies present a number of unique challenges (for example, in sequence analysis and search tools) as well as a number of unique opportunities (standardized numbering schemes, high-quality structural modeling). This chapter will survey a number of these aspects and will attempt to guide the user towards tools and resources that will fulfill these requirements. It is assumed that the reader is familiar with the natural generation of antibody diversity. Concepts of antibody structure will be reviewed only briefly insofar as they influence the rest of the discussion.

5.1.1
Brief Review of Antibody Structure

The four-chain model of antibody structure consisting of two identical light chains and two identical heavy chains, was first proposed by Porter (1959). There are many reviews of antibody structure (Alzari et al. 1988; Padlan 1994; Searle et al. 1994, for example) and of the interactions between antibodies and antigens (Padlan 1977; Mariuzza et al. 1987; Davies et al. 1990; Wilson and Stanfield 1993, for example); the reader is referred to these reviews for more information. In their pioneering work, Wu and Kabat (1970) examined the sequences of the variable domain. They aligned the sequences and generated a "variability plot." While the method used to calculate variability has since been criticized (Valdar and Thornton 2001; Valdar 2002), the trends are remarkably clear and enabled them to identify "hypervariable" regions which they suggested form the actual antigen combining site. They proposed that in the three-dimensional structure, these regions adopt loop conformations supported on a relatively conserved framework. They termed these stretches of hypervariable sequence the "complementarity determining regions" (CDRs).

Electron microscopy revealed the "Y" shape (Valentine and Green 1967) of IgG, the best studied of the immunoglobulin classes. Each arm, or Fab fragment,

Handbook of Therapeutic Antibodies. Edited by Stefan Dübel

ISBN 978-3-527-31453-9

consists of a complete light chain (V_L and C_L domains), and half of a heavy chain (V_H and C_H1 domains). The remainder of each heavy chain (C_H2 and C_H3 domains) pairs to form the stem of the "Y" known as the F_c fragment (IgM and IgE have an additional C_H4 domain). A dimer of V_H and V_L domains is known as the Fv fragment. The first X-ray crystal structure of a Fab fragment, solved by Poljak in 1973 confirmed that CDRs defined by Kabat and Wu corresponded approximately to structural loops which come together to form the antigen-binding site (Poljak et al. 1973).

In the 1980s, Chothia and Lesk showed that the CDRs were much less variable in structure than might have been expected (Chothia and Lesk 1987). They found that for a given length of CDR, if certain amino acids were present at key locations in the CDR and in the structurally adjacent framework, then the conformation of the CDR would be conserved. In general, the amino acids at other positions within the CDRs could be varied freely without any major change in the conformation of the CDR. They defined a set of "canonical conformations" into which the majority of CDRs would be expected to fall. The application of this observation to three-dimensional modeling is discussed later.

The rest of this chapter will be confined to discussion of the Fv fragment (i.e. the V_L and V_H domains) which encompass the variability of antibodies and their ability to interact with antigens.

5.1.2 Conventions Used in this Chapter

In this chapter, the following conventions will be used. The letters "L" and "H" will be used to refer to the light and heavy chains respectively and when referring to a particular residue number, the chain label will be prepended onto the residue number. For example, the 10th residue in the light chain will be referred to as L10. Any of the six CDRs will be referred to using the letters "CDR-" followed by the chain name and the CDR number. For example, the first CDR of the light chain will be referred to as CDR-L1. This serves to avoid confusion with the first residue of the light chain (residue L1). In addition, CDR*n* (e.g. CDR3) is used to refer to both the heavy and light chain CDRs. Similarly the framework regions (those regions outside the CDRs) are termed LFR1, LFR2, LFR3, and LFR4 in the light chain, and HFR1, HFR2, HFR3, and HFR4 in the heavy chain. FR*n* (e.g. FR3) is used to refer to a framework region in both heavy and light chains.

5.2 Numbering Schemes for Antibodies

One of the major advantages of working with antibodies from a bioinformatics viewpoint is the availability of a standardized numbering scheme. This provides a standard way of identifying specific locations within an antibody sequence. In addition, it may provide a standard link between sequence and structure. Unfortunately in the case of antibodies, the idea of having a standard numbering scheme is now so popular that there are at least four different such schemes!

An extremely useful comparison of the different numbering schemes is presented by Honegger (see http://www.biochem.unizh.ch/antibody/).

5.2.1 The Kabat Numbering Scheme

The most commonly used scheme is that introduced by Wu and Kabat when they performed their analysis of sequence variability. This "Kabat numbering scheme," universely recognized by immunologists, was developed purely on the basis of sequence alignment. Insertions in the sequence relative to the standard numbering scheme are indicated using insertion letter codes. For example, residues inserted between residues L27 and L28 are indicated as L27A, L27B, etc. Deletions relative to the standard scheme are simply accommodated by skipping numbers.

Ideally, such schemes are designed in the light of both large amounts of sequence information and multiple structures. Insertion sites (i.e. residue L27A, etc.) are placed only in loop regions (or form β bulges) and have structural meaning such that topologically equivalent residues get the same label.

The numbering scheme for the light and heavy chains is shown in Table 5.1. The residues considered to be part of the six CDRs as defined by Wu and Kabat are CDR-L1: L24–L34; CDR-L2: L50–L56; CDR-L3: L89–L97; CDR-H1: H31–H35B; CDR-H2: H50–H65; CDR-H3: H95–H102.

While the Kabat numbering scheme is the most widely adopted, it was derived from the analysis of a rather limited set of sequence data and, as a result, has some problems.

The numbering adopts a very rigid specification such that the allowed insertions at each position are specified. For example, in CDR-H3, insertions occur between H100 and H101 and the Kabat specification allows insertion letters up to K: (i.e. H100, H100A . . . H100K, H101). This accommodates CDR-H3 loops with lengths up to 19 residues. More than 200 heavy chain sequences are now known in which CDR-H3 is longer than this, some being 30 residues in length and therefore needing insertion letters up to H100U. However, the Kabat standard does not allow insertion letters beyond H100K so there is no agreed way of numbering these very long loops. The Kabat data files place these additional insertions at varying positions. While CDR-H3 is the prime position at which such problems occur, similar situations can arise at other locations.

Even more importantly, when Chothia and Lesk performed their analysis of CDR conformation, they found that the insertion sites within CDR-L1 and CDR-H1 did not correspond to the sites of structural insertions. Therefore, when one looks at the three-dimensional structures, one finds that topologically equivalent residues in these loops are not assigned the same number, leading to the requirement for a structurally correct numbering scheme.

Table 5.1 Kabat numbering scheme.

Light chain										
L0	L1	L2	L3	L4	L5	L6	L7	L8	L9	
L10	L11	L12	L13	L14	L15	L16	L17	L18	L19	
L20	L21	L22	L23	L24	L25	L26	L27			
L27A	L27B	L27C	L27D	L27E	L27F					
								L28	L29	
L30	L31	L32	L33	L34	L35	L36	L37	L38	L39	
L40	L41	L42	L43	L44	L45	L46	L47	L48	L49	
L50	L51	L52	L53	L54	L55	L56	L57	L58	L59	
L60	L61	L62	L63	L64	L65	L66	L67	L68	L69	
L70	L71	L72	L73	L74	L75	L76	L77	L78	L79	
L80	L81	L82	L83	L84	L85	L86	L87	L88	L89	
L90	L91	L92	L93	L94	L95					
					L95A	L95B	L95C	L95D	L95E	L95F
						L96	L97	L98	L99	
L100	L101	L102	L103	L104	L105	L106				
						L106A				
							L107	L108	L109	
Heavy chain										
H0	H1	H2	H3	H4	H5	H6	H7	H8	H9	
H10	H11	H12	H13	H14	H15	H16	H17	H18	H19	
H20	H21	H22	H23	H24	H25	H26	H27	H28	H29	
H30	H31	H32	H33	H34	H35					
					H35A	H35B				
						H36	H37	H38	H39	
H40	H41	H42	H43	H44	H45	H46	H47	H48	H49	
H50	H51	H52								
		H52A	H52B	H52C						
			H53	H54	H55	H56	H57	H58	H59	
H60	H61	H62	H63	H64	H65	H66	H67	H68	H69	
H70	H71	H72	H73	H74	H75	H76	H77	H78	H79	
H80	H81	H82								
		H82A	H82B	H82C						
			H83	H84	H85	H86	H87	H88	H89	
H90	H91	H92	H93	H94	H95	H96	H97	H98	H99	
H100										
H100A	H100B	H100C	H100D	H100E	H100F	H100G	H100H	H100I	H100J	H100K
	H101	H102	H103	H104	H105	H106	H107	H108	H109	
H110	H111	H112	H113							

5.2.1.1 The Chothia Numbering Scheme

The problem of topological equivalents in CDR-L1 and CDR-H1 led Chothia and Lesk to introduce the Chothia numbering scheme. This is identical to the Kabat scheme with the exception of CDR-L1 and CDR-H1, where the insertions are placed at the structurally correct positions such that topologically equivalent residues do get the same label. The extreme variability of conformation in CDR-H3 and the lack of structures with very long CDR-H3 loops means that it has not been possible to assign a definitive numbering scheme with topological equivalence for this CDR.

Unfortunately, Chothia and co-workers confused issues from 1989 (Chothia et al. 1989) when they erroneously changed their numbering scheme such that insertions in CDR-L1 were placed after residue L31 rather than the structurally correct L30. This was corrected in 1997 (Al-Lazikani et al. 1997).

The correct version of the Chothia numbering (as used before 1989 and since 1997) for the light and heavy chains is shown in Table 5.2.

Table 5.2 Chothia numbering scheme.

Light chain										
L0	L1	L2	L3	L4	L5	L6	L7	L8	L9	
L10	L11	L12	L13	L14	L15	L16	L17	L18	L19	
L20	L21	L22	L23	L24	L25	L26	L27	L28	L29	
L30										
L30A	L30B	L30C	L30D	L30E	L30F					
	L31	L32	L33	L34	L35	L36	L37	L38	L39	
L40	L41	L42	L43	L44	L45	L46	L47	L48	L49	
L50	L51	L52	L53	L54	L55	L56	L57	L58	L59	
L60	L61	L62	L63	L64	L65	L66	L67	L68	L69	
L70	L71	L72	L73	L74	L75	L76	L77	L78	L79	
L80	L81	L82	L83	L84	L85	L86	L87	L88	L89	
L90	L91	L92	L93	L94	L95					
					L95A	L95B	L95C	L95D	L95E	L95F
						L96	L97	L98	L99	
L100	L101	L102	L103	L104	L105	L106				
						L106A				
							L107	L108	L109	
Heavy chain										
H0	H1	H2	H3	H4	H5	H6	H7	H8	H9	
H10	H11	H12	H13	H14	H15	H16	H17	H18	H19	
H20	H21	H22	H23	H24	H25	H26	H27	H28	H29	
H30	H31									
	H31A	H31B								
		H32	H33	H34	H35	H36	H37	H38	H39	
H40	H41	H42	H43	H44	H45	H46	H47	H48	H49	
H50	H51	H52								
		H52A	H52B	H52C						
			H53	H54	H55	H56	H57	H58	H59	
H60	H61	H62	H63	H64	H65	H66	H67	H68	H69	
H70	H71	H72	H73	H74	H75	H76	H77	H78	H79	
H80	H81	H82								
		H82A	H82B	H82C						
			H83	H84	H85	H86	H87	H88	H89	
H90	H91	H92	H93	H94	H95	H96	H97	H98	H99	
H100										
H100A	H100B	H100C	H100D	H100E	H100F	H100G	H100H	H100I	H100J	H100K
	H101	H102	H103	H104	H105	H106	H107	H108	H109	
H110	H111	H112	H113							

5.2.2
The IMGT Numbering Scheme

An alternative numbering scheme has been introduced by Lefranc for use with the Immunogenetics (IMGT) database (Lefranc 1997). The advantage of this scheme, based on germline sequences, is that it unifies numbering across antibody lambda and kappa light chains, heavy chains and T-cell receptor alpha, beta, gamma, and delta chains.

However, because the scheme is based on germline sequences it does not span CDR3 or FR4. In addition, insertions and deletions increase in size unidirectionally.

5.2.3
Honegger and Plückthun Numbering Scheme

Yet another numbering scheme was introduced by Honegger and Plückthun in 2001 (Honegger and Plückthun 2001). This takes the same approach as the IMGT scheme, but addresses the problem in the IMGT scheme of being cut short in CDR3. In addition, insertions and deletions, rather than growing unidirectionally, are placed symmetrically around a key position. Furthermore, whereas length variations in CDR1 and CDR2 are accounted for by a single gap position in IMGT, this scheme has two locations at which gaps may be introduced.

In conclusion, while the latter numbering schemes, in particular that of Honneger and Plückthun, have distinct advantages over the earlier schemes, the Kabat scheme in particular is so well established amongst immunologists, it is hard for them to gain acceptance. The Chothia scheme is routinely used by other groups in structural analysis, but these other schemes are rarely used outside the groups who have developed them.

5.3
Definition of the CDRs and Related Regions

The complementarity determining regions (CDRs) were defined by Wu and Kabat from their variability plot. Tips for identifying the location of the CDRs through visual inspection are available on Andrew Martin's website. However, others have also provided definitions of regions related to the CDRs. It is important to note that, in the main, these are not attempts to redefine the CDRs, they are simply alternative subsets of residues, overlapping the CDRs, which are important in different ways. However, confusingly, some authors do use terms such as "Chothia CDRs" although the regions defined by Chothia have a different meaning.

Chothia defined the "structural loops" – those regions likely to vary in conformation between different antibody structures. With the exception of CDR-H1, all the structural loops are contained within the CDRs; in the case of CDR-H1 the structural loop and the CDR overlap one another. Thus the analysis of canonical classes performed by Chothia is based on these structural loops rather than the full CDRs. To make things a little more confusing, the precise boundaries of the structural loops have varied somewhat between different papers from the Chothia group. These differences have been the result of finding changes in conformational variability as new structures have become available and have been added to the analysis. For example, the 1997 paper from Chothia's group (Al-Lazikani et al. 1997) changed from defining the CDR-H2 structural loop as residues H52–H56 to residues H50–H58.

Another region, known as the "AbM loops" was introduced by Martin et al. (1989). These regions are a compromise between the Kabat sequence-variability-defined CDRs and Chothia structural loops. This region is probably the most useful definition to use when trying to generate three-dimensional models of the conformations of the loops likely to interact with antigen.

An analysis of the contact residues from a set of antibody–antigen complexes by MacCallum et al. (1996) introduced the "contact region." Since these are the residues that are most likely to take part in interactions with the antigen, it is likely to be the most useful region for people wishing to perform mutagenesis to modify the affinity of an antibody.

IMGT has introduced another range of residues for the loops which form the combining site and confusingly they do term these CDRs.

These alternative regions are summarized in Table 5.3. Note that when using the Kabat numbering scheme, the C-terminal end of the Chothia structural loop changes residue number depending on the length of the loop.

Table 5.3 Residue ranges for different definitions of regions around the CDRs.

Loop	*Kabat CDR*	*AbM*	*Chothia*	*Contact*	*IMGT*
CDR-L1	L24–L34	L24–L34	L23–L34	L30–L36	L27–L32
CDR-L2	L50–L56	L50–L56	L50–L56	L46–L55	L50–L52
CDR-L3	L89–L97	L89–L97	L89–L97	L89–L96	L89–L97
CDR-H1	H31–H35B (Kabat numbering)	H26–H35B	H26–H32 . . . 34	H30–H35B	H26–H35B
CDR-H1	H31–H35 (Chothia numbering)	H26–H35	H26–H32	H30–H35	H26–H35
CDR-H2	H50–H65	H50–H58	H50–H58	H47–H58	H51–H57
CDR-H3	H95–H102	H95–H102	H95–H102	H93–H101	H93–H102

Note that for the Chothia definition, the C-terminal end of CDR-H1 varies in location under the Kabat numbering scheme depending on the length of the CDR (i.e. if neither H35A nor H35B is present then the loop ends at H32; if only H35A is present it ends at H33 and if both H35A and H35B are present then it ends at H34). It should also be noted that different papers by Chothia use slightly different definitions of the structural loops (for example earlier papers used H52–H56 for the CDR-H2 loop); the most recent definitions are shown.

5.4
Antibody Sequence Data

For those wishing to search and analyze antibody sequences, the standard databanks (Genbank/EMBL for DNA, Genpept/trEMBL/SwissProt for protein data) and search tools are generally not suitable. The sequence databanks make a deliberate effort to avoid including rearranged somatically mutated antibody sequences since the vast number of these can confuse significance statistics and more advanced profile-based search methods. Sequence search tools such as BLAST and FASTA are designed to identify homologs. Of course this is not an issue when comparing antibody sequences as all the sequences are homologous by definition.

Once the standard numbering scheme has been applied to antibody protein sequences, they are effectively multiply aligned. This allows much finer search criteria to be applied with the right tools. For example, one should be able to search a set of antibody sequences to find all examples with a 10-residue long CDR-L1 and a valine at position L30. Tools such as BLAST, FASTA, or SSEARCH are still valuable to search a database of antibody sequences to find the most similar sequence(s) or to rank sequences on the basis of sequence identity. However, the calculation of significance scores (*P*-values and *E*-values) is meaningless when the database contains only closely related homologs. If you wish to use one of these tools to calculate sequence identities with all antibody sequences in a database, you must set an extremely poor *E*-value cutoff (e.g. 10 000) to ensure that all sequences are compared.

5.4.1
Antibody Sequence Databanks

There are two major resources that collect antibody sequence data. Probably the best known, and certainly the oldest, resource is the Kabat database. This collection of data was started by Wu and Kabat in the 1970s when they started their work on analyzing sequence variability. It grew into the book *Sequences of Immunological Interest* (Kabat et al. 1991). The last edition of this appeared in 1991 when it was replaced by an internet-based resource.

The Kabat data have been available as a downloadable resource and as a web-based resource allowing interactive queries. The raw sequence data may be downloaded for local analysis from either the National Center for Biotechnology Information (NCBI) or European Bioinformatics Institute (EBI) FTP sites. The most up-to-date raw data is in the fixlen subdirectory (or in FASTA format in the fasta format subdirectory). The "fixlen" data format contains the sequences with the standard Kabat numbering scheme applied. Unfortunately, these freely available data have not been updated since April 2000 as the Kabat database has now become a paid-for resource. This can be accessed on the Kabat Database website (http://www.kabatdatabase.com/), but requires registration and payment for both commercial and academic use.

The second major resource is IMGT (Lefranc 2001). The data in IMGT are updated regularly and may be downloaded from the EBI FTP site. A huge advantage of IMGT is the adoption of an ontology to describe various features of the data (Giudicelli and Lefranc 1999). The ontology includes terms for species, loci, genes, chains, structure, localization, and specificity amongst numerous other terms. This makes it much easier to perform reliable analyses by allowing direct comparison of sequence characteristics. There are, however, some disadvantages compared with the Kabat data. First the data are only available as EMBL-style files, or as DNA FASTA files (at the time of writing, there are no FASTA files for protein translations). The EMBL-style files have translations that can be extracted, but no standard numbering scheme has been applied. Numbered sequences may be accessed via the web interface at the IMGT database website, but further confusion has been introduced by the IMGT numbering scheme described above. There is also some confusion about the copyright terms on the data. The user manual available from the EBI IMGT data manual FTP site states that the "manual and the database it accompanies may be copied and redistributed freely, without advance permission". On the other hand the IMGT database warranty web page states that "The IMGT software and data are provided as a service to the scientific community to be used only for research and educational purposes. [. . .] Any other use of IMGT material needs prior written permission of the IMGT coordinator and of the legal institutions."

5.4.2 Germline Sequence Databases

The Kabat and IMGT resources collect rearranged, somatically mutated, and expressed antibody sequences (though germline data may also be included). In contrast, Tomlinson's VBase (see the VBase website, http://vbase.mrc-cpe.cam.ac.uk/) is a comprehensive directory of all human germline variable region sequences (Tomlinson et al. 1992). The database was developed over a period of several years, but is now considered to be complete and is no longer updated. The sequences (both DNA and amino acid translation) can be viewed on the site and saved to disk through cut-and-paste. The site includes nucleotide alignments for all functional segments, scale maps of all human V gene loci, DNAPLOT alignment software allowing rearranged genes to be assigned to their closest germline counterparts and various compiled statistics (numbers of functional segments belonging to each V gene family, cuts by different restriction enzymes and polymerase chain reaction (PCR) primers for amplifying rearranged V genes).

Mouse germline data are collected by Almagro (Almagro et al. 1998) in the ABG database (http://www.ibt.unam.mx/vir/V_mice.html). This resource provides access to sequences for the mouse germline sequences together with alignments of murine V_H and V_κ sequences. Like VBase, pseudogene and fragment data are stored, but there is also information on the particular strain of mouse from which the sequence data are obtained. Currently there are no data on murine V_λ sequences, which are relatively rare.

5.4.3 Web Resources for Sequence Analysis

5.4.3.1 Kabat Data

Access to the Kabat data is available from the Kabat Database website. The web interface allows searches of the annotations (ID, name, species, authors, etc.) by keyword, sequence searches using patterns (and allowing for mismatches), selection of specific sequence types, and positional correlations. The website also allows alignment of a light chain sequence against the data and provides an assortment of analyses such as variability, length distribution, and general statistics. However, as stated above, access to these tools requires registration and payment.

The publicly available Kabat protein sequence data (up to April 2000) may also be searched using KabatMan (Martin 1996). This is a specialized database for the analysis of Kabat antibody sequence data which may be queried using a language similar to the standard database query language SQL ("structured query language") using a full web interface, or via a simplified point and click interface both available through Andrew Martin's KabatMan web pages (http://www.bioinf.org.uk/abs/kabatman.html and http://www.bioinf.org.uk/abs/simkab.html). KabatMan is particularly suited to global analysis of the antibody data. It allows searches to specify individual amino acids or the contents of one of the six CDRs. For example, to identify all the antibodies which bind to DNA, but do not contain arginine in CDR-H3 one could use the query:

```
SELECT name, h3
WHERE antigen inc "DNA"
h3 <> " " and
h3 inc "R" not and
```

In this query, the "SELECT" clause specifies which data are to be returned (here, the name of the antibody and the sequence of CDR-H3). The "WHERE" clause specifies that the antigen should be DNA, then requires that the sequence of CDR-H3 should not be blank and that CDR-H3 should not include the letter "R" (i.e. arginine). Detailed examples are given in the KabatMan paper and in the online help.

KabatMan allows selection of antibody name, antigen, CDR sequences and lengths, framework region sequences, light chain class, species, citation, sequences of light and heavy chains, Chothia canonical conformational classes (see below), Kabat identifiers, human subgroups and earliest publication date. All of these properties can be restricted in the "WHERE" clause. While the Kabat data do not provide a link between paired light and heavy chains, KabatMan adds this information and the requirement for a "complete" antibody can be specified in the search.

Almagro (Ramirez-Benitez et al. 2001) has also provided a search interface (VIR-II) to access the public Kabat data (see http://www.ibt.unam.mx/vir/VIR/vir_index.html). This interface allows a subset of sequences to be extracted on the basis of type, gross or fine specificity, sequence completeness, and the presence of paired light and heavy chains. At the time of writing, however, this search facility is not available.

5.4.3.2 IMGT Data

The IMGT data may be accessed via a sequence retrieval service (SRS) interface at the EBI's IMGT website (http://www.ebi.ac.uk/imgt/). The main IMGT database website in France allows searching on the basis of accession number, keywords, name, date, length, species, functionality, specificity, group, subgroup, and reference. The interface to the data is hierarchical in nature, allowing one to home in on a particular sequence. However this is not suited to global analysis of the data. In future, IMGT data will be integrated into KabatMan to enable global analysis.

5.5 Antibody Structure Data

In contrast to the sequence data, structural data for antibodies are stored in the Protein Databank together with other protein structure data (Berman et al. 2000). Therefore, standard resources like the CATH database (http://cathwww.biochem.ucl.ac.uk/) contain information on antibody structures (Orengo et al. 1997). The antibody protein fold falls into the CATH class 2.60.40.10.

In addition, three specialist resources provide summaries of antibody structure data. The earliest of these, SACS (Allcorn and Martin, 2002), may be accessed at the SACS Database website (http://www.bioinf.org.uk/abs/sacs/). This resource is maintained in a fully automated manner with a brief manual check before data are made available. It is updated every 3–6 months. The resource provides names, light chain class, species, antigen, crystal structure details, fragment type and lengths, and sequences of the six CDRs. The data may be sorted on various criteria, including lengths of the CDRs, and the whole dataset may be downloaded. Almagro maintained another similar resource (see VIR Structures, http://www.ibt.unam.mx/vir/structure/structures.html), but at the time of writing this has not been updated since February 2001. Recently, IMGT has introduced a new summary, IMGT/3Dstructure-DB (Kaas et al. 2004) available at the IMGT3D website (http://imgt3d.igh.cnrs.fr/). This allows searches on the basis of species, group, subgroup, gene or allele, CDR length, sequence pattern within the CDRs, specific amino acids, accessible surface area, and backbone conformational details, as well as residue contact information.

5.6
Sequence Families

Ever since the first antibodies were studied, there have been attempts to group them into distinct sets. Initially this enabled estimates of the number of genes and their chromosomal locations, but it also allowed for comparisons between antibodies from different species, generally humans and mice. Groups of sequences are usually defined by analysis of sequence identity at the amino acid or nucleotide level, but have also been defined by antigen specificity and chromosomal location. The term "family" is usually applied to a set of sequences that are defined by protein sequence similarity or identity, and a "subgroup" generally refers to a set defined according to nucleotide sequence identity. In the IMGT-ONTOLOGY a subgroup is defined as a "set of genes which belong to the same group [V, D, J, or C], in a given species, and which share at least 75% identity at the nucleotide level" (Giudicelli and Lefranc 1999). This definition is not universally accepted, however, and "family" and "subgroup" are sometimes used interchangeably, or in a more general sense to denote grouping by any method. The number and definition of families/subgroups in the literature has varied, both because comprehensive data and sufficiently powerful computers have only recently become available, and because the grouping process is necessarily somewhat arbitrary. This arbitrariness is not a problem if the groupings are useful for some particular purpose.

5.6.1
Families and Subgroups

Inspection of amino acid sequences suggests that the variable region genes are arranged in subgroups, which evolved by gene duplication, and gene diversity is partly due to the expansion and contraction of the subgroups. DNA sequence analysis has confirmed the existence of subgroups for V_L and V_H genes and at least some of the hypervariability in CDR-H1 and CDR-H2 is present in the germline V_H genes, suggesting that an evolutionary process has caused substitution in the hypervariable regions.

Subdivision at the gene level can be made by cross-hybridization of different V genes with cDNA probes. There is not always a correspondence between groups defined on the basis of protein and nucleotide sequences, but the correlation is often good (Rechavi et al. 1983).

The V gene subgroups are multigene families that have been maintained throughout evolution; a subgroup preserves some characteristics in the noncoding segments that differ in other subgroups. It is difficult to relate subgroups from different species based on amino acid similarity; the germline structure, including the noncoding regions, provides a better understanding of the evolutionary relationship of V subgroups within, and subsequently between, species. Each cluster of V genes that constitutes a multigene family is likely to undergo concerted evolution, with preservation of the sequence characteristics of the subgroup (Kroemer et al. 1991; Rechavi et al. 1983).

5.6.2
Human Family Chronology

6.6.2.1 Human Heavy Chain Variable Genes (V_H)

Kabat et al. (1991) classified V_H genes into three groups (I, II, and III) according to amino acid sequence identity; this grouping was confirmed by analysis at the nucleotide level (Kodaira et al. 1986). An additional family, homologous to a mouse family, was defined by Lee et al. (1987), and a further small family, with an unusual DNA sequence, was determined by Shen et al. (1987). Berman et al. (1988) defined six families; Kabat's subgroup II was divided into two families, one of which corresponds to Lee's family (named IV), and the fifth group (V) corresponds to Shen's family. The new sixth family (VI), which has 70% identity with family IV, was determined with Southern blot analysis and nucleotide identity.

A set of V_H sequences, previously classified as members of family I, differ at a clustered region and were proposed as a seventh family (VII) by Schroeder et al. (1990). This family can be considered as a subfamily of I, or as a family in transition to independence – either way, the classification has generally been accepted (Cook and Tomlinson 1995). Membership of a family is generally defined by >80% sequence identity at the nucleotide level, and this definition is supported by phylogenetic analysis (Honjo and Matsuda 1995).

Kabat et al. (1991) noted that the threshold of 80% nucleotide identity is somewhat arbitrary, and in the 1994 version of the Kabat database, the sequences are divided into families based on amino acid identity, where members of a family differ by 12 amino acids or less (Déret et al. 1995). This criterion creates 14 V_H families, but the classification into seven families described above is more generally used.

5.6.2.2 Human Light Chain Variable Genes (V_κ and V_λ)

In 1984, human V_κ sequences were classified into four families (I–IV) on the basis of amino acid sequence similarity (Pech et al. 1984), and Kabat et al. (1991) continued to use this grouping. Kroemer et al. (1991) found that analysis of nucleotide sequence identity largely paralleled this classification, but some sequences could not be assigned to a family using a threshold of 80% identity. Four sequences were grouped into three additional families, partly based on nucleotide identity and partly based on similarity to mouse V_κ families. A phylogenetic analysis of human V_κ genes showed four major clusters, and three groups with a single sequence (Sitnikova and Nei 1998), corresponding to the seven families defined by sequence similarity.

Based on >75% nucleotide sequence identity, human V_λ genes were placed into 10 subgroups (Frippiat et al. 1995), a result which agrees with later phylogenetic analysis (Williams et al. 1996).

In the 1994 version of the Kabat database, V_κ and V_λ sequences are grouped into six and nine families, respectively.

5.6.3 Mouse Family Chronology

5.6.3.1 Mouse Heavy Chain Variable Genes (V_H)

Amino acid sequence similarity defined seven families of mouse V_H sequences, each family having a different specificity. However, this grouping was not proposed as definitive, since the analyzed data did not constitute a representative sample (Dildrop 1984). These families corresponded with those determined by Southern blot analysis, and all nucleotide sequences available at that time had >80% sequence identity to members of one of these seven families (Brodeur and Riblet 1984). Two additional families were later defined (Dildrop et al. 1985).

Kabat divided the three subgroups that he had previously defined to correspond with the nine groups based on amino acid sequences (Kabat et al. 1991). As with the human V_H sequences, a more stringent rule was applied in the 1994 version of the Kabat database, creating a somewhat unwieldy 27 groups.

The number of V_H families was subsequently revised further, based on family members sharing >80% nucleotide sequence identity, and nonfamily members having 70–75% identity: first by Strohal et al. (1989) to 11, then by Honjo and Matsuda (1995) to 14, and finally by Mainville et al. (1996) to 15. This grouping into 15 subgroups is now recognized as standard and is used by the IMGT database (Lefranc 2001).

5.6.3.2 Mouse Light Chain Variable Genes (V_κ and V_λ)

Classification of mouse V_κ sequences into subgroups based on nucleotide sequence identity is not as unambiguous as it is with V_H sequences (Strohal et al. 1989). Using an 80% threshold resulted in 16 subgroups, though the existence of more was predicted (D'Hoostelaere and Klinman 1990); and the figure was increased to 19 by Kofler and Helmberg (1991). However, some subgroups shared >75% identity, members of different subgroups sometimes shared >80% identity, and there was not always a correspondence between these subgroups and V_κ protein groups. These awkward results suggest that while the arbitrary threshold of 80% might be useful, it is not necessarily significant in evolutionary terms (Kroemer et al. 1991).

In 1991, Kabat arranged the V_κ sequences into seven families (Kabat et al. 1991), with an additional "miscellaneous" family for sequences that were problematic. These groupings were revised in the 1994 database, creating 26 families. Lambda chains represent only around 5% of the total murine light chains, and the three genes can be classified into two families (Sitnikova and Su 1998), but such a classification is not particularly useful.

5.6.4
Correspondence Between Human and Mouse Families

5.6.4.1 Heavy Chain Variable Genes (V_H)

Genes of the same family in different species can be more alike than genes of different families in the same species (Lee et al. 1987). Such interspecies similarities could be explained by evolution from common ancestral genes or by shared requirements for structure and diversity (Berman et al. 1988).

As described above, early work by Kabat classified human V_H sequences into subgroups I, II, and III. More early work by Rechavi et al. (1982) suggested that the human V_H III subgroup underwent a significant gene expansion compared to the equivalent mouse subgroup. Surprisingly, however, they found through comparison of amino acid sequences, that the large set of human genes correspond to a small subset of mouse genes. Analysis of human and mouse germline V_H regions suggests that V_H families developed before speciation, and that they have been conserved by selection at the protein level (Brodeur and Riblet 1984).

More recent work has expanded the number of families based on analysis of complete genomes. Excluding pseudogenes and rearranged gene sequences, mouse V_H genes are generally classified into 15 families (Mainville et al. 1996) (termed "subgroups" in IMGT), while human V_H genes are classified into seven families (Schroeder et al. 1990).

However, human V_H genes have historically been clustered using more liberal criteria than those applied to mouse V_H genes, so to allow better comparisons mouse V_H sequences have been grouped into sets using the same criteria normally used for humans (de Bono et al. 2004). This produced eight rather than 15 "sets," each containing between one and five of the 15 conventional V_H subgroups (Table 5.4). Three of these sets have a one-to-one match with human V_H families; in two cases two mouse sets correspond to a single human family and in one case two human families correspond to a single mouse set. Phylogenetic analysis by de Bono et al. (2004) has confirmed the human family–mouse set relationships established by sequence comparison.

5.6.4.2 Light Chain Variable Genes (V_κ and V_λ)

As with heavy chains, V_κ sequences from corresponding families of different species can show greater similarity than sequences of different families within a species, indicating that V_κ family genes were fixed prior to mammalian speciation (Kroemer et al. 1991). Using nucleotide sequence identity, equivalent human and mouse families were ascertained, though the complexity of the mappings varied. The single gene in human family V_κ IV has at least 10 equivalents in mice, spread across three families; and a small mouse family is related to two human V_κ families, estimated to contain at least three times as many genes. A single member human family and large mouse family do not correspond to a family from the other species. Despite this variation, the maintenance of family specificity over 5070 million years suggests some degree of environmental selection pressure.

Table 5.4 Correspondence between mouse V_H sets (de Bono et al. 2004), the standard 15 families (Mainville et al. 1996) or subgroups as used by IMGT and the seven human V_H families (Schroeder et al. 1990).

Mouse V_H set	*IMGT subgroup*	*Human V_H family*
1	1	I
	14	
2	8	II
3a	4	III
	11	
3b	5	III
	6	
	7	
	10	
	13	
4a	3	IV
	12	
4b	2	IV
N/A	–	V
N/A	–	VI
7	9	VII
8	15	–

Data from de Bono et al. (2004).

It is difficult to establish interspecies correspondence for V_λ genes, since the number of mouse V_λ genes is small. The high diversity of mouse V_κ genes which compensates for this means that some of those families may have equivalents in the human V_λ families.

5.6.5 Tools for Assigning Subgroups

The Kabat Database website allows one to submit a sequence to determine the subgroup as specified in the 1991 Kabat book (Kabat et al. 1991).

Déret's SUBIM program (Déret et al. 1995) which may be downloaded from Paris Institute of Mineralogy FTP site (ftp://ftp.lmcp.jussieu.fr/pub/sincris/software/protein/subim/), allows the assignment of the variability subgroup of human sequences by comparison of the N-terminal 15 residues with consensus sequences determined by Kabat et al. (1991). The subgroup assignment function of Déret's program is also accessible via a web interface available on Andrew Martin's Subgroup website (http://www.bioinf.org.uk/abs/hsubgroup.html). The assignment algorithm is also built into the KabatMan software discussed above.

5.7
Screening new antibody sequences

Given a new antibody sequence, one can assign families and subgroups using the tools described above. In addition it may be interesting to look for unusual features in your sequence. These may indicate cloning artifacts, errors in the sequencing, or residues that are critical to the binding of this particular antibody. A server which compares your sequence against sequences in the Kabat database and reports amino acids occurring in less than 1% of the data in the database is available on Andrew Martin's Seqtest web page (http://www.bioinf.org.uk/abs/seqtest.html) and a typical sequence does have one or two "unusual" residues. To use the server, you simply enter the amino acid sequence of your Fv fragment (one or both chains). Optionally you may include the whole Fab fragment, but only the Fv portion will be tested. The method is described in detail on the web pages.

A new server from Raghavan and Martin on Andrew Martin's SHAB web page (http://www.bioinf.org.uk/abs/shab/) assesses the "humanness" of an antibody. This may be of value in selecting non-human antibodies which can be used successfully as chimerics in human therapy or for *in vivo* diagnostics. The method compares the antibody sequence with all known human sequences and calculates an average sequence identity. This is then plotted onto a distribution of scores achieved for every human antibody sequence in the Kabat database and a *Z*-score is assigned. Positive *Z*-scores indicate sequences that are more typically human than average while negative scores indicate sequences with less than average scores. Using this tool, it is observed that many mouse antibodies have positive *Z*-scores, indicating that they are more typically "human" than some human antibodies.

5.8
Antibody Structure Prediction

Since the 1980s there has been much interest in modeling the three-dimensional structure of antibodies. In part this is because antibodies provide a unique opportunity for protein modelers. The framework is so conserved that the problem is essentially reduced to one of modeling the six CDRs and Chothia's analysis has determined rules that make it straightforward to obtain generally accurate models for all but CDR-H3 in at least 75% of cases. However, modeling can also be valuable in understanding the binding of antibody and antigen with a view to modifying affinity or specificity, or in understanding cross-reactivity and autoimmune disease.

The problem of modeling the three-dimensional structure of an antibody can be broken down into two major steps: building the framework and building the CDRs.

5.8.1
Build the framework

Antibody crystal structures with the most similar light and heavy chains are identified from the Protein Databank (PDB, see RCSB website, http://www.rcsb.org/pdb/). Light and heavy chains may be identified separately and then must be combined. The combination of chains will inherit the packing between V_L and V_H from one of the parent structures. Thus, if one selects light chain, L*a* (paired with H*a*) and heavy chain, H*b* (paired with L*b*) then one can choose the V_L/V_H packing based on either antibody *a* or antibody *b*. To inherit the packing from antibody *b*, the structure of L*a* is fitted to L*b* and chains L*a* and H*b* are retained, discarding H*a* and L*b*. To inherit the packing from antibody *a*, H*b* is fitted to H*a*, and H*b* and L*a* are retained discarding H*a* and L*b*. Currently the choice of packing is arbitrary and it may be worth constructing two models. The sidechains of the framework are then replaced using automated processes available in molecular graphics programs, or software such as SCWRL (Canutescu et al. 2003).

5.8.2
Build the CDRs

Typically CDRs are built using the canonical classes described by Chothia. A server which will assign canonical classes automatically given the sequence of an antibody is available on Andrew Martin's Chothia web page (http://www.bioinf.org.uk/abs/chothia.html). The website also has a summary of the key residues described in the various papers by Chothia et al. and those resulting from an automated classification by Martin and Thornton (1996). Once canonical classes have been identified, examples from the Protein Databank are selected (typically on the basis of maximum sequence identity throughout the CDR) and fitted onto the framework.

In general, canonical classes can be identified for 4 or 5 of the 6 CDRs. CDR-H3 is too variable to be classified into canonical classes at the same level of detail although some work has been done classifying its conformations into groups (Martin and Thornton 1996; Shirai et al. 1996; Morea et al. 1998; Oliva et al. 1998).

CDRs which cannot be built using canonicals must be built using another modeling method. This can be conformational search using software such as CONGEN (Bruccoleri and Karplus 1987), by searching the PDB for loops of the same length and with similar distance between the attachment points to the framework, or by combined methods such as CAMAL (Martin et al. 1989).

5.8.3
Automated Modeling Tools

A number of automated tools are available for general protein modeling. These include MODELLER (http://www.salilab.org/modeller/) and SwissModel (http://

swissmodel.expasy.org/). They may both be used to generate a model in a quick and simple manner, but they do not take advantage of the special properties of antibodies and therefore are unlikely to produce a model of the quality that can be generated if a specialist program is used.

The commercial program AbM, designed specifically for antibody modeling, is, unfortunately, no longer available. However, a modified version of the software known as WAM (Whitelegg and Rees 2000, 2004) is accessible over the web (http://antibody.bath.ac.uk/). This server automates the generation of antibody models using the methods described above. Canonicals are used to build the hypervariable loops where possible and the remaining loops are built using the CAMAL method of Martin et al. (1989).

Another automated approach, ABGEN, has been proposed by Mandal et al. (1996), but is no longer available as a server.

5.9 Summary

This chapter has briefly reviewed antibody structure and discussed the different numbering schemes that have been used for antibody sequences. While the latest schemes may have their advantages, the Kabat and Chothia schemes are unlikely to be replaced by immunologists. The chapter then went on to look at the different definitions of regions around the CDRs which have been used by different groups for different purposes. Antibody sequence and structure databases were presented together with the tools and web resources that can be used to access them. The plethora of schemes for classifying antibody sequences were discussed highlighting some recent information on the equivalence between human and mouse groupings. Finally, tools for dealing with a new antibody sequence, both at the level of sequence analysis and structure prediction were presented.

It is the nature of the web that resources come and go or move location. Links to all the services mentioned here will be maintained on Andrew Martin's web pages.

References

Al-Lazikani, B., Lesk, A.M., Chothia, C. (1997) Standard conformations for the canonical structures of immunoglobulins. *J Mol Biol* 273: 927–948.

Allcorn, L.C., Martin, A.C.R. (2002) SACS – self-maintaining database of antibody crystal structure information. *Bioinformatics* 18: 175–181.

Almagro, J.C., Hernández, I., Ramírez, M. C., Vargas-Madrazo, E. (1998) Structural differences between the repertoires of mouse and human germline genes and their evolutionary implications. *Immunogenetics* 47: 355–363.

Alzari, P.M., Lascombe, M.-B., Poljak, R.J. (1988) Three-dimensional structure of antibodies. *Annu Rev Immunol* 6: 555–580.

Berman, J., Mellis, S., Pollock, R., Smith, C., Suh, H., Heinke, B., Kowal, C., Surti, U., Chess, L., Cantor, C., F.W., A. (1988)

Content and organization of the human Ig V_H locus: definition of three new V_H families and linkage to the Ig C_H locus. *EMBO J*, 7: 727–738.

Berman, H.M., Westbrook, J., Feng, Z., Gilliland, G., Bhat, T. N., Weissig, H., Shindyalov, I. N., Bourne, P. E. (2000) The Protein Data Bank. *Nucleic Acids Res* 28: 235–242.

Brodeur, P., Riblet, R. (1984) The immunoglobulin heavy chain variable region (IgH-V) locus in the mouse. I. One hundred IgH-V genes comprise seven families of homologous genes. *Eur J Immunol* 14: 922–930.

Bruccoleri, R.E., Karplus, M. (1987) Prediction of the folding of short polypeptide segments by uniform conformational sampling. *Biopolymers*, 26: 137–168.

Canutescu, A.A., Shelenkov, A.A., Dunbrack, R.L. (2003) A graph-theory algorithm for rapid protein side-chain prediction. *Protein Sci* 12: 2001–2014.

Chothia, C., Lesk, A.M. (1987) Canonical structures for the hypervariable regions of immunoglobulins. *J Mol Biol* 196: 901–917.

Chothia, C., Lesk, A.M., Tramontano, A., Levitt, M., Smith-Gill, S.J., Air, G., Sheriff, S., Padlan, E.A., Davies, D., Tulip, W.R., Colman, P.M., Spinelli, S., Alzari, P.M., Poljak, R.J. (1989) Conformations of immunoglobulin hypervariable regions. *Nature (Lond)* 342: 877–883.

Cook, G., Tomlinson, I. (1995) The human immunoglobulin V_H repertoire. *Immun Today* 16: 237–242.

Davies, D.R., Padlan, E.A., Sheriff, S. (1990) Antibody-antigen complexes. *Annu Rev Biochem* 59: 439–473.

de Bono, B., Madera, M., Chothia, C. (2004) V_H gene segments in the mouse and human genome. *J Mol Biol* 342: 131–143.

Déret, S., Maissiat, C., Aucouturier, P., Chomilier, J. (1995) SUBIM: A program for analysing the Kabat database and determining the variability subgroup of a new immunoglobulin sequence. *Comput Appl Biosci* 11: 435–439.

D'Hoostelaere, L., Klinman, D. (1990) Characterization of new mouse v groups. *J Immunol* 145: 2706–2712.

Dildrop, R. (1984) A new classification of mouse V_H sequences. *Immunol Today* 5: 85–86.

Dildrop, R., Krawinkel, U., Winter, E., Rajewsky, K. (1985) V_H-gene expression in murine lipopolysaccharide blasts distributes over the nine known V_H-gene groups and may be random. *Eur J Immunol* 15: 1154–1156.

Frippiat, J.-P., Williams, S., Tomlinson, I., Cook, G., Cherif, D., Le Paslier, D., Collins, J., Dunham, I., Winter, G., Lefranc, M.-P. (1995) Organization of the human immunoglobulin lambda light-chain locus on chromosome 22q11.2. *Hum Mol Genet* 4: 983–991.

Giudicelli, V., Lefranc, M.-P. (1999) Ontology for immunogenetics: the IMGT-ONTOLOGY. *Bioinformatics* 15: 1047–1054.

Honegger, A., Plückthun, A. (2001) Yet another numbering scheme for immunoglobulin variable domains: an automatic modeling and analysis tool. *J Mol Biol* 309: 657–670.

Honjo, T., Matsuda, F. (1995) *Immunoglobulin Genes*, 2nd edn. London: Academic Press, pp. 145–171.

Kaas, Q., Ruiz, M., Lefranc, M.-P. (2004) IMGT/3Dstructure-DB and IMGT/StructuralQuery, a database and a tool for immunoglobulin, T cell receptor and MHC structural data. *Nucleic Acids Res* 32: D208–D210.

Kabat, E.A., Wu, T.T., Perry, H.M., Gottesman, K.S., Foeller, C. (1991). *Sequences of Proteins of Immunological Interest*, 5th edn. Bethesda, MD: US Department of Health and Human Services, National Institutes for Health.

Kodaira, M., Kinashi, T., Umemura, I., Matsuda, F., Noma, T., Ono, Y., Honjo, T. (1986) Organization and evolution of variable region genes of the human immunoglobulin heavy chain. *J Mol Biol* 190: 529–541.

Kofler, R., Helmberg, A. (1991) A new Igk-V family in the mouse. *Immunogenetics* 34: 139–140.

Kroemer, G., Helmberg, A., Bernot, A., Auffray, C., Kofler, R. (1991) Evolutionary relationship between human and mouse immunoglobulin kappa light chain variable region genes. *Immunogenetics* 33: 42–49.

Lee, K., Matsuda, F., Kinashi, T., Kodaira, M., Honjo, T. (1987) A novel family of variable region genes of the human immunoglobulin heavy chain. *J Mol Biol* 195: 761–768.

Lefranc, M.P. (1997) Unique database numbering system for immunogenetic analysis. *Immunol Today* 18: 509–509.

Lefranc, M.-P. (2001) Imgt, the international immunogenetics database. *Nucleic Acids Res* 29: 207–209.

MacCallum, R.M., Martin, A.C., Thornton, J.M. (1996) Antibody-antigen interactions: Contact analysis and binding site topography. *J Mol Biol* 262: 732–745.

Mainville, C., Sheehan, K., Klaman, L., Giorgetti, C., Press, J., Brodeur, P. (1996) Deletional mapping of fifteen mouse V_H gene families reveals a common organization for three Igh haplotypes. *J Immunol* 156: 1038–1046.

Mandal, C., Kingery, B.D., Anchin, J.M., Subramaniam, S., Linthicum, D.S. (1996) ABGEN: a knowledge-based automated approach for antibody structure modeling. *Nat Biotechnol* 14: 323–328.

Mariuzza, R.A., Phillips, S.E.V., Poljak, R.J. (1987) The structural basis of antigen-antibody recognition. *Annu Rev Biophys Bioeng* 139–159.

Martin, A.C.R. (1996) Accessing the Kabat antibody sequence database by computer. *Proteins Struct Funct Genet* 25: 130–133.

Martin, A.C.R., Thornton, J.M. (1996) Structural families in homologous proteins: automatic classification, modelling and application to antibodies. *J Mol Biol* 263: 800–815.

Martin, A.C.R., Cheetham, J.C., Rees, A.R. (1989) Modelling antibody hypervariable loops: A combined algorithm. *Proc Natl Acad Sci USA* 86: 9268–9272.

Morea, V., Tramontano, A., Rustici, M., Chothia, C., Lesk, A.M. (1998) Conformations of the third hypervariable region in the V_H domain of immunoglobulins. *J Mol Biol* 275: 269–294.

Oliva, B., Bates, P.A., Querol, E., Avilés, F. X., Sternberg, M.J. (1998) Automated classification of antibody complementarity determining region 3 of the heavy chain (H3) loops into canonical forms and its application to protein structure prediction. *J Mol Biol* 279: 1193–1210.

Orengo, C.A., Michie, A.D., Jones, S., Jones, D.T., Swindells, M.B., Thornton, J.M. (1997) CATH – a hierarchic classification of protein domain structures. *Structure* 5: 1093–1108.

Padlan, Eduardo, A. (1994) Anatomy of the antibody molecule. *Mol Immunol* 31: 169–217.

Padlan, E.A. (1977) The structural basis for the specificity of antibody-antigen reactions and structural mechanisms for the diversification of antigen-binding specificities. *Quant Rev Biophys* 10: 35–65.

Pech, M., Jaenichen, H., Pohlenz, H., Neumaier, P., Klobeck, H., Zachau, H. (1984) Organization and evolution of a gene cluster for human immunoglobulin variable regions of the kappa type. *J Mol Biol* 176: 189–204.

Poljak, R.J., Amzel, L.M., Avey, H.P., Chen, B.L., Phizackerley, R.P., Saul, F. (1973) The three-dimensional structure of the Fab fragment of a human immunoglobulin at 2.8Å resolution. *Proc Natl Acad Sci USA* 70: 3305–3310.

Porter, R.R. (1959) The hydrolysis of rabbit-globulin and antibodies with crystalline papain. *Biochem J* 73: 119–127.

Ramirez-Benitez, M.d.C., Moreno-Hagelsieb, G., Almagro, J.C. (2001) VIR.II: a new interface with the antibody sequences in the Kabat database. Biosystems, 61: 125–131.

Rechavi, G., Bienz, B., Ram, D., Ben-Neriah, Y., Cohen, J., Zakut, R., Givol, D. (1982) Organization and evolution of immunoglobulin V_H gene subgroups. *Proc Natl Acad Sci USA* 79: 4405–4409.

Rechavi, G., Ram, D., Glazer, L., Zakut, R., Givol, D. (1983) Evolutionary aspects of immunoglobulin heavy chain variable region (V_H) gene subgroups. *Proc Natl Acad Sci USA* 90: 855–859.

Schroeder, H.J., Hillson, J., Perlmutter, R. (1990) Structure and evolution of mammalian V_H families. *Int Immunol* 2: 41–50.

Searle, S.J., Pedersen, J.T., Henry, A.H., Webster, D.M., Rees, A.R. (1994) Antibody structure and function. In: Borreback, C.

A.K. (ed.) *Antibody Engineering*. Oxford: Oxford University Press, pp. 3–51.

Shen, A., Humphries, C., Tucker, P., Blattner, F. (1987) Human heavy-chain variable region gene family nonrandomly rearranged in familial chronic lymphocytic leukemia. *Proc Natl Acad Sci USA* 84: 8563–8567.

Shirai, H., Kidera, A., Nakamura, H. (1996) Structural classification of CDR-H3 in antibodies. *FEBS Lett* 399: 1–8.

Sitnikova, T., Nei, M. (1998) Evolution of immunoglobulin kappa chain variable region genes in vertebrates. *Mol Biol Evol* 15: 50–60.

Sitnikova, T., Su, C. (1998) Coevolution of immunoglobulin heavy- and light-chain variable-region gene families. *Mol Biol Evol* 15: 617–625.

Strohal, R., Helmberg, A., Kroemer, G., Kofler, R. (1989) Mouse V_κ gene classification by nucleic acid sequence similarity. *Immunogenetics* 30: 475–493.

Tomlinson, I. M., Walter, G., Marks, J. D., Llewelyn, M. B., Winter, G. (1992) The repertoire of human germline V_H sequences reveals about fifty groups of V_H segments with different hypervariable loops. *J Mol Biol* 227: 776–798.

Valdar, W.S., Thornton, J.M. (2001) Protein-protein interfaces: Analysis of amino acid conservation in homodimers. *Proteins Struct Funct Genet* 42: 108–124.

Valdar, W.S.J. (2002) Scoring residue conservation. *Proteins Struct Funct Genet* 48: 227–241.

Valentine, R.C., Green, N.M. (1967) Electron microscopy of an antibody-hapten complex. *J Mol Biol* 27: 615–617.

Whitelegg, N.R., Rees, A.R. (2000) WAM: an improved algorithm for modelling antibodies on the WEB. *Protein Eng* 13: 819–824.

Whitelegg, N., Rees, A.R. (2004) Antibody variable regions: Toward a unified modeling method. *Methods Mol Biol* 248: 51–91.

Williams, S., Frippiat, J.-P., Tomlinson, I., Ignatovich, O., Lefranc, M.-P., Winter, G. (1996) Sequence and evolution of the human germline V lambda repertoire. *J Mol Biol* 264: 220–232.

Wilson, I.A., Stanfield, R.L. (1993) Antibodyantigen interactions. *Curr Opin Struct Biol* 3: 113–118.

Wu, T.T., Kabat, E.A. (1970) An analysis of the sequences of the variable regions of Bence Jones proteins and myeloma light chains and their implications for antibody complementarity. *J Exp Med* 132: 211–250.

Websites

ABG Database: http://www.ibt.unam.mx/vir/V_mice.html

Andrew Martin: http://www.bioinf.org.uk/abs/

Andrew Martin Chothia: http://www.bioinf.org.uk/abs/chothia.html

Andrew Martin Kabatman: http://www.bioinf.org.uk/abs/kabatman.html and http://www.bioinf.org.uk/abs/simkab.html

Andrew Martin Seqtest: http://www.bioinf.org.uk/abs/seqtest.html

Andrew Martin SHAB: http://www.bioinf.org.uk/abs/shab/

Andrew Martin Subgroup: http://www.bioinf.org.uk/abs/hsubgroup.html

Annemarie Honegger: http://www.biochem.unizh.ch/antibody/

CATH Database: http://cathwww.biochem.ucl.ac.uk/

EBI IMGT: http://www.ebi.ac.uk/imgt/

EBI IMGTdata: ftp://ftp.ebi.ac.uk/pub/databases/imgt/ligm/

EBI IMGT data manual: ftp://ftp.ebi.ac.uk/pub/databases/imgt/ligm/userman_doc.html

EBI Kabat data: ftp://ftp.ebi.ac.uk/pub/databases/kabat/

IMGT3D: http://imgt3d.igh.cnrs.fr/

IMGT Database: http://imgt.cines.fr/

IMGT Database Warranty: http://imgt.cines.fr/textes/Warranty.html

Kabat Database: http://www.kabatdatabase.com/

MODELLER: http://www.salilab.org/modeller/
NCBI Kabat data: ftp://ftp.ncbi.nlm.nih.gov/repository/kabat/
Paris Institute of Mineralogy FTP site: ftp://ftp.lmcp.jussieu.fr/pub/sincris/software/protein/subim/
RCSB: http://www.rcsb.org/pdb/
SACS Database: http://www.bioinf.org.uk/abs/sacs/
SwissModel: http://swissmodel.expasy.org/
VBase Database: http://vbase.mrc-cpe.cam.ac.uk/
VIR Database: http://www.ibt.unam.mx/vir/VIR/vir_index.html
VIR Structures: http://www.ibt.unam.mx/vir/structure/structures.html
WAM: http://antibody.bath.ac.uk/

Note added in proof

The search and analysis tools are no longer available on the Kabat database website at http://www.kabatdatabase.com/ but the data and tools may be obtained for a fee.

A new compilation of germline sequences in human and mouse (soon to be extended to other species) has become available at http://www.vbase2.org/

6
Molecular Engineering I: Humanization

José W. Saldanha

6.1
Introduction

Humanization, also referred to as reshaping, complementarity determining region (CDR) grafting, veneering, resurfacing, specificity determining residue (SDR) transfer or DeImmunization, comprises strategies for reducing the immunogenicity of monoclonal antibodies (mAbs) from animal sources and for improving their activation of the human immune system. There are now many humanized mAbs in late-phase clinical trials and several have been given approval to be used as biopharmaceuticals (www.fda.gov). The source of the donor antibodies is usually mouse or rat, but rabbit (Steinberger et al. 2000) and chicken (Tsurushita et al. 2004) have also been used, the former because their CDR-H3 length is closer to human than mouse, the latter because they are useful for raising antibodies against conserved mammalian antigens. Although the mechanics of producing engineered mAbs using the techniques of molecular biology are relatively straightforward, the design of the humanized antibody sequence is critical for reproducing the affinity, specificity, and function of the original molecule whilst minimizing human anti-mouse antibody (HAMA) responses elicited in patients. In some cases, humanization has even led to an increase in the affinity of the antibody (Kolbinger et al. 1993; Brams et al. 2001; Luo et al. 2003).

There are many strategies leading to the design of the humanized variable regions (Fvs) and thus various choices are open to the antibody designer. These strategies and choices are the subject matter of this chapter. However, it is worth noting that some animal mAbs have proved difficult to humanize using current protocols (Pichla et al. 1997) and there is a need to experimentally verify the various approaches. The design and engineering of humanized mAbs are still interesting areas of research, as much for the light they shed on protein structure and function as for the potential therapeutic and diagnostic benefits.

Handbook of Therapeutic Antibodies. Edited by Stefan Dübel

ISBN 978-3-527-31453-9

6.2 History of Humanization

Over a century ago, Paul Ehrlich proposed that antibodies could be used as magic bullets to target and destroy human diseases. This vision is still being pursued today since antibodies combine the properties of specificity and affinity with the ability to recruit effector functions of the immune system such as complement-dependent cytolysis (CDC) and antibody-dependent cell-mediated cytotoxicity (ADCC). Alternatively, a toxic payload (such as a radioactive isotope, protein, or small molecule toxin) attached to the antibody can be accurately delivered to the target. Historically, antibodies have been produced from the serum of animals containing a cocktail of polyclonals, but the advent of hybridoma technology (Kohler and Milstein 1975) allowed monoclonals to become useful research and diagnostic tools, even though their use as therapeutics has been hindered by the elicitation of the HAMA response. Despite this problem, several animal mAbs have been approved by the US Food and Drug Administration (FDA). The obvious solution to this problem would be to raise human mAbs to the therapeutic targets, but this is difficult both practically and ethically using hybridoma technology. Nevertheless, production of fully human mAbs from transgenic mice and phage display has been possible since the early 1990s and Humira (adalimumab), isolated by phage display, was approved for rheumatoid arthritis in 2002. However, it does incur immunogenicity in 12% of the treated population when used alone (Hwang and Foote 2005).

Scientists are now using the techniques of molecular biology to design, engineer, and express mAbs from hybridoma technology to produce humanized mAbs. These approaches are suitable because of the domain structure of antibody molecules that allows functional domains carrying antigen binding or effector functions to be exchanged (Fig. 6.1). The first step was to produce a chimeric antibody (Morrison et al. 1984; Boulianne et al. 1984) where the xenogeneic variable light (V_L) or variable heavy (V_H) and human constant (Fc) domains were constructed by linking together the genes encoding them and expressing the engineered, recombinant antibodies in myeloma cells. In particular, the Fc was chosen to provide an isotype relevant to the desired biological function. However, when these antibodies were used therapeutically in humans, some still generated human antichimeric antibody (HACA) responses directed against the V regions. Since the level of HACA varies depending on the chimeric antibody, several have still been approved by the FDA.

6.3 CDR Grafting

The next step was to replace only the antigen-binding site from the human antibody by that of the source antibody. The first reported CDR graft was performed using the heavy-chain CDRs of a donor antihapten antibody B1-8 from a murine

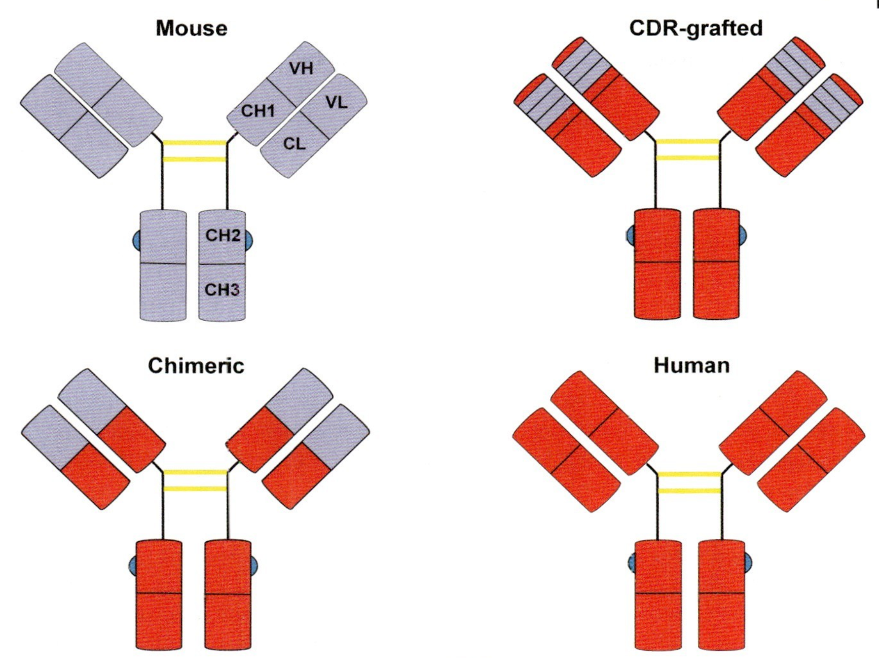

Fig. 6.1 Antibody engineering for chimeric and CDR-grafted mAbs. Blue – mouse protein domain or CDRs; red – human protein domain; green – carbohydrate; yellow – disulfide bridge.

source grafted into human acceptor V_H NEWM frameworks (Jones et al. 1986) to determine "whether the frameworks represent a simple beta-sheet scaffold on which new binding sites may be built and whether the structure of the CDRs (and antigen binding) is therefore independent of the framework context." Although the affinity of the (hemi-) CDR-grafted antibody was 2/3-fold lower than the mouse, proof of principle was established. However, the assumption was made that mutations in the frameworks do not affect the conformation of the CDR loops. This first experiment was followed a couple of years later by a similar CDR graft using the heavy-chain CDRs of murine antilysozyme antibody D1.3 (Verhoeyen et al. 1988). The results were considered remarkable, despite the binding being 10-fold less than the source antibody, given that CDR loops are not standalone structures and framework mutations actually do affect their conformation.

In the same year, the complete CDR graft of the first antibody of therapeutic interest was reported (Riechmann et al. 1988). All six CDRs from both the V_L and V_H of rat antibody Campath-1R were grafted into the V_L frameworks of human antibody REI and the V_H frameworks of human antibody NEWM. These frameworks were chosen since structural data was available for these human antibodies. The light chain of NEW was not used because there is a deletion at the beginning of the third framework region. In addition, framework reversions (backmutations from human to rat) were made, for the first time, in the V_H domain to enhance the affinity. The reshaped antibody showed 3-fold lower affinity, but comparable activity to the source antibody in functional assays. This CDR-grafted antibody is now on the market as Campath (alemtuzumab) for the treatment of B-cell chronic lymphocytic leukemia (B-CLL).

The following year, the first completely CDR-grafted mouse antibody was reported (Queen et al. 1989). All six CDRs were grafted into human EU frameworks (V_H and V_L). In this case, the human V_H framework was chosen based on similarity to the mouse, and the complementary V_L from the same antibody was also used. Several backmutations were introduced, based on analysis of a computer model of the mouse Fv. The engineered antibody had 3-fold lower affinity but is now marketed as Zenapax (daclizumab) for the prophylaxis of acute organ rejection in patients receiving renal transplants. Later work in cynomolgus monkeys (Schneider et al. 1993) showed the immune response to the CDR-grafted anti-Tac antibody was mainly directed against idiotopes rather than to the modified human framework regions. In general, chimerization reduces the immunogenicity of therapeutic mAbs, and CDR grafting reduces it further (Hwang and Foote 2005).

These early experiments illustrated that choices were necessary in the design of CDR-grafted antibodies, for instance in the selection of human frameworks.

6.4 The Design Cycle

The design of CDR-grafted antibodies often involves an iterated approach where sequence designs are generated and tested in binding and/or functional assays. An outline of the general approach to this design cycle is presented below:

1. Analyze the source donor amino acid sequences.
2. Construct a 3D computer model of the Fv.
3. Find suitable human acceptor framework sequences.
4. Identify putative backmutations (reversions) in the chosen frameworks.
5. Reconsider the framework choice and design the humanized antibody sequence.
6. Construct humanized and (if possible) chimeric antibody sequences.
7. Test constructs (humanized light chain can be tested independently by combining with heavy chimeric chain (hemi-chimeric) and vice versa).
8. Success? If "No" then continue, If "Yes" consider how to reduce backmutations by returning to "Identify putative backmutations . . .".
9. Return to "Find suitable human acceptor . . .".

6.4.1
Analysis of the Source (Donor) Sequence

6.4.1.1 Complementarity Determining Regions (CDRs)

These are six highly variable regions in the Fv, three in V_L, three in V_H. It is worth noting that the preponderance of backmutations at position 73 in V_H suggests that the loop encompassing this residue may be a fourth CDR. The CDRs contain the residues most likely to bind antigen and are therefore usually retained in the humanized antibody. They can be defined by sequence according to Kabat (Wu and Kabat 1970; Kabat et al. 1987, 1991) or by structure according to Chothia (Chothia and Lesk 1987; Chothia et al. 1989) (see Chapter 5). These definitions may be mixed in a humanization experiment. The advantage of using the Chothia definitions is that the CDRs are shorter and therefore the humanized antibody should have less xenogeneic fragments in it. Since Kabat et al. (1987, 1991) place CDR-H1 from positions 31–35, whereas crystal structures show the loop to be from positions 26–32 (Chothia et al. 1986), and murine residues 28–30 have been reported to exacerbate the immunogenic response in humans (Tempest et al. 1995), this advantage may be true.

However, the experience of Rodrigues et al. (1992) has shown that the shorter Chothia definition of CDR-H2 required several backmutations, mainly in the region covered by the Kabat definition. Conversely, the Kabat definition of CDR-H1 often requires backmutations in the region covered by the Chothia definition. Therefore, some strategies have combined the Kabat and Chothia definitions of CDR-H1, increasing its length (Presta et al. 1997; Vajdos et al. 2002). Others have kept some CDR residues human (Presta et al. 1993; Hsiao et al. 1994), tried to match CDR lengths between the source and human frameworks (Sims et al. 1993) and even found the mouse and human CDRs to be identical in sequence (as was the case with CDR-L2, Park et al. 1996). Pulito et al. (1996) tried to reduce the number of murine residues in the humanized antibody OKT4A, but found that CDR residues that do not contact antigen directly are also essential for antigen binding. In fact, Vajdos et al. (2002) subsequently discovered, through alanine scanning, that residues in the CDRs that contribute to antigen binding fall into two groups: solvent exposed residues that make direct contact with the antigen, and buried sidechains that can pack against other CDR residues and act as a scaffold. The ideal situation would be to have the crystal structure of the source antibody in complex with its antigen so that only the antigen-binding CDRs would be grafted. The structure of antibody 26-2F FAB was solved with its antigen (Chavali et al. 2003) and the authors predict the humanization would require the graft of only four of the six CDRs, since they are solely in contact with the antigen.

6.4.1.2 Canonical Residues

Originally defined by Chothia and Lesk (1987), the canonical residues have now been revised by Martin and Thornton (1996). The web page www.bioinf.org.uk/abs/chothia.html allows input of variable region sequences and automatic identification of the canonical structure class and important residues. Canonical residues are key residues in the CDR and/or framework that determine the conformation of the CDR loop. They can be hydrophobic residues that pack in the body of the loop, polar residues which form important stabilizing hydrogen bonds, or residues which can assume abnormal phi/psi conformations. Canonical residues should be retained in the humanized antibody if they are different to those in the human acceptor frameworks. However, it should also be noted that sometimes, backmutation of canonical residues has no effect or even decreases binding (see Section 6.4.4). The cause of these anomalies might be (1) the particular amino acid at that canonical position in the donor sequence is unimportant or (2) the acceptor residue at that position is better – a case of *in vitro* affinity maturation.

6.4.1.3 **Interface Packing Residues**

Originally investigated by Novotny and colleagues (Novotny et al. 1983; Novotny and Haber 1985) and defined by Chothia et al. (1985) (Table 6.1), interface packing residues occur at the interface between the V_L and V_H domains. These residues govern the packing of the variable domains, thus affecting the binding site. The main reason for the selection of human frameworks for V_L and V_H from the same antibody clone is to maintain the integrity of the interface between the variable domains (Section 6.4.3.2). Generally, unusual packing residues should be retained in the humanized antibody if they differ from those in the human frameworks. Their influence on affinity is illustrated in the humanization of 1B4 (Singer et al. 1993). Interestingly, their importance might also be functional. In antibody KM966, Nakamura et al. (2000) retained murine residues at V_L/V_H packing positions 38 and 40 in V_H. These residues had little effect on binding, but did improve the CDC of the humanized antibody.

6.4.1.4 **Rare Framework Residues**

Rare or atypical residues can be located by determining the Kabat subgroup (Kabat et al. 1987, 1991) and identifying the residue positions which differ from the consensus sequence (Gussow and Seemann 1991). These donor-specific differences may point to somatic mutations that enhance activity, atypical residues near to the binding site possibly contacting antigen (e.g. antibody BMA 031, Shearman et al. 1991). Humanizing these rare residues can cause the loss of binding affinity (Chan et al. 2001). However, if they are not important for binding, then it is desirable to get rid of them because they may create immunogenic neoepitopes in the humanized antibody. Note that unusual residues in the donor sequence are sometimes actually common residues in the acceptor (Queen et al. 1989). Atypical residues in the acceptor frameworks are not desirable because of the possibility of immunogenicity, unless of course they correspond to unusual residues in the donor and thus may be important functionally. Rarely occurring amino acids in the human frameworks have been mutated to human consensus residues (Co et al. 1991; Baker et al. 1994).

Table 6.1 Residues at the V_L/V_H interface[a].

	Kabat number[b]	*Mouse*[c]	*Human*[c]
V_K	34	H678 N420 A408 Y147 E114	A531 N147 D66
	36	Y1653 F198 L96	Y748 F80
	38	Q1865 H47	Q799 H22
	44(+)	P1767 V132 I40	P839 L5
	46	L1381 R374 P97	L760 V37
	87	Y1457 F448	Y795 F41
	89	Q1170 L206 F144	Q687 M107
	91	W376 S374 G356 Y295 H182	Y404 R115 S105 A84
	96(+)	L537 Y380 W285	L134 Y215 F78 W73 I71
	98(+)	F1724	F654
V_H	35	H1001 N636 S402	E184 S527 H340 G167 A143
	37	V2336 I200	V1037 I477 L27
	39	Q2518 K67	Q1539 R16
	45(+)	L2636 P16	L1531 P24
	47	W2518 L64 Y50	W1534 Y21
	91	Y2149 F479	Y1429 F116
	93	A2202 T222 V102	A1346 T90 V71
	95	Y399 G375 S340 D340 R226	D268 G266 R109 E100
	100k(+)	F1285 M450	F540 M109 L33
	103(+)	W1469	W323

a The positions of interdomain residues were as defined by Chothia et al. (1985).

b Numbering is according to Kabat et al. (1991). Residues underlined are in the framework, other residues are in the CDRs. (+) residues are the six that form the core of the V_L/V_H interface according to Chothia et al. (1985).

c The number following the one-letter amino acid code is the frequency taken from the Kabat database (November 1997 dataset).

6.4.1.5 *N*- or *O*-Glycosylation Sites

Potential *N*-glycosylation sites are specific to the consensus pattern asparagine-X-serine/threonine. It must be noted that the presence of the consensus tripeptide is not sufficient to conclude that an asparagine residue is glycosylated, due to the fact that the folding of the protein plays an important role in the regulation of *N*-glycosylation. It has been shown that the presence of proline between asparagine and serine/threonine will inhibit *N*-glycosylation and about 50% of the sites that have a proline C-terminal to serine/threonine are not glycosylated. It should also be noted that there are a few reported cases of glycosylation sites with the pattern asparagine-X-cysteine. Potential *N*-glycosylation sites can be located with the ExPASy (Expert Protein Analysis System) proteomics server of the Swiss Institute of Bioinformatics (SIB) (www.expasy.ch). It was expected that addition, removal or modification of glycosylation sites in the humanized antibody might affect the binding or immunogenicity. However, removal of potential *N*-glycosylation sites in either the V_L or V_H domain has not destroyed the binding affinity of a humanized antibody thus far (Leger et al. 1997) and in the case of M195 (Co et al. 1993) and LL2 (Leung et al. 1995) it actually increased. In another example, an *N*-glycosylation site was found at canonical residue 30 in V_H, although its removal through backmutation did not influence binding (Sato et al. 1996). Glycosylation sites can also be used as conjugation sites for drug or radionuclides as has been the case in antibody constant domains (Qu et al. 1998).

O-Glycosylation sites are usually found in helical segments, meaning that they are uncommon in the beta-sheet structure of antibodies. They have no consensus pattern. Couto et al. (1994) ruled out the removal of an *N*-linked glycosylation site in their humanized antibody (BrE-3) as being responsible for increased binding, but were open to the possibility of differences in *O*-linked glycosylation.

6.4.2
Three-Dimensional Computer Modeling of the Antibody Structure

Early humanizations by CDR grafting utilized human frameworks for which the crystal structures were available (Riechmann et al. 1988). Analysis of these structures was useful in the design of the humanized antibody. Although some solely used sequence analysis (Poul et al. 1995), later approaches relied on a carefully built model of the Fv regions of the source antibody and in some cases also the humanized antibody (Nakatani et al. 1994; Hsiao et al. 1994). Superposition of the source and humanized antibody models and analysis of size, charge, hydrophobicity, or hydrogen bond potential between equivalent residues highlighted important residues in the frameworks for maintaining the conformation of the CDRs or contacting antigen. The identification of these residues was useful in suggesting putative backmutations (Section 6.4.4).

In some cases, where possible, a model of the antigen was also built (Nisihara et al. 2001). A model of the source antibody docked to the antigen would be ideal for the design of the humanized antibody, in the absence of a crystal structure for the complex, and has been achieved in some cases by computer-guided docking (Zhang et al. 2005). The first reported use of a model in the design of a humanized antibody was by Queen et al. (1989) where a molecular model of the anti-Tac Fv was constructed with the ENCAD (Levitt 1983) computer program and examined with the MIDAS (Ferrin et al. 1991) visualization software on an IRIS (Silicon Graphics Inc., Mountain View, CA, USA) graphics workstation. This model was used to identify several framework amino acids in the mouse antibody that might interact with the CDRs or directly with the antigen.

Using a computer model of the donor mouse Fv, Kettleborough et al. (1991) analyzed the influence of certain framework residues in antigen binding. This model was built using QUANTA (Accelrys, San Diego, CA, USA) and the CHARMm (Brooks et al. 1983) force field on a Silicon Graphics IRIS workstation.

Carter et al. (1992) constructed a model using seven Fab crystal structures from the Protein Data Bank (PDB) (Deshpande et al. 2005). Their modeling program was Discover & Insight (Accelrys, San Diego, CA, USA) with the AMBER force field (Weiner et al. 1984). They acknowledged the crucial role of molecular modeling illustrated by the designed antibody binding antigen 250-fold more tightly than the simple CDR loop swap.

The modeling of the source or humanized antibody usually begins with a search over the PDB (Deshpande et al. 2005) to find template structures on which to build the model. The search is performed with standard packages (BLASTP (Altschul et al. 1997) or FASTA (Pearson 2000)) and the selection of template structures for V_L and V_H takes into account such parameters as sequence identity, overall sequence conservation and the resolutions of the structures. If different structures are chosen for V_L and V_H, then invariant residues at the V_L/V_H interface (Novotny and Haber 1985) must be superposed to model the interaction between the protein domains. Sometimes, several structures are used as templates (Carter et al. 1992) and structurally conserved regions (SCRs) are determined. The model can be built by splicing together regions with the highest sequence identity between the SCRs and the sequence to be modeled or modeling on average α-carbon coordinates in the SCRs. Identical residues in the framework regions are retained while nonidentical residues are substituted with the modeling package.

Modeling of the CDR loop regions poses a greater challenge. Searches can be performed to find an antibody structural loop of the same length and similar stem or overall canonical structure as tabulated by Martin and Thornton (1996). The matching structural loop can then be grafted into the model using the chosen modeling package. For loop regions of unknown canonical structure, a search can be performed using the positions of residues flanking the loop as anchors, over all known structures to find a loop region of the same length and similar base structure. The best match can be grafted into the model. Alternatively, *ab initio* methods can be employed using a conformational search program such as CONGEN (Bruccoleri and Karplus 1987), or a combination of database search for the base stem structure of the loop and *ab initio* conformational search for the central portion of the loop (Martin et al. 1989).

Of particular importance is the CDR-H3 loop. Shirai et al. (1996, 1999) showed that in many cases these loops exhibit “kinked” or “extended” C-terminal regions predicted by sequence-based rules. These rules can be applied to determine additional features of CDR-H3, thus aiding the modeling of its conformation.

The entire model is finally energy minimized to relieve unfavorable atomic contacts and to optimize nonbonded interactions. Stereochemical verification of the model is generally performed using ProCheck (Laskowski et al. 1993), while VERIFY3D (Eisenberg et al. 1997) and PROSA-II (Sippl 1993) can be used to measure model quality in terms of packing and solvent exposure.

Nowadays, it is entirely possible to build a model completely automatically using programs such as Swiss PDB Viewer and academic servers such as Swiss-Model (Guex and Peitsch 1997). However, the danger of allowing a computer to make all the decisions is highlighted in the humanization of antibody AT13/5 (Ellis et al. 1995) where the interaction between residues at positions 29 and 78 in the heavy chain was not modeled correctly. The experience of an expert in protein structure modeling is always welcomed. Additionally, it may be better to also model the constant regions of the antibody, since interactions at the variable/constant domain interface are likely to affect the affinity and/or activity of the molecule. Landolfi et al. (2001) found that altering framework position 11 in the V_H although only slightly affecting affinity, severely diminished the activity of humanized antibody AF2. This residue had been identified as being involved in a “ball-and-socket” joint between the V and C domains of the immunoglobulin Fab (Lesk and Chothia 1988).

Modeling can only be an interim measure on the way to determining the structure of the source antibody by X-ray crystallography or NMR. Increasingly, antibody structures are being determined, sometimes in complex with antigen, which can help the design process. Redwan et al. (2003) used the structure of a cocaine-binding antibody to humanize it using real structure-based design to incorporate human residues that would not affect the binding pocket or key cocaine-contacting residues. Yazaki et al. (2004) used the program VAST (Gibrat et al. 1996) to identify a human (or humanized Fv) acceptor for the CDR graft of antibody T84.66, whose crystal structure had already been determined. Interestingly, Herceptin (trastuzumab) (Eigenbrot et al. 1993), a successfully humanized antibody, was chosen for its high degree of overlap of α-carbon atoms and overall sequence identity.

6.4.3 Choice of Human Framework Sequences

This is the most critical area of the humanization design cycle, yet there are no hard-and-fast rules for choosing the human acceptor frameworks into which to graft the donor CDRs. This is because the benefits of the various choices in terms of recovery of affinity, specificity, and activity with the fewest backmutations, and also immunogenicity in the patient, have not been clearly proven in the clinic. Therefore, there are only sets of approaches that need to be combined with the collective experience of previous humanizations, although the antibodies gaining FDA approval are giving clues to which approach may be best (Table 6.2).

6.4.3.1 Fixed Frameworks or Best Fit?

Some groups prefer to use fixed frameworks (usually NEW for the heavy chain and REI for the light chain since their structures are solved, sometimes KOL for the heavy chain) for all their humanized antibodies. This was the case for the first therapeutically CDR-grafted antibody, Campath-1 (Riechmann et al. 1988) and Actemra (tocilizumab) (Sato et al. 1993) that has also reached the market. Other groups try to use the most similar frameworks to their donor sequence (homology matching, also called “best-fit” by Gorman et al. 1991) by searching over large sequence databases – nowadays the nonredundant (nr) database at the

Table 6.2 CDR-grafted biopharmaceuticals approved by the Food and Drug Administration (FDA).

Drug name	*Year FDA*	*Active ingredient*	*Therapeutic area*[a]	*Antigen*[b]	*Model*[c]	*Frameworks*[d]	*CDRs*[e]	*Back*[f]	*Affinity with source mAb*	*% patients with AAR*[g]	*Reference*
Zenapax	1997	daclizumab	immune	CD25	m	best-fit	kab	yes	3-fold less	8.4	Queen et al. 1989
Herceptin	1998	trastuzumab	oncology	HER2	m/h	consen	mix	yes	3-fold more	0.1	Carter et al. 1992
Synagis	1998	palivizumab	infectious disease	RSVF	m	germ V_L expr V_H	kab	yes	similar	0.7–1.8	Johnson et al. 1997
Mylotarg	2000	gemtuzumab, ozogamicin	oncology	CD33	?	fixed	mix	yes	?	2.9†	Hamann et al. 1999
Campath	2001	alemtuzumab	oncology	CD52	strs	fixed	kab	yes	3-fold less	2–63	Riechmann et al. 1988
Raptiva	2003	efalizumab	immune	CD11a	m/h	consen	chot	yes	similar	6.3	Werther et al. 1996
Xolair	2003	omalizumab	immune	IgE	m/h	consen	mix	yes	similar	<0.1	Presta et al. 1993
Avastin	2004	bevacizumab	oncology	VEGF	m/h	consen	mix	yes	2-fold less	?	Presta et al. 1997

a Immune, immune disorders.
b VEFG, vascular endothelial growth factor; HER2, human epidermal growth factor receptor 2; RSV F, respiratory syncytial virus F protein; IgE, immunoglobulin E.
c m, mouse; h, human; strs, structures available for human frameworks.
d Consen, human consensus; fixed, fixed framework approach; germ, human germline; expr, functionally expressed human antibody; best-fit, best-fit framework approach.
e Mix, mixture of Kabat sequence (kab) and Chothia structure (chot) definitions.
f Rodent backmutations incorporated into human frameworks in either V_L, V_H or both.
g AAR, anti-antibody response (data from Tabrizi 2005, except † Hwang and Foote 2005).

National Center for Biotechnology Information (www.ncbi.nlm.nih.gov). This was the approach of Queen et al. (1989) for the V_H of the anti-Tac antibody, but the V_L frameworks were chosen to match the V_H (i.e. the same human antibody for both chains). Others used V_L/V_H frameworks from different human antibodies (Singer et al. 1993). The best-fit strategy requires other choices to be made. Should the sequence similarity extend over the whole V-region, only the framework regions, or a mix of matching CDR lengths and framework identity? A subtle comparison of the fixed frameworks and homology matched (best-fit) methods, in terms of the ease of producing a functional humanized antibody, can be found in the humanization of antibody M22 (Graziano et al. 1995). The preferential choice appears to be the latter, where the more homologous KOL frameworks gave better binding than NEWM.

Hamilton et al. (1997) questioned the best-fit approach, arguing that the advantage of choosing homologous frameworks might be outweighed by the advantage of using fixed frameworks for which a database of experience had been assembled. Certainly, if the sequences of the fixed frameworks have a low homology to the original source sequences, then there is increased likelihood of low binding (Sha and Xiang 1994) but knowledge of which backmutations to make should restore binding. The crystal structures of two humanized forms of antibody AF2, which differed in the sequence identity of the donor heavy chain to the human frameworks, provided evidence supporting the best-fit approach (Bourne et al. 2004). The humanized form with the greater identity between donor V_H and human frameworks was significantly more structurally similar to the mouse antibody.

Note that it is possible to mix and match frameworks. Qu et al. (1999) for antibody Immu31 mixed human EU frameworks 1, 2, and 3, with NEW framework 4 in an essentially fixed framework approach. In the best-fit approach, it is possible to choose frameworks 1–4 from different human antibodies according to sequence similarity. Additionally, Ono et al. (1999) ensured that all the framework regions came from naturally processed human antibodies. In their case framework 3 of the heavy chain came from a different human antibody to frameworks 1 and 2, and framework 4 was the human JH6 germline.

6.4.3.2 V_L/V_H Frameworks from the Same or Different Clone?

In general, light and heavy chains from the same antibody are more likely to associate to form a functional binding site than light and heavy chains from different antibodies. Frameworks from the same clone were used for antibodies FD79 and FD138-80 by Co et al. (1991) to reduce the possibility of incompatibility in assembly of the heavy and light chains. However, since the interface between the chains is so well conserved, this is not usually a problem. A comparison of the two approaches was made with antibody anti-B4 (Roguska et al. 1994, 1996). Once again, the preferential choice appears to be the latter, full restoration of antigen-binding affinity being achieved when the most identical, but clonally different, human V_L and V_H acceptor frameworks were selected.

6.4.3.3 Human Subgroup Consensus or Expressed Framework?

Being limited to expressed frameworks from particular human antibodies runs the risk of their somatic mutations creating immunogenic epitopes, even though the frameworks are human. An alternative approach is to use the frameworks from human consensus sequences where idiosyncratic somatic mutations will have been removed – first suggested by Shearman et al. (1991). The two approaches have been compared; in one case showing no difference in binding affinities (Kolbinger et al. 1993) and in the other case showing better binding with individual frameworks (Sato et al. 1994). The fixed framework approach versus using consensus human frameworks was tested by Maeda et al. (1991). The fixed framework humanized antibodies showed loss of binding whilst the consensus framework antibodies did not. However, since no model was used in this case, judicious choice of backmutations might have led to different results. In several instances, homologous human frameworks have been chosen, only to then change some residue positions to their consensus amino acid (e.g. Hakimi et al. 1993) (see also Section 6.4.1.4).

6.4.3.4 Germline Frameworks

Consensus sequences are artificial and although they have no idiosyncratic residues, may create unnatural sequence motifs that are immunogenic. An alternative is to use human germline sequences that have been compiled in various databases (IMGT (Lefranc et al. 2005), VBASE2 (Retter et al. 2005)). Originally suggested by Shearman et al. (1991), several groups have reported using germline frameworks (Rosok et al. 1996; Caldas et al. 2003; DiJoseph et al. 2005). In fact, in our laboratory consensus sequences were first considered because it was reasoned that they would most closely resemble the germline sequences when the latter were still unavailable. Also, somatic mutations from the germline may indicate residues that contact antigen. This was the case in antibody MOC-31 (Beiboer et al. 2000) where modeling predicted that one of the germline mutations might bind antigen. Germline frameworks can be chosen based on similarity to the donor sequences, including amino acid identity at important residue positions (Brams et al. 2001) or multiple germline frameworks can be used (Gonzales et al. 2004).

Tan et al. (2002) matched the canonical templates of the donor antibody with human genomic V gene sequences and CDR grafted into these frameworks. No backmutations were included, and the resultant "superhumanized" antibody showed a 30-fold reduction in affinity but maintained biological activity. A later experiment with anti-lysozyme murine antibody D1.3 showed only a 6-fold reduction in affinity (Hwang et al. 2005).

6.4.3.5 Database Search

Having decided on an approach to take in order to choose the human frameworks; which particular human antibody, consensus or germline sequence should be used? This is simple in the fixed framework approach since the choice is always

NEW for the heavy chain and REI for the light. The approach generally taken is to perform a search for the most similar human acceptor sequence over the appropriate database. Choice of the particular human frameworks for the light and heavy chain variable regions should be made by trying to match the length of the CDRs, the canonical residues and the interface packing residues (Section 6.4.1) as well as trying to find the highest percentage identity between the donor and acceptor sequences. Try to find human frameworks that are similar (in terms of percentage identity) to the source sequences and also require the least number of backmutations. The consensus approach utilizes V_L and V_H frameworks derived from the most common amino acid found at each position within a given human subgroup from the Kabat database and can be chosen based on similarity (Kettleborough et al. 1991) to the mouse V-region sequences using search programs as above. Another approach is to use the same human consensus frameworks for each design, regardless of sequence similarity. The most abundant human subgroups are V_H subgroup III, V_L kappa subgroup I. These have been used in several humanization experiments (Carter et al. 1992; Presta et al. 1993; Presta et al. 1997; Luo et al. 2003; Adams et al. 2005).

6.4.4
Identify Putative Backmutations

Although straight CDR grafts can recover both the affinity and specificity of the source antibody (Thomas et al. 1996), usually backmutations (reversions) to the donor residues are required in the human acceptor frameworks. This is the most difficult and unpredictable procedure in any humanization strategy, sometimes requiring many different versions of the humanized antibody to be made (Presta et al. 1993). It is also the area that throws much light on protein structure and function. A solid body of data for helping to identify strategic alterations is available on the "Humanization bY Design" website (www.cryst.bbk.ac.uk/~ubcg07s). Riechmann et al. (1988) were the first to employ backmutations reverting two residues in the heavy chain at positions 27 and 30, and found an increase in binding to antigen. It is from this work that we gain the term "reshaped" to indicate a CDR-grafted antibody with backmutations. Queen et al. (1989) incorporated seven backmutations in the heavy chain and two in the light chain that either influenced CDR conformation or interacted directly with antigen for their anti-Tac humanization. This work also muddied the waters of what a backmutation actually was, since they additionally changed some human framework residues to mouse, arguing that the human residues were unusual and the mouse residues were more typically human. Taken together this meant that the humanized anti-Tac antibody had 13 "backmutations" in the heavy chain and three in the light.

By introducing ordered steps of additional backmutations, Tempest et al. (1991) minimized the number of changes required to restore affinity and specificity. One simple approach to identifying backmutations is to keep all source residues within four positions of every CDR (Shearman et al. 1991). Alternatively, the structural model can be used to analyze residues within 5Å (Graves et al. 1999) of any CDR residue. This will ensure the integrity of the Vernier zone (Foote and Winter 1992) (a platform on which the CDRs rest) and identify residues in the framework that may bind antigen.

However, the main method for identifying backmutations is to study the differences between donor and acceptor frameworks and analyze them on the structural model. Most of the differences are not important, lying on the surface and far from the CDRs. Those not on the surface and/or close to the CDRs are worthy of greater attention. Putative point reversions from the acceptor residue back to the original donor residue will already have been identified from the analysis of the sequences for canonical residues, interchain packing residues and rare residues. Experience has shown that it is especially important to retain the source's canonical and interchain packing residues, though not in all cases. In antibody HMFG1 (Krauss et al. 2004) position 71 in the heavy chain, although being a canonical residue, was found to stabilize the scFV while having only a minor effect on the binding. Sometimes, position 71 was found to have no effect (Hsiao et al. 1994; Tempest et al. 1995) while Sato et al. (1993) reported that the backmutation at position 71 actually caused worse binding.

Due to the extreme variability in sequence and length of CDR-H3, there are no canonical residues defined for this loop although certain positions are known to interact and maintain some conformational stability. For instance, a salt-bridge between the positively charged position 94 and the usually invariant aspartic acid at 101 in CDR-H3 is seen in many antibody structures (Tempest et al. 1991) and arginine 94 also forms interactions with positions 31 and 74 in the heavy chain (Chuntharapai et al. 2001). In addition, residue 101 is known to form a cooperative (possibly indirect) interaction with position 49 in the light chain of an anti-CD40 antibody (Wu et al. 1999). Therefore, special attention should be paid to this loop, analyzing the structural model for residues that may potentially affect its conformation.

Backmutations are not transferable between different antibodies, even if they have high sequence homology and similar antigen specificity (Rosok et al. 1996). Ohtomo et al. (1995) introduced a new method to identify important backmutations when the first version of their humanized V_L did not bind. They constructed "hybrid variable" regions, joining together mouse and CDR-grafted framework regions. In this way, they found that proline 46 in framework 2 was required to recreate the functional binding site, and this was not even among the five backmutations that they had originally considered.

Having decided on the residues to backmutate, the question of human acceptor frameworks should be reconsidered. It is not unlikely that an overlooked human framework may actually contain the backmutations that are to be retained. If this is the case, then there is no need to introduce residues from the source sequence, thus making the humanized antibody more "human." However, not all backmutations are necessary, and there is a need for experimental validation. In some cases, the acceptor residue can be better than the donor (Gonzales et al. 2003; Caldas et al. 2000) either decreasing the immunogenicity or increasing the affinity of a humanized antibody. Backmutations can also have an effect on antibody expression, a 5-fold increase being found with a backmutation at position 75 in V_H of antibody ABL364 (Co et al. 1996). There are also surprises in store. Caldas et al. (2003) through a systematic analysis of related structures unexpectedly identified position 37 in the light chain as a putative site for backmutation. This led to a more effective humanized antibody in cell binding assays, although in this case the affinity was not measured.

6.5 Other Approaches to Antibody Humanization

6.5.1 Resurfacing/Veneering

The backmutations required in CDR-grafted antibodies may introduce new antigenic epitopes or lose the advantage over chimeric antibodies if a great many are required. A solution to this is to maintain the core and CDRs of the murine variable regions, but replace the surface residues with those from a human sequence by a strategy known as "resurfacing." This strategy originates from a systematic analysis of known antibody structures to determine the relative solvent accessibility distributions of residues in human and mouse variable regions (Pedersen et al. 1994). A description of the differences in the presentation of surface residues in a small number of mouse and human antibody variable regions had already been published in a process known as "veneering" (Padlan 1991). The analysis showed that the sequence alignment positions of surface residues were conserved 98% of the time between the two species. Also, the pattern of amino acid substitution was conserved within a species, but not between the species (i.e. no mouse framework displayed the exact pattern of surface residues found in any human framework). Thus, it was possible to convert a murine surface pattern to that of human with relatively few mutations. However, a choice was still required for selecting a characteristic human surface pattern.

Two methods were compared for two different murine antibodies (Roguska et al. 1994). For anti-B4 (an anti-CD19 mAb) a database of clonally derived human V_L/V_H sequence pairs was used. For N901 (an anti-CD56 mAb), sequences for V_L and V_H were independently selected from the Kabat database. Both resurfaced antibodies presented apparent affinities for their antigens identical to those of their source murine antibodies. Further versions of these antibodies (where the number of murine surface residues was reduced) were compared with CDR-grafted versions (Roguska et al. 1996) (Section 6.4.3.2). The goal of generating humanized antibodies that retained the affinity and specificity of the source murine antibody, but with as few murine residues as possible in the variable domain framework was achieved by selecting the most similar human V_L and V_H frameworks, without regard to clonal origin. This was the case for both CDR-grafted and resurfaced antibodies. However, the CDR-grafted anti-B4 antibodies had more murine residues at surface positions than the resurfaced antibodies, and were more difficult to engineer requiring seventeen attempts. The conclusion was that resurfaced antibodies are easier to produce and are conceivably less immunogenic, although this ignores the possibility of T-cell epitopes presented from the murine core.

Resurfacing technology has been developed by Immunogen (Cambridge, MA, USA) and two antibodies humanized using this strategy are currently in clinical trials.

6.5.2 SDR Transfer

CDR-grafting of a xenogeneic antibody does not necessarily eliminate the immunogenicity of the molecule because of idiotypic responses directed against the xenogeneic CDRs, particularly when given in multiple doses. CDR-grafted mAbs have been shown to be immunogenic in both primate animal models (Schneider et al. 1993; Stephens et al. 1995) and in humans (Richards et al. 1999; Ritter et al. 2001). Antigen binding usually involves between 20% and 33% of the CDR residues (Padlan 1994) that have been labeled "specificity determining residues" (SDRs). Padlan et al. (1995) using the PDB determined the boundaries of the potential SDRs in different CDRs and called these segments "abbreviated CDRs." The SDRs are commonly located at positions of high variability and are possibly unique to each mAb. However, they can be identified by site-directed mutation or determination of the 3D structure of the Fv or, in the absence of this information, the variability of positions within the abbreviated CDRs can be used to suggest which residues are SDRs.

Transfer of SDRs has only been used successfully in the humanization of anti-carcinoma mAb CC49 which specifically recognizes tumor-associated glycoprotein (TAG)-72 (Tamura et al. 2000). SDR transfer has also been utilized in the humanization of murine mCOL-1 that specifically recognizes carcinoembryonic antigen (CEA). In this case, the SDR-transfer antibody had comparable binding activity to the CDR-grafted equivalent and significantly higher activity compared with the abbreviated CDR-grafted antibody. It also showed decreased reactivity for anti-V-region antibodies present in the sera of patients treated with mCOL-1 (Gonzales et al. 2004).

6.5.3 DeImmunization Technology

This technology applied by Biovation (an associate of Merck KGaA, Germany) combines veneering (based on Padlan's approach) to effectively humanize surface residues (thus removing B-cell epitopes) with the identification and removal of potential helper T-cell epitopes from antibody biopharmaceuticals. Helper T-cell epitopes are short peptide sequences within proteins that bind to MHC class II molecules. These epitopes can be created by somatic mutations occurring naturally in human antibodies or by the veneering process. The peptide – MHC class II complexes are recognized by T cells and trigger the activation and differentiation of helper T cells, thus stimulating a cellular immune response. Helper T cells initiate and maintain immunogenicity by interacting with B cells, resulting in the production of antibodies that bind specifically to the administered antibody. In DeImmunization, helper T cell epitopes are identified within the primary sequence of the antibody using prediction software and these sequences are altered by amino acid substitution to avoid recognition by T cells.

The prediction software is principally based on modeling work with the crystal structures of MHC class II allotypes combined with a database search of known T-cell epitopes. As a result, the modified antibody should no longer trigger T cell help. In this way immunogenicity may be eliminated or substantially reduced. However, particular peptides are not necessarily processed and presented by MHC class II, so some unnecessary epitope deletion is possible. Furthermore, there is the issue of tolerance, and this is handled by ignoring peptides present in human immunoglobulin sequences. Currently, two products are in clinical trials following DeImmunization.

6.5.4 Phage Libraries

Since the relative importance of backmutations varies between different mAbs, identifying important positions and determining the optimal amino acid at those positions has proven difficult. With the advent of bacterial expression and phage display of antibody fragments (Fvs, Fabs, etc.) (McCafferty et al. 1990) combined with efficient screening methods, large numbers of variants can be rapidly characterized for activity. This permits antibody function to be optimized or even evolved *in vitro*, as opposed to using successive iterations in the design cycle. Thus, combinatorial antibody libraries have been used for the humanization of mAbs from murine, chicken, or rabbit sources. Rosok et al. (1996) grafted murine CDRs into homologous human germline frameworks and, keeping surface residues human, determined buried positions to be randomized with all possible combinations of murine/human amino acids in a phage library. Thus, the screening simultaneously selected the best binders out of many different humanized Fabs, differing only in their backmutations. Baca et al. (1997) took a similar approach, but differed in details. They CDR grafted into consensus human frameworks and only selected those positions empirically found to be important for antigen binding. Thus the randomized set of backmutations were almost entirely different. This approach can also be used to optimize CDR residues (Wu et al. 1999), combining humanization and *in vitro* affinity maturation in the same procedure.

A different strategy termed guided selection or chain shuffling has been used to isolate human mAbs from phage display libraries in a two-step process. In the first stage, the source V_H is paired with a repertoire of human V_Ls. The resulting Fabs are displayed on filamentous phage and the selected human V_L isolated from the screening process is paired in the second stage with a human V_H repertoire. Thus the source variable domains are sequentially replaced by human variable domains to derive high affinity human (Jespers et al. 1994) or mainly human (Rader et al. 1998) mAbs. The approach of Rader et al. (1998) has also been called framework shuffling where the CDR-L3 of the source antibody is grafted into human V_L domains and the selected V_L paired in the second step with a library of human V_H domains grafted with the source CDR-H3. This process thus obviates the necessity of information from a structural model, while the resulting humanized mAbs can maintain the specificity and affinity of the original source mAb.

References

Adams, C.W., Allison, D.E., Flagella, K., Presta, L., Clarke, J., Dybdal, N., McKeever, K., Sliwkowski, M.X. (2005) Humanization of a recombinant monoclonal antibody to produce a therapeutic HER dimerization inhibitor, pertuzumab. *Cancer Immunol Immunother* 55: 717–727.

Altschul, S.F., Madden, T.L., Schaffer, A.A., Zhang, J., Zhang, Z., Miller, W., Lipman, D.J. (1997) Gapped BLAST and PSI-BLAST: a new generation of protein database search programs. *Nucleic Acids Res* 25: 3389–3402.

Baca, M., Presta, L.G., O'Connor, S.J., Wells, J.A. (1997) Antibody humanization using monovalent phage display. *J Biol Chem* 272: 10678–10684.

Baker, T.S., Bose, C.C., Caskey-Finney, H. M., King, D.J., Lawson, A.D., Lyons, A., Mountain,A., Owens, R.J., Rolfe, M.R., Sehdev, M., Yarranton, G.T., Adair, J.R. (1994) Humanization of an anti-mucin antibody for breast and ovarian cancer therapy. *Adv Exp Med Biol* 353: 61–82.

Beiboer, S.H., Reurs, A., Roovers, R.C., Arends, J.W., Whitelegg, N.R., Rees, A.R., Hoogenboom, H.R. (2000) Guided selection of a pan carcinoma specific antibody reveals similar binding characteristics yet structural divergence between the original murine antibody and its human equivalent. *J Mol Biol* 296: 833–849.

Boulianne, G.L., Hozumi, N., Shulman, M.J. (1984) Production of functional chimaeric mouse/human antibody. *Nature* 312: 643–646.

Bourne, P.C., Terzyan, S.S., Cloud G., Landolfi, N.F., Vasquez, M., Edmundson, A.B. (2004) Three-dimensional structures of a humanized anti-IFN-gamma Fab (HuZAF) in two crystal forms. *Acta Crystallogr D Biol Crystallogr* 60: 1761–1769.

Brams, P., Black, A., Padlan, E.A., Hariharan, K., Leonard, J., Chambers-Slater, K., Noelle, R.J., Newman, R. (2001) A humanized anti-human CD154 monoclonal antibody blocks CD154-CD40 mediated human B cell activation. *Int Immunopharmacol* 1: 277–294.

Brooks, B.R., Bruccoleri, R.E., Olafson, B.D., States, D.J., Swaminathan, S., Karplus, M. (1983) CHARMm: A program for macromolecular energy, minimization, and dynamics calculations. *J Comp Chem* 4: 187–217.

Bruccoleri, R.E., Karplus, M. (1987) Prediction of the folding of short polypeptide segments by uniform conformational sampling. *Biopolymers* 26: 137–168.

Caldas, C., Coelho, V.P., Rigden, D.J., Neschich, G., Moro, A.M., Brigido, M.M. (2000) Design and synthesis of germline-based hemi-humanized single-chain Fv against the CD18 surface antigen. *Protein Eng* 13: 353–360.

Caldas, C., Coelho V., Kalil J., Moro, A.M., Maranhao, A.Q., Brigido, M.M. (2003) Humanization of the anti-CD18 antibody 6.7: an unexpected effect of a framework residue in binding to antigen. *Mol Immunol* 39: 941–952.

Carter, P., Presta, L., Gorman, C.M., Ridgway, J.B., Henner, D., Wong, W.L., Rowland, A.M., Kotts, C., Carver, M.E., Shepard, H.M. (1992) Humanization of an anti-p185HER2 antibody for human cancer therapy. *Proc Natl Acad Sci USA* 89: 4285–4289.

Chan, K.T., Cheng, S.C., Xie, H., Xie, Y. (2001) A humanized monoclonal antibody constructed from intronless expression vectors targets human hepatocellular carcinoma cells. *Biochem Biophys Res Commun* 284: 157–167.

Chavali, G.B., Papageorgiou, A.C., Olson, K.A., Fett, J.W., Hu, G., Shapiro, R., Acharya, K.R. (2003) The crystal structure of human angiogenin in complex with an antitumor neutralizing antibody. *Structure* 11: 875–885.

Chothia, C., Lesk, A.M. (1987) Canonical structures for the hypervariable regions of immunoglobulins. *J Mol Biol* 196: 901–917.

Chothia, C., Novotny, J., Bruccoleri, R., Karplus, M. (1985) Domain association in immunoglobulin molecules. The packing of variable domains. *J Mol Biol* 186: 651–663.

Chothia, C., Lesk, A.M., Levitt, M., Amit, A. G., Mariuzza, R.A., Phillips, S.E., Poljak, R.J. (1986) The predicted structure of immunoglobulin D1.3 and its comparison with the crystal structure. *Science* 233: 755–758.

Chothia, C., Lesk, A.M., Tramontano, A., Levitt, M., Smith-Gill, S.J., Air, G., Sheriff, S., Padlan, E.A., Davies, D., Tulip, W.R., Colman, P.M., Spinelli, S., Alzari, P.M., Poljak, R.J. (1989) Conformations of immunoglobulin hypervariable regions. *Nature* 342: 877–883.

Chuntharapai, A., Lai, J., Huang, X., Gibbs, V., Kim, K.J., Presta, L.G., Stewart, T.A. (2001) Characterization and humanization of a monoclonal antibody that neutralizes human leukocyte interferon: a candidate therapeutic for IDDM and SLE. *Cytokine* 15: 250–260.

Co, M.S., Deschamps, M., Whitley, R.J., Queen, C. (1991) Humanized antibodies for antiviral therapy. *Proc Natl Acad Sci USA* 88: 2869–2873.

Co, M.S., Scheinberg, D.A., Avdalovic, N.M., McGraw, K., Vasquez, M., Caron, P.C., Queen, C. (1993) Genetically engineered deglycosylation of the variable domain increases the affinity of an anti-CD33 monoclonal antibody. *Mol Immunol* 30: 1361–1367.

Co, M.S., Baker, J., Bednarik, K., Janzek, E., Neruda, W., Mayer, P., Plot, R., Stumper, B., Vasquez, M., Queen, C., Loibner, H. (1996) Humanized anti-Lewis Y antibodies: in vitro properties and pharmacokinetics in rhesus monkeys. *Cancer Res* 56: 1118–1125.

Couto, J.R., Blank, E.W., Peterson, J.A., Kiwan R., Padlan, E.A., Ceriani, R.L. (1994) Engineering of antibodies for breast cancer therapy: construction of chimeric and humanized versions of the murine monoclonal antibody BrE-3. *Adv Exp Med Biol* 353: 55–59.

Deshpande, N., Addess, K.J., Bluhm, W.F., Merino-Ott, J.C., Townsend-Merino, W., Zhang, Q., Knezevich, C., Xie, L., Chen, L., Feng, Z., Green, R.K., Flippen-Anderson, J.L., Westbrook, J., Berman, H.M., Bourne, PE. (2005) The RCSB Protein Data Bank: a redesigned query system and relational database based on the mmCIF schema. *Nucleic Acids Res* 33(Database issue): D233–237.

DiJoseph, J.F., Popplewell, A., Tickle, S., Ladyman, H., Lawson, A., Kunz, A., Khandke, K., Armellino, D.C., Boghaert, E.R., Hamann, P., Zinkewich-Peotti, K., Stephens, S., Weir, N., Damle, N.K. (2005) Antibody-targeted chemotherapy of B-cell lymphoma using calicheamicin conjugated to murine or humanized antibody against CD22. *Cancer Immunol Immunother* 54: 11–24.

Eigenbrot, C., Randal, M., Presta, L., Carter, P., Kossiakoff, A.A. (1993) X-ray structures of the antigen-binding domains from three variants of humanized anti-p185HER2 antibody 4D5 and comparison with molecular modeling. *J Mol Biol* 229: 969–995.

Eisenberg, D., Luthy, R., Bowie, J.U. (1997) VERIFY3D: assessment of protein models with three-dimensional profiles. *Methods Enzymol* 277: 396–404.

Ellis, J.H., Barber, K.A., Tutt, A., Hale, C., Lewis, A.P., Glennie, M.J., Stevenson, G.T., Crowe, J.S. (1995) Engineered anti-CD38 monoclonal antibodies for immunotherapy of multiple myeloma. *J Immunol* 155: 925–937.

Ferrin, T.E., Couch, G.S., Huang, C.C., Pettersen, E.F., Langridge, R. (1991) An affordable approach to interactive desktop molecular modeling. *J Mol Graph* 9: 27–32, 37–38.

Foote, J., Winter, G. (1992) Antibody framework residues affecting the conformation of the hypervariable loops. *J Mol Biol* 224: 487–499.

Gibrat, J.F., Madej, T., Bryant, S.H. (1996) Surprising similarities in structure comparison. *Curr Opin Struct Biol* 6: 377–385.

Gonzales, N.R., Padlan, E.A., De Pascalis, R., Schuck, P., Schlom, J., Kashmiri, S.V. (2003) Minimizing immunogenicity of the SDR-grafted humanized antibody CC49 by genetic manipulation of the framework residues. *Mol Immunol* 40: 337–349.

Gonzales, N.R., Padlan, E.A., De Pascalis, R., Schuck, P., Schlom, J., Kashmiri, S.V. (2004) SDR grafting of a murine antibody using multiple human germline templates to minimize its immunogenicity. *Mol Immunol* 41: 863–872.

Gorman, S.D., Clark, M.R., Routledge, E.G., Cobbold, S.P., Waldmann, H. (1991) Reshaping a therapeutic CD4 antibody. *Proc Natl Acad Sci USA* 88: 4181–4185.

Graves, S.S., Goshorn, S.C., Stone, D.M., Axworthy, D.B., Reno, J.M., Bottino, B., Searle, S., Henry, A., Pedersen, J., Rees, A. R., Libby, R.T. (1999) Molecular modeling and preclinical evaluation of the humanized NR-LU-13 antibody. *Clin Cancer Res* 5: 899–908.

Graziano, R.F., Tempest, P.R., White, P., Keler, T., Deo, Y., Ghebremariam, H., Coleman, K., Pfefferkorn, L.C., Fanger, M. W., Guyre, P.M. (1995) Construction and characterization of a humanized anti-gamma-Ig receptor type I (Fc gamma RI) monoclonal antibody. *J Immunol* 155: 4996–5002.

Guex, N., Peitsch MC. (1997) SWISS-MODEL and the Swiss-PdbViewer: an environment for comparative protein modeling. *Electrophoresis* 18: 2714–2723.

Gussow, D., Seemann, G. (1991) Humanization of monoclonal antibodies. *Methods Enzymol* 203: 99–121.

Hakimi, J., Ha, V.C., Lin, P., Campbell, E., Gately, M.K., Tsudo, M., Payne, P.W., Waldmann, T.A., Grant, A.J., Tsien, W.H., Schneider, W.P. (1993) Humanized Mik beta 1, a humanized antibody to the IL-2 receptor beta-chain that acts synergistically with humanized anti-TAC. *J Immunol* 151: 1075–1085.

Hamann, P.R., Hinman, L., Hollander, I., Holcomb, R., Hallett, W., Tsou, H-R., Weiss, M.J. (1999) Process for preparing conjugates of methyltrithio antitumor agents. U.S. patent 5, 877, 296.

Hamilton, A.A., Manuel, D.M., Grundy, J.E., Turner, A.J., King, S.I., Adair, J.R., White, P., Carr, F.J., Harris, W.J. (1997) A humanized antibody against human cytomegalovirus (CMV) gpUL75 (gH) for prophylaxis or treatment of CMV infections. *J Infect Dis* 176: 59–68.

Hsiao, K.C., Bajorath, J., Harris, L.J. (1994) Humanization of 60.3, an anti-CD18 antibody; importance of the L2 loop. *Protein Eng* 7: 815–822.

Hwang, W.Y., Foote, J. (2005) Immunogenicity of engineered antibodies. *Methods* 36: 3–10.

Hwang, W.Y., Almagro, J.C., Buss, T.N., Tan, P., Foote, J. (2005) Use of human germline genes in a CDR homology-based approach to antibody humanization. *Methods* 36: 35–42.

Jespers, L.S., Roberts, A., Mahler, S.M., Winter, G., Hoogenboom, H.R. (1994) Guiding the selection of human antibodies from phage display repertoires to a single epitope of an antigen. *Biotechnology* 12: 899–903.

Johnson, S., Oliver, C., Prince, G.A., Hemming, V.G., Pfarr, D.S., Wang, S.C., Dormitzer, M., O'Grady, J., Koenig, S., Tamura, J.K., Woods, R., Bansal, G., Couchenour, D., Tsao, E., Hall, W.C., Young, J.F. (1997) Development of a humanized monoclonal antibody (MEDI-493) with potent in vitro and in vivo activity against respiratory syncytial virus. *J Infect Dis* 176: 1215–1224.

Jones, P.T., Dear, P.H., Foote, J., Neuberger, M.S., Winter, G. (1986) Replacing the complementarity-determining regions in a human antibody with those from a mouse. *Nature* 321: 522–525.

Kabat, E.A., Wu, T.T., Reid-Miller, M., Perry, H., Gottesman, K. (1987) *Sequences of Proteins of Immunological Interest*, 4th edn. US Govt. Printing Off. No. 165-492.

Kabat, E.A., Wu, T.T., Perry, H., Gottesman, K., Foeller, C. (1991) *Sequences of Proteins of Immunological Interest*, 5th edn. NIH Publication No. 91-3242.

Kettleborough, C.A., Saldanha, J., Heath, V. J., Morrison, C.J., Bendig, M.M. (1991) Humanization of a mouse monoclonal antibody by CDR-grafting: the importance of framework residues on loop conformation. *Protein Eng* 4: 773–783.

Kohler, G., Milstein, C. (1975) Continuous cultures of fused cells secreting antibody of predefined specificity. *Nature* 256: 495–497.

Kolbinger, F., Saldanha, J., Hardman, N., Bendig, M.M. (1993) Humanization of a mouse anti-human IgE antibody: a potential therapeutic for IgE-mediated allergies. *Protein Eng* 6: 971–980.

Krauss, J., Arndt, M.A., Zhu, Z., Newton, D.L., Vu, B.K., Choudhry, V., Darbha, R., Ji X., Courtenay-Luck, N.S., Deonarain, M.P., Richards, J., Rybak, S.M. (2004) Impact of antibody framework residue V_H-71 on the stability of a humanised anti-MUC1 scFv and derived immunoenzyme. *Br J Cancer* 90: 1863–1870.

Landolfi, N.F., Thakur, A.B., Fu, H., Vasquez, M., Queen, C., Tsurushita, N. (2001) The integrity of the ball-and-socket joint between V and C domains is essential for complete activity of a humanized antibody. *J Immunol* 166: 1748–1754.

Laskowski, R.A., MacArthur, M.W., Moss, D.S., Thornton, J.M. (1993) PROCHECK: a program to check the stereochemical quality of protein structures *J Appl Cryst* 26: 283–291.

Lefranc, M.P., Giudicelli, V., Kaas, Q., Duprat, E., Jabado-Michaloud, J., Scaviner, D., Ginestoux, C., Clement, O., Chaume, D., Lefranc, G. (2005) IMGT, the international ImMunoGeneTics information system. *Nucleic Acids Res* 33: D593–D597.

Leger, O.J., Yednock, T.A., Tanner L., Horner, H.C., Hines, D.K., Keen, S., Saldanha, J., Jones, S.T., Fritz, L.C., Bendig, M.M. (1997) Humanization of a mouse antibody against human alpha-4 integrin: a potential therapeutic for the treatment of multiple sclerosis. *Hum Antibodies* 8: 3–16.

Lesk, A.M., Chothia, C. (1988) Elbow motion in the immunoglobulins involves a molecular ball-and-socket joint. *Nature* 335: 188–190.

Leung, S.O., Goldenberg, D.M., Dion, A.S., Pellegrini, M.C., Shevitz, J., Shih, L.B., Hansen, H.J. (1995) Construction and characterization of a humanized, internalizing, B-cell (CD22)-specific, leukemia/lymphoma antibody, LL2. *Mol Immunol* 32: 1413–1427.

Levitt, M. (1983) Molecular dynamics of native protein. I. Computer simulation of trajectories. *J Mol Biol* 168: 595–617.

Luo, G.X., Kohlstaedt, L.A., Charles, C.H., Gorfain, E., Morantte, I., Williams, J.H., Fang, F. (2003) Humanization of an anti-ICAM-1 antibody with over 50-fold affinity and functional improvement. *J Immunol Methods* 275: 31–40.

Maeda, H., Matsushita, S., Eda, Y., Kimachi, K., Tokiyoshi, S., Bendig, M.M. (1991) Construction of reshaped human antibodies with HIV-neutralizing activity. *Hum Antibodies Hybridomas* 2: 124–134.

Martin, A.C., Cheetham, J.C., Rees, A.R. (1989) Modeling antibody hypervariable loops: a combined algorithm. *Proc Natl Acad Sci USA* 86: 9268–9272.

Martin, A.C., Thornton, J.M. (1996) Structural families in loops of homologous proteins: automatic classification, modelling and application to antibodies. *J Mol Biol* 263: 800–815.

McCafferty, J., Griffiths, A.D., Winter, G., Chiswell, D.J. (1990) Phage antibodies: filamentous phage displaying antibody variable domains. *Nature* 348: 552–554.

Morrison, S.L., Johnson, M.J., Herzenberg, L.A., Oi, V.T. (1984) Chimeric human antibody molecules: mouse antigen-binding domains with human constant region domains. *Proc Natl Acad Sci USA* 81: 6851–6855.

Nakamura, K., Tanaka, Y., Fujino, I., Hirayama, N., Shitara, K., Hanai, N. (2000) Dissection and optimization of immune effector functions of humanized anti-ganglioside GM2 monoclonal antibody. *Mol Immunol* 37: 1035–1046.

Nakatani, T., Lone, Y.C., Yamakawa, J., Kanaoka, M., Gomi, H., Wijdenes, J., Noguchi, H. (1994) Humanization of mouse anti-human IL-2 receptor antibody B-B10. *Protein Eng* 7: 435–443.

Nisihara, T., Ushio, Y., Higuchi, H., Kayagaki, N., Yamaguchi, N., Soejima, K., Matsuo, S., Maeda, H., Eda, Y., Okumura, K., Yagita, H. (2001) Humanization and epitope mapping of neutralizing anti-human Fas ligand monoclonal antibodies: structural insights into Fas/Fas ligand interaction. *J Immunol* 167: 3266–3275.

Novotny, J., Haber, E. (1985) Structural invariants of antigen binding: comparison of immunoglobulin V_L-V_H and V_L-V_L domain dimers. *Proc Natl Acad Sci USA* 82: 4592–4596.

Novotny, J., Bruccoleri, R., Newell, J., Murphy, D., Haber, E., Karplus, M. (1983) Molecular anatomy of the antibody binding site. *J Biol Chem* 258: 14433–14437.

Ohtomo, T., Tsuchiya, M., Sato, K., Shimizu, K., Moriuchi, S., Miyao, Y., Akimoto, T., Akamatsu, K., Hayakawa, T., Ohsugi, Y. (1995) Humanization of mouse ONS-M21 antibody with the aid of hybrid variable regions. *Mol Immunol* 32: 407–416.

Ono, K., Ohtomo, T., Yoshida, K., Yoshimura, Y., Kawai, S., Koishihara, Y., Ozaki, S., Kosaka, M., Tsuchiya, M. (1999) The humanized anti-HM1.24 antibody effectively kills multiple myeloma cells by human effector cell-mediated cytotoxicity. *Mol Immunol* 36: 387–395.

Padlan, E.A. (1991) A possible procedure for reducing the immunogenicity of antibody variable domains while preserving their ligand-binding properties. *Mol Immunol* 28: 489–498.

Padlan, E.A. (1994) Anatomy of the antibody molecule. *Mol Immunol* 31: 169–217.

Padlan, E.A., Abergel C., Tipper, J.P. (1995) Identification of specificity-determining residues in antibodies. *FASEB J* 9: 133–139.

Park, S.S., Ryu, C.J., Gripon, P., Guguen-Guillouzo, C., Hong, H.J. (1996) Generation and characterization of a humanized antibody with specificity for preS2 surface antigen of hepatitis B virus. *Hybridoma* 15: 435–441.

Pearson, W.R. (2000) Flexible sequence similarity searching with the FASTA3 program package. *Methods Mol Biol* 132: 185–219.

Pedersen, J.T., Henry, A.H., Searle, S.J., Guild, B.C., Roguska, M., Rees, A.R. (1994) Comparison of surface accessible residues in human and murine immunoglobulin Fv domains. Implication for humanization of murine antibodies. *J Mol Biol* 235: 959–973.

Pichla, S.L., Murali, R., Burnett, RM. (1997) The crystal structure of a Fab fragment to the melanoma-associated GD2 ganglioside. *J Struct Biol* 119: 6–16.

Poul, M.A., Ticchioni, M., Bernard, A., Lefranc, M.P. (1995) Inhibition of T cell activation with a humanized anti-beta 1 integrin chain mAb. *Mol Immunol* 32: 101–116.

Presta, L.G., Lahr, S.J., Shields, R.L., Porter, J.P., Gorman, C.M., Fendly, B.M., Jardieu, P.M. (1993) Humanization of an antibody directed against IgE. *J Immunol* 151: 2623–2632.

Presta, L.G., Chen, H., O'Connor, S.J., Chisholm, V., Meng, Y.G., Krummen, L., Winkler, M., Ferrara, N. (1997) Humanization of an anti-vascular endothelial growth factor monoclonal antibody for the therapy of solid tumors and other disorders. *Cancer Res* 57: 4593–4599.

Pulito, V.L., Roberts, V.A., Adair, J.R., Rothermel, A.L., Collins, A.M., Varga, S. S., Martocello, C., Bodmer, M., Jolliffe, L.K., Zivin, R.A. (1996) Humanization and molecular modeling of the anti-CD4 monoclonal antibody. OKT4A. *J Immunol* 156: 2840–2850.

Qu, Z., Sharkey, R.M., Hansen, H.J., Shih, L.B., Govindan, S.V., Shen J., Goldenberg, D.M., Leung, S.O. (1998) Carbohydrates engineered at antibody constant domains can be used for site-specific conjugation of drugs and chelates. *J Immunol Methods* 213: 131–144.

Qu, Z., Losman, M.J., Eliassen, K.C., Hansen, H.J., Goldenberg, D.M., Leung, S.O. (1999) Humanization of Immu31: an alpha-fetoprotein-specific antibody. *Clin Cancer Res* 5: 3095s–3100s.

Queen, C., Schneider, W.P., Selick, H.E., Payne, P.W., Landolfi, N.F., Duncan, J.F., Avdalovic, N.M., Levitt M., Junghans, R.P., Waldmann, T.A. (1989) A humanized antibody that binds to the interleukin 2 receptor. *Proc Natl Acad Sci USA* 86: 10029–10033.

Rader, C., Cheresh, D.A., Barbas, C.F. III (1998) A phage display approach for rapid antibody humanization: designed combinatorial V gene libraries. *Proc Natl Acad Sci USA* 95: 8910–8915.

Redwan, el-, R.M., Larsen, N.A., Zhou, B., Wirsching, P., Janda, K.D., Wilson, I.A. (2003) Expression and characterization of a humanized cocaine-binding antibody. *Biotechnol Bioeng* 82: 612–618.

Retter, I., Althaus, H.H., Munch, R., Muller, W. (2005) VBASE2: an integrative V gene database. *Nucleic Acids Res* 33: D671–D674.

Richards, J., Auger, J., Peace, D., Gale, D., Michel, J., Koons, A., Haverty, T., Zivin, R., Jolliffe, L., Bluestone, J.A. (1999) Phase I evaluation of humanized OKT3: toxicity and immunomodulatory effects of hOKT3gamma4. *Cancer Res* 59: 2096–2101.

Riechmann, L., Clark, M., Waldmann, H., Winter, G. (1988) Reshaping human antibodies for therapy. *Nature* 332: 323–327.

Ritter, G., Cohen, L.S., Williams, C, Jr., Richards, E.C., Old, L.J., Welt, S. (2001) Serological analysis of human anti-human antibody responses in colon cancer patients treated with repeated doses of humanized monoclonal antibody A33. *Cancer Res* 61: 6851–6859.

Rodrigues, M.L., Shalaby, M.R., Werther, W., Presta, L., Carter, P. (1992) Engineering a humanized bispecific F(ab')2 fragment for improved binding to T cells. *Int J Cancer Suppl* 7: 45–50.

Roguska, M.A., Pedersen, J.T., Keddy, C.A., Henry, A.H., Searle, S.J., Lambert, J.M., Goldmacher, V.S., Blattler, W.A., Rees, A. R., Guild, B.C. (1994) Humanization of murine monoclonal antibodies through variable domain resurfacing. *Proc Natl Acad Sci USA* 91: 969–673.

Roguska, M.A., Pedersen, J.T., Henry, A.H., Searle, S.M., Roja, C.M., Avery, B., Hoffee, M., Cook S., Lambert, J.M., Blattler, W.A., Rees, A.R., Guild, B.C. (1996) A comparison of two murine monoclonal antibodies humanized by CDR-grafting and variable domain resurfacing. *Protein Eng* 9: 895–904.

Rosok, M.J., Yelton, D.E., Harris, L.J., Bajorath, J., Hellstrom, K.E., Hellstrom, I., Cruz, G.A., Kristensson, K., Lin, H., Huse, W.D., Glaser, S.M. (1996) A combinatorial library strategy for the rapid humanization of anticarcinoma BR96 Fab. *J Biol Chem* 271: 22611–22618.

Sato, K., Tsuchiya, M., Saldanha, J., Koishihara, Y., Ohsugi, Y., Kishimoto, T., Bendig, M.M. (1993) Reshaping a human antibody to inhibit the interleukin 6-dependent tumor cell growth. *Cancer Res* 53: 851–856.

Sato, K., Tsuchiya M., Saldanha J., Koishihara Y., Ohsugi Y., Kishimoto T., Bendig MM. (1994) Humanization of a mouse anti-human interleukin-6 receptor antibody comparing two methods for selecting human framework regions. *Mol Immunol* 31: 371–381.

Sato, K., Ohtomo, T., Hirata, Y., Saito, H., Matsuura, T., Akimoto, T., Akamatsu, K., Koishihara, Y., Ohsugi, Y., Tsuchiya, M. (1996) Humanization of an anti-human IL-6 mouse monoclonal antibody glycosylated in its heavy chain variable region. *Hum Antibodies Hybridomas* 7: 175–183.

Schneider, W.P., Glaser, S.M., Kondas, J.A., Hakimi, J. (1993) The anti-idiotypic response by cynomolgus monkeys to humanized anti-Tac is primarily directed to complementarity-determining regions H1: H2: and L3. *J Immunol* 150: 3086–3090.

Sha, Y., Xiang, J. (1994) A heavy-chain grafted antibody that recognizes the tumor-associated TAG72 antigen. *Cancer Biother* 9: 341–349.

Shearman, C.W., Pollock, D., White, G., Hehir, K., Moore, G.P., Kanzy, E.J., Kurrle, R. (1991) Construction, expression and characterization of humanized antibodies directed against the human alpha/beta T cell receptor. *J Immunol* 147: 4366–4373.

Shirai, H., Kidera, A., Nakamura, H. (1996) Structural classification of CDR-H3 in antibodies. *FEBS Lett* 399: 1–8.

Shirai, H., Kidera, A., Nakamura, H. (1999) H3-rules: identification of CDR-H3 structures in antibodies. *FEBS Lett* 455: 188–197.

Sims, M.J., Hassal, D.G., Brett, S., Rowan, W., Lockyer, M.J., Angel, A., Lewis, A.P., Hale, G., Waldmann, H., Crowe, J.S.

(1993) A humanized CD18 antibody can block function without cell destruction. *J Immunol* 151: 2296–2308.

Singer, I.I., Kawka, D.W., DeMartino, J.A., Daugherty, B.L., Elliston, K.O., Alves, K., Bush, B.L., Cameron, P.M., Cuca, G.C., Davies, P., Forrest, M.J., Kazazis, D.M., Law, M-F., Lenny, A.B., MacIntyre, D.E., Meurer, R., Padlan, E.A., Pandya, S., Schmidt, J.A., Seamans, T.C., Scott, S., Silberklang, M., Williamson, A.R., Mark, G.E. (1993) Optimal humanization of 1B4: an anti-CD18 murine monoclonal antibody, is achieved by correct choice of human V-region framework sequences. *J Immunol* 150: 2844–2857.

Sippl, M.J. (1993) Recognition of errors in three-dimensional structures of proteins. *Proteins* 17: 355–362.

Steinberger, P., Sutton, J.K., Rader, C., Elia, M., Barbas, C.F. III (2000) Generation and characterization of a recombinant human CCR5-specific antibody. A phage display approach for rabbit antibody humanization. *J Biol Chem* 275: 36073–36078.

Stephens, S., Emtage, S., Vetterlein, O., Chaplin, L., Bebbington, C., Nesbitt, A., Sopwith, M., Athwal, D., Novak, C., Bodmer, M. (1995) Comprehensive pharmacokinetics of a humanized antibody and analysis of residual anti-idiotypic responses. *Immunology* 85: 668–674.

Tabrizi, M. (2005) Immunogenicity of therapeutics: monoclonal antibodies (mAbs). Presentation at the IBC Antibody Engineering Conference, San Diego, USA, December 2005.

Tamura, M., Milenic, D.E., Iwahashi, M., Padlan, E., Schlom, J., Kashmiri, S.V. (2000) Structural correlates of an anticarcinoma antibody: identification of specificity-determining residues (SDRs) and development of a minimally immunogenic antibody variant by retention of SDRs only. *J Immunol* 164: 1432–1441.

Tan, P., Mitchell, D.A., Buss, T.N., Holmes, M.A., Anasetti, C., Foote, J. (2002) 'Superhumanized' antibodies: reduction of immunogenic potential by complementarity-determining region grafting with human germline sequences: application to an anti-CD28. *J Immunol* 169: 1119–1125.

Tempest, P.R., Bremner, P., Lambert, M., Taylor, G., Furze, J.M., Carr, F.J., Harris, W.J. (1991) Reshaping a human monoclonal antibody to inhibit human respiratory syncytial virus infection in vivo. *Biotechnology (NY)* 9: 266–271.

Tempest, P.R., White, P., Buttle, M., Carr, F.J., Harris, W.J. (1995) Identification of framework residues required to restore antigen binding during reshaping of a monoclonal antibody against the glycoprotein gB of human cytomegalovirus. *Int J Biol Macromol* 17: 37–42.

Thomas, T.C., Rollins, S.A., Rother, R.P., Giannoni, M.A., Hartman, S.L., Elliott, E.A., Nye, S.H., Matis, L.A., Squinto, S.P., Evans, M.J. (1996) Inhibition of complement activity by humanized anti-C5 antibody and single-chain Fv. *Mol Immunol* 33: 1389–1401.

Tsurushita, N., Park, M., Pakabunto, K., Ong, K., Avdalovic, A., Fu, H., Jia, A., Vasquez, M., Kumar, S. (2004) Humanization of a chicken anti-IL-12 monoclonal antibody. *J Immunol Methods* 295: 9–19.

Vajdos, F.F., Adams, C.W., Breece, T.N., Presta, L.G., de Vos, A.M., Sidhu, S.S. (2002) Comprehensive functional maps of the antigen-binding site of an anti-ErbB2 antibody obtained with shotgun scanning mutagenesis. *J Mol Biol* 320: 415–428.

Verhoeyen, M., Milstein, C., Winter, G. (1988) Reshaping human antibodies: grafting an antilysozyme activity. *Science* 239: 1534–1536.

Weiner, S.J., Kollman, P.A., Case, D.A., Singh, U.C., Ghio, C., Alagona, G., Profeta, S., Weiner, P. (1984) A new force field for molecular mechanical simulation of nucleic acids and proteins. *J Am Chem Soc* 106: 765–784.

Werther, W.A., Gonzalez, T.N., O'Connor, S. J., McCabe, S., Chan, B., Hotaling, T., Champe, M., Fox, J.A., Jardieu, P.M., Berman, P.W., Presta, LG. (1996) Humanization of an anti-lymphocyte function-associated antigen (LFA)-1 monoclonal antibody and reengineering of the humanized antibody for binding to rhesus LFA-1. *J Immunol* 157: 4986–4995.

Wu, T.T., Kabat, EA. (1970) An analysis of the sequences of the variable regions of

Bence Jones proteins and myeloma light chains and their implications for antibody complementarity. *J Exp Med* 132: 211–250.
Wu, H., Nie, Y., Huse, W.D., Watkins, J.D. (1999) Humanization of a murine monoclonal antibody by simultaneous optimization of framework and CDR residues. *J Mol Biol* 294: 151–162.
Yazaki, P.J., Sherman, M.A., Shively, J.E., Ikle, D., Williams, L.E., Wong, J.Y., Colcher, D., Wu, A.M., Raubitschek, A.A. (2004) Humanization of the anti-CEA T84.66 antibody based on crystal structure data. *Protein Eng Des Sel* 17: 481–489.
Zhang, W., Feng, J., Li, Y., Guo, N., Shen, B. (2005) Humanization of an anti-human TNF-alpha antibody by variable region resurfacing with the aid of molecular modeling. *Mol Immunol* 42: 1445–1451.

7
Molecular Engineering II: Antibody Affinity

Lorin Roskos, Scott Klakamp, Meina Liang, Rosalin Arends, and Larry Green

7.1
Introduction

Antibody affinity is a quantitative measurement of the strength of association between a single antibody-binding site and a single antigen-binding site. For any given concentration of antibody and antigen, the affinity determines the number of antigen molecules that will be bound by antibody. Affinity, therefore, is a critical variable affecting the potency of an antibody therapeutic. Potency will affect the dose and possibly the dosing interval. Since therapeutic doses (Roskos et al. 2004) and costs of goods of antibodies are frequently high, an adequate affinity might ultimately determine the clinical and commercial success of an antibody therapeutic. In this chapter the process of affinity maturation, the relationship between affinity and potency, and analytical approaches to the measurement of affinity are reviewed.

7.2
Affinity Maturation

7.2.1
Maturation *In Vivo*

Antibody diversity, which allows the antibodies of the primary repertoire to bind and to modulate a vast number of different antigens and antigen epitopes, is generated through sequential processes that occur at different stages of B cell development (Wu et al. 2003). VDJ rearrangement of gene segments in the immunoglobulin heavy chain locus, VJ rearrangements in the κ and λ immunoglobulin light chain loci, addition and deletion of nucleotides at the joints between the gene segments, and random pairing of the rearranged V_HDJ_H and $V_κJ_κ$ and $V_λJ_λ$ segments leads to enormous diversity in the order of 10^{11} B-cell receptors (BCRs) expressed by naive cells in humans. However, antigen binding affinity to

Handbook of Therapeutic Antibodies. Edited by Stefan Dübel

ISBN 978-3-527-31453-9

the BCRs of the primary repertoire is usually weak. With appropriate signaling upon encountering antigen, additional antibody diversity is introduced by class switch DNA recombination, which changes the antibody constant region and alters antibody effector functions, and by somatic hypermutation, which drives the affinity maturation process.

The clonal selection of B cells producing high-affinity antibodies to an antigen is an essential part of the humoral response to an antigen challenge. Affinity maturation is the result of two processes of positive selection that take place in different compartments of the germinal center in secondary lymphoid organs (Wabl et al. 1999; Defrance et al. 2002; Wu et al. 2003). Somatic V(D)J hypermutation occurs by introduction of point mutations, with occasional insertions and deletions, in the variable region gene of antigen-activated B cells residing in the dark zone of the germinal center (Defrance et al. 2002). The somatic mutation rate in these cells is about one million times higher than the spontaneous somatic mutation rate of the genome at large (Wu et al. 2003). Only those cells expressing BCRs with sufficiently high affinity to compete with antigen trapped as IgM immune complexes on follicular dendritic cells receive a survival signal by antigen binding to the BCR. The antigen-activated B cells receive a second survival signal in the light zone of the germinal center, where positive selection of mutant B cell clones with high antigen binding affinity relies on efficient binding and presentation of antigen to helper T cells (Defrance et al. 2002). If the hypermutation mechanism creates a BCR that has lost antigen binding or has become self-reactive, a negative selection process can trigger either receptor editing, a "last-ditch" effort of reactivated V gene rearrangement to rescue antigen binding, or apoptosis. This natural process of affinity maturation, hypermutation combined with positive and negative selection for improvements in antibody affinity, is iterative, and can lead to one million fold or greater improvements in monoclonal antibody (mAb) affinity compared to the primary mAb with the germline V(D)J sequence from the naive repertoire.

Foote and Eisen have proposed a ceiling for affinities that can be obtained through *in vivo* selection (Foote and Eisen 1995, 2000). The basis for the affinity ceiling is that the association rate of the antigen with the BCR will eventually be rate limited by the diffusion rate to the BCR (rather than the association rate constant for binding), and that the dissociation rate of the antigen bound to BCR cannot be slower than the internalization rate of the antigen–BCR complex into the B cell (the net dissociation rate constant will be the sum of the internalization rate constant and the antigen–BCR dissociation rate constant). The maturation ceiling for the equilibrium dissociation constant has been estimated to be approximately 10^{-10} mol L^{-1}. Although the kinetic basis for an affinity ceiling is sound, somatic hypermutation is a stochastic process; therefore, there would be large variation in antibody affinity on both sides of a mean affinity ceiling. Three conditions can allow for the *in vivo* generation of antibodies with affinities exceeding the theoretical ceiling: (1) iterative rounds of positive selection generate BCRs that can evolve towards the theoretical affinity plateau; (2) the hypermutation process stochastically generates a BCR that exceeds the plateau for selection, evolved either from a BCR near the ceiling or from a "jackpot" rearrangement or mutation; and (3) the absence of negative selection against BCRs with affinity $<10^{-10}$ mol L^{-1} allows for the continued survival of the B cell. Thus, though B cells expressing a BCR with an affinity $<10^{-10}$ mol L^{-1} would have no more or no less advantage than those with an affinity of 10^{-10} mol L^{-1}, they can be created and maintained. Indeed, rigorous screening of large pools of diverse mAbs recovered from hyperimmunized mice can yield antibodies with picomolar and even femtomolar affinity, which is many fold higher than the proposed *in vivo* ceiling (Rathanaswami et al. 2005).

Somatic hypermutation has evolved to favor mutations in the complementarity determining regions (CDRs) of antibodies; the framework regions (FRs) are mutated at a lower frequency (Wu et al. 2003). The FR mutations are usually less important for affinity and specificity than the CDR mutations. The V_H CDRs in general, and the V_H and V_L CDR3 in particular, tend to dominate the antibody–antigen interactions (Sundberg and Mariuzza 2002). For these reasons, random mutations frequently exist in the FRs and CDRs that do not contribute to antibody affinity and specificity. In the generation of therapeutic antibodies, some antibody generation groups routinely identify the nonessential mutations (usually in the FRs) by alanine scanning mutagenesis and convert these mutations back to the germline sequence (a process called "germlining"). The rationale is that nonessential, random mutations may increase the risk of immunogenicity relative to an antibody with the germline sequence. To date, the risk of immunogenicity as related to the number of CDR and FR mutations remains only a theoretical concern.

7.2.2
Maturation *In Vitro*

In vitro antibody display technologies can be used to generate high-affinity, antigen-specific antibodies (Hoogenboom 2005). Generally, DNA for rearranged V_H and either V_κ or V_λ derived either from naive B cells or selected V gene framework regions with synthetic CDRs are molecularly coupled and then expressed recombinantly in systems such as bacteriophage, yeast, or via translation *in vitro*. Because the protein product is engineered to remain physically linked to the nucleic acid encoding it, *in vitro* screening for binding of the displayed variable region repertoire to the antigen allows for recovery of the DNA or RNA encoding the antibody. Antibodies with moderate affinities on the order of 10^{-8}–10^{-9} are usually recovered after 2–3 rounds of standard selection, enrichment and screening for antigen binding. In theory, the *in vitro* nature of the process allows for generation of libraries of V_H/V_L combinations on the order of 10^{10}–10^{13}, sizes that greatly exceed the size of the B-cell compartment *in vivo*, which would therein provide superior diversity in the antibody repertoire. In practice, however, the individual dynamics of the various systems for *in vitro* display often can bias the repertoire and diminish the theoretical diversity by orders of magnitude.

Because most therapeutic mAbs would require affinities better than those of antibodies recovered directly from *in vitro* antibody display systems, large efforts have been undertaken to develop efficient *in vitro* means to mimic the *in vivo* affinity maturation process. Typically, the processes currently in use employ error prone processes such as polymerase chain reaction (PCR) and/or reverse transcription focused in and adjacent to the CDRs of the V_H and V_L coupled with stringent selection for higher affinity binding *in vitro*. The screening procedures may be manipulated to recover variable regions with selective improvements in k_{on} and/or in k_{off}, something not feasible with affinity maturation *in vivo*. As typical for the *in vivo* process, iterative rounds of *in vitro* mutagenesis and selection are employed to recover and incrementally improve selected variants. In this way, antibodies with affinities $<10^{-9}$–10^{-10} mol L^{-1} and occasionally greater can be routinely generated through *in vitro* processes. These processes, and with advancements such as utilization of combinations of *in vitro* procedures that complement deficiencies inherent in each (phage plus either ribosome display or yeast display) and improvements in operational efficiencies that allow for both parallel-processing of multiple starting V region templates and screening of larger pools of variants, have yielded antibodies with picomolar and femtomolar affinities (Schier et al. 1995; Boder et al. 2000; Hanes et al. 2000; Zahnd et al. 2004; Hoet et al. 2005; Rathanaswami et al. 2005).

7.3
Effect of Affinity on Antigen Binding and Antibody Potency

Affinity describes the strength of reversible association between antibody and antigen:

$$\mathrm{Ab} + \mathrm{Ag} \underset{k_{off}}{\overset{k_{on}}{\rightleftarrows}} \mathrm{Ab} \cdot \mathrm{Ag}$$

The strength of antibody-antigen binding is enhanced by a fast association rate, which is proportional to the *association rate constant* (k_{on} or k_a), and by a slow dissociation rate, which is proportional to the *dissociation rate constant* (k_{off} or k_d). The value of affinity is most frequently described by the *equilibrium dissociation constant* (K_D). The K_D, which is readily calculated by k_{off} divided by k_{on}, is the concentration of antibody-binding sites that will bind 50% of the antigen-binding sites when the concentration of antigen is much less than the K_D. This simple definition of K_D assumes that all antibody-binding sites are accessible to all antigen-binding sites and that no avid interactions occur. *Avidity* reflects the strength of binding when multivalent binding results in a cooperative antigen–antibody interaction. An example of an avid interaction is when both antibody-binding sites simultaneously bind an antigen on a surface, or form cyclic or lattice immune complexes. In such cases, the avidity may be much stronger (by several orders of magnitude) than reflected by the 1:1 site binding K_D. The values of K_D,

k_{on}, and k_{off} can be determined experimentally; the most robust methods are reviewed later in this chapter.

An understanding of the basic relationship between affinity, antibody concentration, antigen concentration, and the fraction of antigen bound is essential to the understanding of the relationship between antibody affinity and potency. The fraction of antigen bound ($F_{b,Ag}$) can be readily calculated as a function of the K_D, the antibody-binding site concentration (Ab_s), and the antigen binding site concentration (Ag_s):

$$F_b^{Ag} = \frac{Ab_s + Ag_s + K_D - \sqrt{Ab_s^2 + Ag_s^2 + K_D^2 - 2Ab_S Ag_s + 2Ab_S K_D + 2Ag_s K_D}}{2Ag_s}$$

This relationship is illustrated by simulation in Fig. 7.1. The fraction of antigen bound is plotted as a function of the ratio of antibody concentration (each mole of antibody is assumed to bind 2 moles of antigen) to the K_D for different multiples of antigen concentration relative to K_D (antigen concentration varying from $K_D/100$ to $100K_D$). When the concentration of antigen is less than or equal to $K_D/10$, 50% of the antigen is bound when the antibody concentration is one-half the K_D (i.e. when the total antibody-binding site concentration is equal to the K_D). Under these conditions, the binding is said to be *K_D-dependent*. When the con-

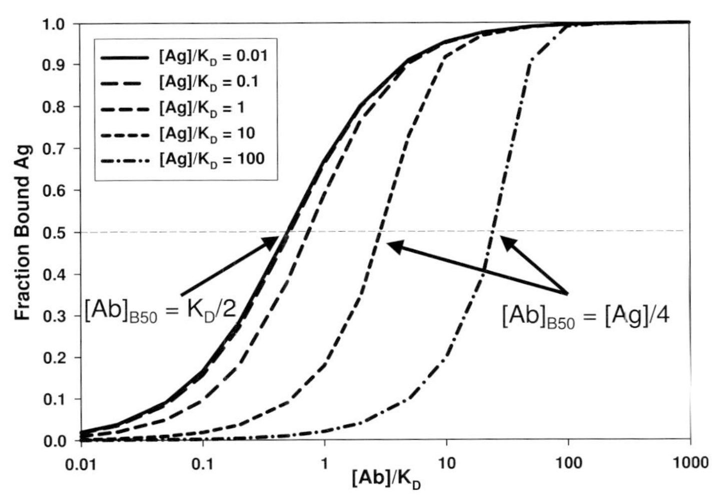

Fig. 7.1 Theoretical fraction of antigen bound by antibody as a function of antigen concentration (Ag) and antibody concentration (Ab) relative to the K_D. One antibody molecule was assumed to be capable of binding two antigen molecules. When Ag is 1/10th the K_D or less, the concentration of antibody required to bind 50% of Ag ($[Ab]_{B50}$) is K_D-dependent. When Ag is 10 times or more greater than the K_D, $[Ab]_{B50}$ is stoichiometric (antigen concentration-dependent).

centration of antigen exceeds the K_D by a multiple of 10 or more, then 50% of the antigen is bound when the antibody concentration is a quarter the antigen concentration (i.e., the antibody-binding site concentration is equal to one-half the antigen concentration). Under these conditions, the binding is said to be *stoichiometric,* since antigen is bound in approximately a 1:1 molar ratio to the available antibody-binding sites, and the binding is not dependent on the K_D. Likewise, for any fixed concentration of antigen, improvements in antibody affinity will eventually result in a transition from K_D-dependent binding conditions to stoichiometric conditions.

From these kinetic observations, a simple relationship between affinity and binding potency emerges. *For any given antigen concentration, an antibody affinity exists beyond which further improvements in affinity will not enhance antigen binding.* This potency ceiling for affinity occurs when the K_D of the antibody falls to approximately 1/10th the antigen concentration. As shown in the following section, this relationship holds *in vitro* and *in vivo.*

7.3.1
Binding and Potency *In Vitro*

Cell-based bioassays are routinely used to compare the functional potency of various mAbs to a given target. By varying antibody concentration in the presence of a fixed concentration of antigen (cell membrane or soluble antigen), the antibody potency can be expressed as a maximum effect in presence of a large excess of antigen (I_{max} or E_{max}) and the concentration of antibody producing the half-maximal effect (IC_{50} or EC_{50}). Since the affinity of the antibody under K_D-dependent conditions affects the fraction of antigen bound, the affinity might be a very influential variable in determination of the IC_{50} and EC_{50}. But interpretation of the results must be taken in context of the antibody affinities and experimental and *in vivo* antigen concentrations. In many cases, supraphysiological antigen concentrations may be needed in bioassays to allow for adequate analytical quantitation limits.

Examples of the effect of affinity on antibody potency are illustrated in Fig. 7.2. In Fig.7.2a, the IC_{50} for neutralizing mAbs to a soluble cytokine (present in cell culture at a 4 pmol L^{-1} concentration, reasonably reflective of *in vivo* concentrations) generated by immunization of XenoMouse animals (Mendez et al. 1997) is plotted as a function of antibody affinity, ranging from 100 fmol L^{-1} to 10 nmol L^{-1}. Since the concentration of antigen was less than the antibody K_D (with the exception of the femtomolar affinity antibody), a strong correlation existed between potency and affinity. In Fig. 7.2b, a supraphysiological concentration (2 nmol L^{-1}) of another soluble cytokine was required for the bioassay. In this case, the affinities of all the antibodies (ranging from 1 to 200 pmol L^{-1}), also generated in XenoMouse animals, were less than 1/10th the antigen concentration; thus the assay was conducted under stoichiometric rather than K_D-dependent conditions. As expected, no relationship existed between affinity and the IC_{50}. Variation in the IC_{50} reflected the intrinsic variability of assay. In this case, rank ordering

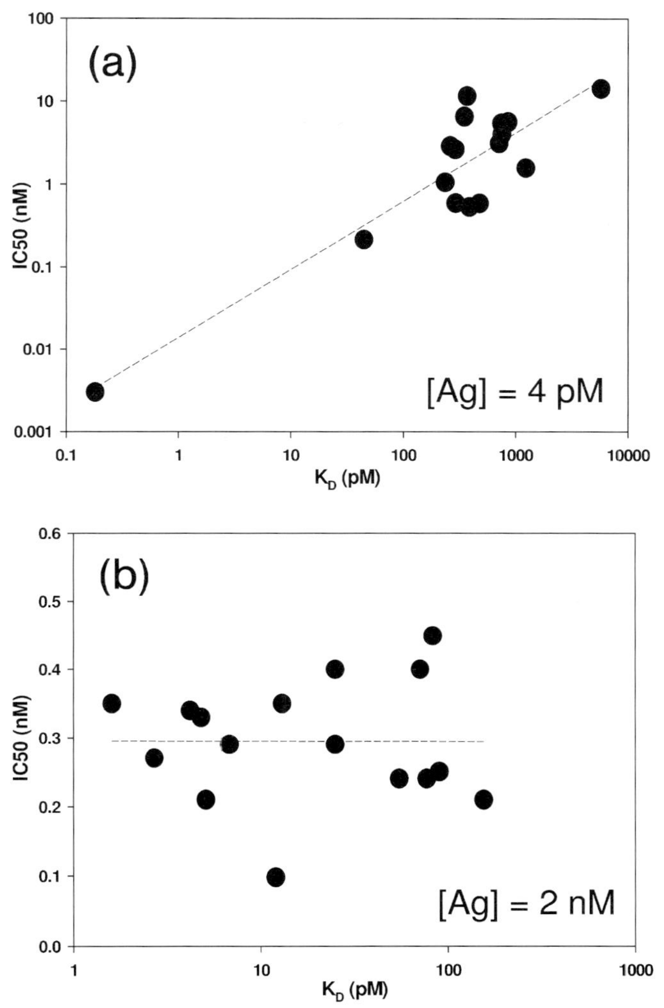

Fig. 7.2 Effect of antibody affinity and antigen concentration on antibody potency *in vitro* for two panels of monoclonal antibodies targeting cytokines. (a) In an experiment conducted under K_D-dependent binding conditions ([Ag] << K_D of most antibodies), a strong correlation of IC_{50} with K_D was observed. (b) In an experiment conducted under stoichiometric conditions ([Ag] >> K_D for all antibodies), no correlation of IC_{50} with K_D was observed because IC_{50} is dependent on the antigen concentration, not the K_D.

antibodies by IC_{50} might erroneously prioritize an antibody with low affinity and low *in vivo* potency.

In vitro assays should also be conducted under equilibrium conditions. In these assays, antigen and antibody are added prior to conduct of the assay, with a pre-incubation to allow antibody and antigen to reach equilibrium. However, if the

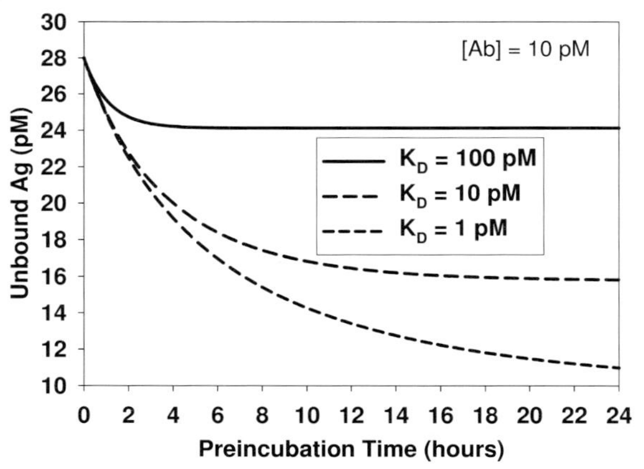

Fig. 7.3 Simulated effect of affinity on time to equilibrium for an antibody–antigen mixture. Under the particular conditions of this *in vitro* experiment ([Ag] = 28 pmol L^{-1}, [Ab] = 10 pmol L^{-1}), the time to reach equilibrium would be over 18 h for a 1 pmol L^{-1} affinity antibody.

k_{off} and the antigen concentration are very low, a long preincubation time may be required to reach equilibrium. If equilibrium for the high-affinity antibodies is not achieved during preincubation, the observed differences in potency might be diminished. Figure 7.3 demonstrates a simulation that was conducted to determine the optimum preincubation time for a bioassay comparing the potency of a panel of antibodies of known affinity in the presence of a fixed concentration of antigen. The time-course of unbound antigen concentration is shown as a function of incubation time and affinity. Over the range of antibody affinities, a minimum preincubation time of 18 h was determined to be necessary for binding to approach equilibrium when the antibody affinity was high. The simulations were supported by potency assays run with 1-h and 18-h preincubation times; potency differentiation between the antibodies was observed only after the 18-h preincubation.

7.3.2
Binding and Potency *In Vivo*

Most *in vitro* potency experiments are conducted under conditions where there is negligible production/input and degradation of antibody and antigen during the course of the experiment. In the *in vivo* situation, antibody is eliminated continuously and is usually dosed repeatedly, and antigen is continuously produced and degraded. Immune complexation of antigen will usually alter the pharmacokinetics of the antibody and/or the antigen (Tabrizi et al. 2006). When a soluble antigen binds to an antibody *in vivo*, the bound antigen will generally take on the elimination kinetics of the antibody. If an antibody binds to a cell membrane antigen, the bound antibody can be eliminated at the internalization

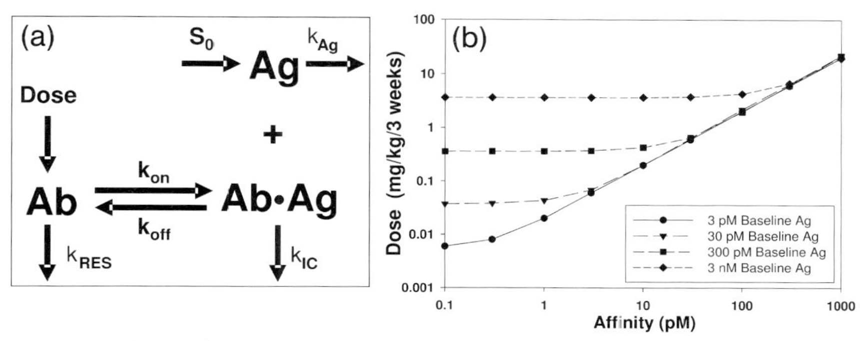

Fig. 7.4 Simulation of the maintenance dose of antibody required to suppress unbound concentrations of a soluble antigen in serum *in vivo* by 90% at steady-steady state prior to the next maintenance dose. (a) Kinetic model of Ag–Ab interaction *in vivo* used to simulate the interaction. S_0, production rate of Ag; k_{Ag} (rate constant for Ag elimination); k_{RES}, rate constant antibody elimination by the reticuloendothelial system; k_{IC}, rate constant for elimination of the immune complex. (b) The simulation, conducted as a function of affinity and [Ag], predicts that increased affinity will improve antibody potency until K_D falls below 1/10th the baseline [Ag].

rate of the antibody–antigen complex; if the antigen is widely expressed and the internalization rate is rapid, the antigen may create a saturable sink for antibody elimination.

Despite the complexity of the antibody and antigen kinetics *in vivo*, the effect of affinity on antibody potency is similar to that observed *in vitro*. Figure 7.4a illustrates a simple kinetic model of antibody and antigen interaction within the plasma pool. For simplicity, a "one-compartment" pharmacokinetic model is assumed for antibody and antigen. The model in Fig. 7.4a can be described by differential equations for unbound antibody, unbound antigen, and the antibody–antigen complex. In Fig. 7.4B, these equations have been used to calculate the dose of antibody, administered every three weeks, needed to decrease unbound concentrations of antigen at predose, steady-state levels of antibody, by 90%. Simulations were conducted assuming a soluble, intermediate clearance rate antigen with steady-state baseline concentrations ranging from $3\,\text{pmol}\,\text{L}^{-1}$ to $3\,\text{nmol}\,\text{L}^{-1}$ in plasma. Clearance of the immune complex was assumed to equal the reticuloendothelial clearance of antibody in absence of an antigen interaction. As illustrated in Fig. 7.4b, a point is reached where further improvements in affinity do not produce additional improvements in potency. Similar to the *in vitro* results described previously, this potency ceiling occurs when the affinity is reduced to about 1/10th the concentration of antigen. Therefore, for any antibody design goal aimed at maximizing the binding potency of a therapeutic antibody *in vivo*, the pathophysiological concentrations of antigen in the relevant biophase should be considered.

When a saturable antigen sink is present, high-affinity antibodies, under certain conditions, can be cleared at a faster rate than low-affinity antibodies. To further illustrate the importance of the antigen concentration when considering the required affinity of the antibody for the antigen, a kinetic model was established (similar to the one illustrated in Fig. 7.4a) that described the bimolecular interaction of an mAb with a membrane-bound antigen, where the antibody exhibits two-compartment distribution kinetics and linear elimination through the reticuloendothelial system (k_{res}) and nonlinear elimination (k_{IC}) through a sink provided by a cell membrane antigen. Simulations were conducted after administering a single intravenous dose of antibody, and the model assumed a rate constant k_{int} of 0.017 min^{-1} (internalization half-life of 40 min) for receptor internalization and a total receptor concentration of 1 nmol L^{-1}. The model was used to simulate antibody pharmacokinetics (Fig. 7.5a) and the concentration of unbound antigen (Fig. 7.5b) in serum for antibodies with different affinities. As seen in Fig. 7.5a, the serum half-life of the high-affinity antibodies becomes shorter than that of the low-affinity antibody when the suppression of unbound antigen is less than ~90% (around 1.5 days after administration of the single dose). However, the high-affinity antibodies produced greater suppression of unbound antigen when the antibody was present at saturating levels. As expected, the unbound antibody and unbound antigen profiles became nearly identical when the K_D of the antibody fell to 1/10th the antigen concentration or below. If subsaturating concentrations of antibody are required clinically, as might occur for some agonist antibodies or antibodies with a dose-limiting toxicity, then a lower affinity antibody might theoretically present a more favorable pharmacokinetic/pharmacodynamic profile. Under conditions of multiple dosing achieving saturating levels of antibody, high-affinity antibodies are generally expected to be advantageous with respect to dose potency.

7.4 High-Throughput Selection of Hybridomas Secreting High-Affinity Antibodies

The immunization and hybridoma selection strategies for antibodies derived from hyperimmunized animals vary depending on antigen type and antibody design criteria. A commonly used antibody screening process is described below for soluble and cell surface antigens.

7.4.1 Soluble Antigens

For soluble protein and peptide antigens, animals are immunized with the antigen or keyhole limpet hemocyanin (KLH)-conjugated antigen; hybridomas are generated and screened using an enzyme-linked immunosorbent assay (ELISA). The antigen is captured on the surface of ELISA microtiter plates. Antibodies in hybridoma supernatants bound to the antigen on the plates are detected with horseradish peroxidase (HRP)-conjugated antibodies against the immunoglobulin constant region. The hybridomas secreting antibodies that generate a

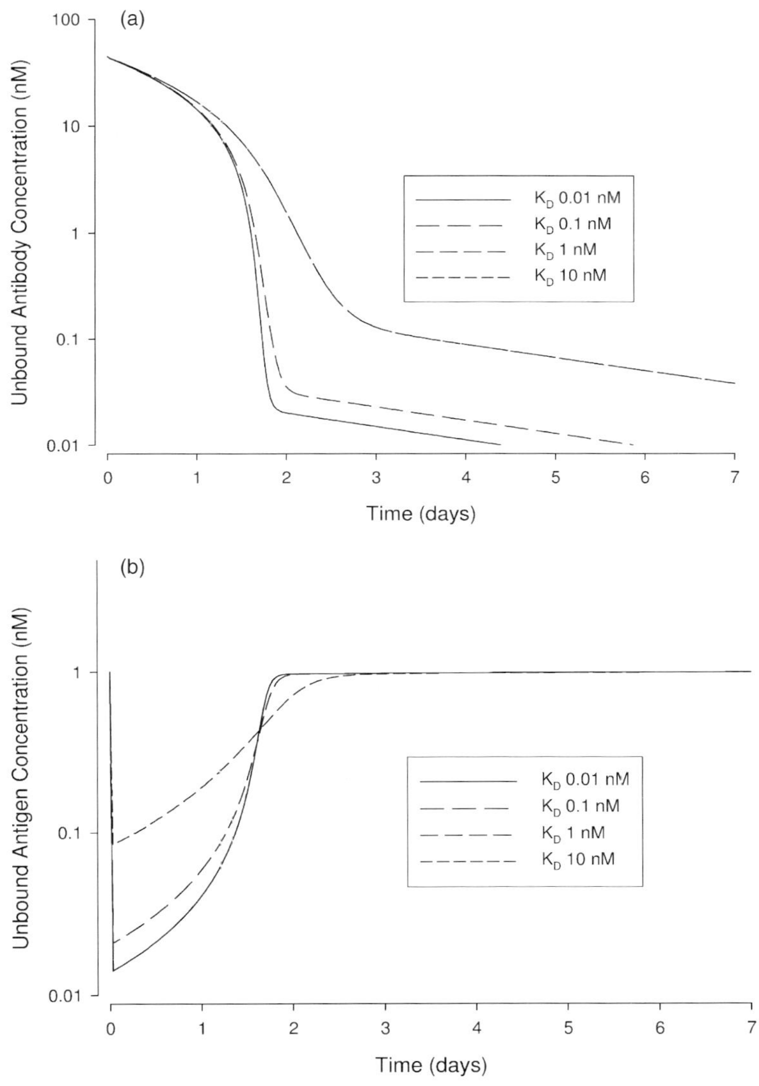

Fig. 7.5 Simulated effect of antibody affinity on (a) antibody pharmacokinetics and (b) unbound antigen levels for an antibody targeting an internalizing, cell membrane receptor. The simulations predict, under the conditions of these simulations, that higher affinity antibodies will have a shorter half-life at subsaturating concentrations, but higher receptor occupancy at saturating concentrations.

positive OD signal in ELISA are cultured continuously and the supernatants are tested again using the same ELISA format to confirm the antibody secretion. In addition, an ELISA assay using a closely related protein or an irrelevant protein, such as bovine serum albumin (BSA), is also performed in order to assess the specificity of antigen binding. Antibodies that bind to the antigen only, not to other proteins, are considered specific for the antigen.

A sophisticated immunization procedure and a significantly improved hybridoma technology can routinely result in few hundreds of antigen-specific, high-affinity mAbs. Therefore, more *in vitro* assays are needed to select the top antibodies for affinity measurement using Biacore or KinExA technologies (discussed in the next section) and ultimately to select the lead antibody candidates for *in vivo* efficacy determination. Typically, epitope binning, affinity ranking, and activity-based assays are utilized for this purpose.

Antibodies with different antigen-binding epitopes may show different potency and specificity profiles *in vivo*. Therefore, it is advantageous to select the mAb candidates that represent various antigen-binding specificities for *in vivo* testing. The conventional method to determine the antigen-binding epitope for an antibody is epitope mapping. However, this method is very time-consuming. To overcome this drawback, an epitope-binning methodology was developed to efficiently group antibodies with distinct antigen-binding specificity (Jia et al. 2004). This method utilizes Luminex bead-based multiplexing technology (Fulton et al. 1997) to detect antibody competition for antigen binding and enables sorting of a large panel of mAbs into different bins based on cross-competition. In this assay, each mAb is captured by an anti-IgG mAb coupled to a unique spectrally encoded bead from 100 commercially available Luminex beads. A Luminex 100 instrument quantified the extent of binding competition of any given mAb to an antigen against up to 99 other antibodies in a single assay. Although this assay does not identify the amino acid sequence in an antigen for an antibody binding as determined by epitope mapping, it provides valuable information to ensure selection of antibodies with different antigen-binding specificities for further testing.

Although Biacore and KinExA assays provide accurate affinity measurements, the complexity of these assays limits the number of antibodies that can be tested. To overcome these hurdles, an affinity ranking methodology using Luminex technology was developed. In this assay, three concentrations of a biotinylated antigen are each captured on a unique streptavidin-coupled spectrally encoded Luminex bead. MAb bound at four concentrations to each of these antigen beads are labeled with phycoerythrin (PE)-conjugated antibody against the IgG constant region, and the fluorescence associated with each of the beads is quantified with a Luminex 100 instrument. The relative fluorescence on each bead is proportional to the amount of bound antibodies to the antigen in the linear range of the binding curve. For a panel of mAbs tested in the same assay, the rank order of the amount of bound antibodies at a selected concentration in the linear range of the curve to a given concentration of antigen is generally correlated to the rank order of their affinities. This method provides a medium-throughput procedure to rank a large panel of mAbs based on their affinities and enables a quick selection of the antibody candidates for affinity measurement by Biacore or KinExA.

7.4.2 Cell Surface Antigens

A number of cell surface proteins have proven to be valid therapeutic targets. However, antibody generation for this class of proteins, especially membrane multi-spanning proteins, can be more challenging than soluble antigens. Challenges exist for both immunogens and hybridoma selection processes. For this type of protein, immunogens can be a purified extracellular domain of an antigen, cell membrane preparations, or whole cells. A purified extracellular domain (ECD) of an antigen is an ideal immunogen to use; however, the ECD is not always available and may not have the native conformation. Although cell membranes or whole cells generally present cell surface proteins in their native conformations, specificity issues may arise with the use of membranes and cells as immunogens due to the existence of other membrane proteins. If the protein of interest was not the most abundant or the most immunogenic on the cell surface, generation of antibodies against these antigens may be difficult.

A conventional hybridoma screening method, ELISA, requires a purified ECD of the antigen, which may not be available or may not be conformationally similar to the cell membrane antigen. Therefore, a conventional ELISA might not be optimal for hybridoma selection for cell surface antigens. An ELISA assay using cell membranes or whole cells often gives high background, high coating variability, and low signal, resulting in a suboptimal assay. Fluorescence-activated cell sorting (FACS) analysis provides advantages compared to ELISA for detecting antibody binding to cell surface antigens. However, this methodology requires multiple washing steps and a complicated data collection/analysis process, not suitable for analyzing large number of hybridomas in a high-throughput manner. To overcome these drawbacks, a high-throughput cell-based antibody–antigen binding assay using fluorometric microvolume assay technology (FMAT) (Miraglia et al. 1999) has been developed. This assay provides high detection sensitivity that is comparable to FACS and is amenable to high-throughput hybridoma screening.

In the FMAT assay, antibodies in hybridoma supernatants bound to an antigen on the surface of the antigen-expressing cells are detected by a Cy5-conjugated F(ab′)2 anti-IgG, Fc-specific antibody. A FMAT 8200 HTS instrument is used to measure the fluorescence associated with the cells. In order to distinguish antibodies that bind to the antigen from those that bind nonspecifically to cells, the antibody binding is assessed using both antigen-expressing and parental cells. The antibodies binding specifically to antigen are identified as those that bind to antigen-expressing cells only, and nonspecific antibodies are identified as those that bind to both antigen-expressing and parental cells.

The key factor for a good FMAT assay is the antigen-expression level on the cells. Although FMAT assay can detect antibody binding to cells expressing an antigen at 9000 copies per cell, the assay performance is improved with a cell line expressing an antigen at 30 000 copies per cell or above. One important step in FMAT assay optimization is to reduce the background. A high background is generally due to low cell viability or nonspecific internalization leading to antibody trapping inside cells. Cell viability can be improved by modifying cell culture and cell detaching conditions. Internalization can be minimized by treatment of cells with sodium azide.

A hook effect is observed with FMAT assay due to high concentration of antibodies in solution saturating Cy5-anti-IgG antibodies. This can potentially cause failure to identify antigen-binding antibodies when their concentrations in supernatants are extremely high. A strategy to address this issue is to test supernatants in more than one dilution. However, due to the broad, detectable range of antibody concentrations by a FMAT assay, the false negative rate estimated using a panel of known positive antibodies is very low and might not be a significant issue in hybridoma screening.

Incorporation of cell-based antibody binding screening using FMAT in a hybridoma selection process for cell-surface antigens has demonstrated significant improvement in finding antibodies that show antigen neutralization. For multispanning membrane proteins, FMAT generally identifies significantly more neutralizing antibodies than ELISA.

One disadvantage observed with FMAT assays is a higher false positive rate compared with FACS due to a no wash format of FMAT assays. To eliminate these false positives, the positive hybridomas identified by FMAT are confirmed by a FACS assay. Hybridoma supernatants are incubated with antigen-expressing cells and parental cells. The cell-associated antibodies are detected with Cy5-conjugated anti-IgG antibody. The fluorescence associated with the cells is analyzed with a FACSCalibur instrument. Antigen-specific antibodies are identified as those that only bind to antigen-expressing cells, but not to parental cells. Nonspecific antibodies are those that bind to both antigen-expressing and parental cells.

Antibodies that bind to cell surface antigen specifically are then assessed for their affinities using a FACS-based affinity determination methodology described in the next section. In addition, the functional activities of these antibodies are also determined in a panel of activity assays. Functional activity and affinity *in vitro* are used to select the top antibodies for *in vivo* testing.

7.5 Kinetic and Equilibrium Determinations of Antibody Affinity

7.5.1 Biacore Technology

Biacore technology is based on a surface plasmon resonance biosensor and is the premier biophysical method currently used for measuring antigen–antibody affinity. With Biacore one reactant is covalently immobilized to a biosensor surface and the other binding partner is flowed across the surface, and the binding of the reactants is followed in real time by surface plasmon resonance (Karlsson and Falt 1997; Morton and Myszka 1998; Drake et al. 2004). Frequently, Biacore is criticized for not giving accurate or "solution-phase" rate constants and hence an inaccurate equilibrium constant for a biomolecular interaction since one binding partner is immobilized to a biosensor surface while the other reactant is flowed across the surface. This criticism may be warranted if the most optimal and advanced experimental design and data-processing techniques are not utilized in the conduct of a Biacore experiment, which is often the case observed in the Biacore literature. Many published articles do not give the reader enough information to reproduce the given experiment or even to critically evaluate whether the experiment was conducted appropriately (Myszka, 1999b; Rich and Myszka 2000, 2005). However, as nicely pointed out by Van Regenmortel (2003), Biacore results are often artifactual because the scientists conducting the experiment do not use the most advanced experimental design and processing techniques to start with. As a result, the Biacore instrument is blamed for giving inaccurate kinetic results when the real cause of the erroneous data is the "unprofessional" use of the technology.

One of the pitfalls of performing a Biacore experiment results from the highly automated nature and user-friendliness of the Biacore instrument, which has a tendency to lull users into never learning or forgetting the basic principles of physical chemistry that apply to the chemical binding reaction taking place in the flow cell (Van Regenmortel 2003). Understanding the basic principles of physical chemistry and how they relate specifically to the particular biophysical technique being undertaken is not unique to Biacore, but really applies to any physiochemical methodology, even KinExA technology as will be discussed later. In addition, the physical chemistry of the particular biomolecular interaction, in this case a bivalent monoclonal antibody and a monovalent or multivalent antigen, also needs to be considered in setting up a Biacore or KinExA experiment in order to yield accurate kinetic and thermodynamic results. Biacore does, indeed, give accurate kinetic rate constants and equilibrium constants that are very similar to values derived from solution-phase kinetic and equilibrium methods when optimal and advanced Biacore data collection and processing techniques are implemented (Day et al. 2002; Drake et al. 2004).

Biacore also is useful in screening mAbs for relative and absolute affinities even from crude mixtures like hybridoma supernatants (Canziani et al. 2004). The K_D values resulting from these Biacore experiments, where monoclonal antibody was captured on the biosensor surface from hybridoma supernatant and one concentration of purified antigen was flowed over the surface, were accurate and reliable because the same kinetics were observed with purified monoclonal antibody using standard Biacore methodology (Canziani et al. 2004).

In general, many therapeutic antibodies have equilibrium dissociation constants (K_D) that are less than 100 pmol L^{-1}. Measurement of on-rates (k_{on}), off-rates (k_d), and K_D values for mAbs possessing very high affinities (<100 pmol L^{-1}) can be challenging for three unique reasons: (1) the time for the antigen–antibody mixture to reach equilibrium can be very long, on the order of days; (2) usually, the k_d for such a tight complex is extremely low, requiring long periods of data collection to discern enough information to predict complex stability; and (3) in cases where the k_d is easily measurable ($>5 \times 10^{-4}\,s^{-1}$), the k_a can be very fast, greater than 1×10^7 ($L\,mol^{-1}$) s^{-1}. When performing a Biacore experiment with any multivalent molecule like a mAb it is important to immobilize the bivalent antibody to the surface and to flow the monovalent antigen as shown in Fig. 7.6a. Many times Biacore experiments are performed in the incorrect orientation shown in Fig. 7.6b where antigen is immobilized to the biosensor surface and bivalent monoclonal antibody is flowed. This orientation quite often leads to sensorgrams that are described by complex kinetic interaction models.

In designing a Biacore experiment, it is absolutely crucial to stay away from orientations that may introduce complexity into the sensorgram data. Interpretation of complex sensorgram data and selection of the correct complex interaction mechanism that describes the complex kinetic data can be painstakingly difficult and take an inordinate amount of time and resources. The third scenario shown in Fig. 7.6c is a plausible Biacore orientation with which to measure monoclonal antibody binding kinetics, but as observed in reality quite frequently, it is difficult to achieve a low enough surface capacity of antigen to completely rid the sensorgrams of complexity and still observe a reasonable binding signal when antibody is flowed.

One interesting dilemma that occurs when measuring antigen–antibody kinetics by Biacore is when the antigen is multivalent like the antibody. What is the best Biacore experimental orientation for a situation like this? When both binding reactants are multivalent, the chance of observing complex kinetic sensorgrams increases dramatically owing to the possible multiple binding steps involved in the interaction as shown in Fig. 7.6b. The best way to optimize the orientation for a Biacore experiment involving two multivalent molecules is to set the experiment up as given in Fig. 7.6c. As discussed earlier, immobilizing a small amount

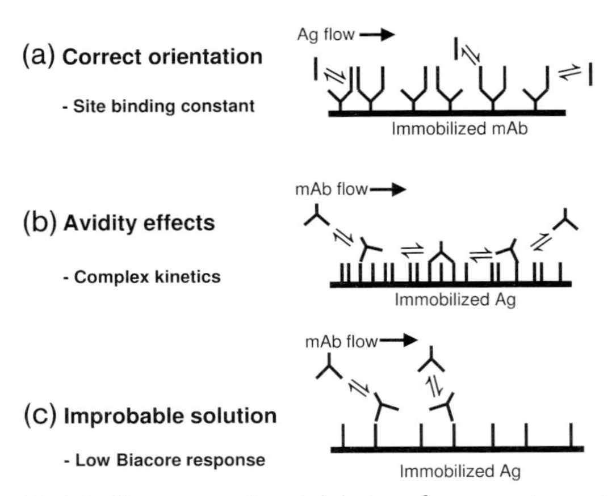

Fig. 7.6 Biacore experimental designs for measuring antigen–antibody interactions.

of multivalent antigen, or for that matter even bivalent monoclonal antibody, to the biosensor surface in no way necessarily guarantees sensorgrams that can be described by a simple 1:1 binding mechanism when both binding partners are multivalent. However, a low surface capacity greatly hedges the probability of collecting sensorgrams that fit a basic 1:1 model. Even in an experiment involving two multivalent binding partners, like in the case of a multivalent antigen and a bivalent antibody, in which sensorgrams are collected that fit a 1:1 interaction model, it is still ambiguous whether the K_D determined corresponds to a stoichiometric (macroscopic) binding constant or an intrinsic site (microscopic) binding constant. For example, with two bivalent molecules two binding sites of one molecule binding to two binding sites of the other molecule fit a 1:1 binding model potentially as well as one site of one molecule binding to one site of the other. The same argument applies to higher order complexes too (i.e. a bivalent antibody binding to a trivalent antigen, etc.).

From a biophysical standpoint, it is always best to determine the site binding dissociation equilibrium constant since this is an intrinsic binding constant that is inherent to the complex. Macroscopic or stoichiometric constants do not pertain to any particular binding site, but rather only provide information from a thermodynamic viewpoint. If the flowing bivalent binding partner is crosslinking two different multivalent immobilized ligands on the biosensor surface, then the stoichiometric binding constant really becomes an avidity constant. Obviously, avidity binding constants can change as a function of the amount of immobilized ligand on the biosensor surface, which in a sense yields an equilibrium "inconstant" since this so-called binding "constant" potentially could change with different surface densities. While it is attractive to argue from a biological standpoint that the avidity constant is the more meaningful equilibrium constant to measure since it may mimic the *in vivo* interaction, this argument breaks down quickly when scrutinized. For example, how is it possible to know experimentally that the biosensor surface created has the same antigen molecular density as found on the target tissue *in vivo*? It would appear that any avidity constant that was desired could be measured by making a biosensor surface with varying multivalent antigen capacities.

One other type of complex, in contrast to the "crosslinked" variety mentioned above, could occur between a multivalent antigen (immobilized on the surface) with antibody flowing wherein both sites of the antibody bind to one distinct antigen molecule. This type of interaction would result in a more legitimate avidity constant in a biophysical sense, though not necessarily from a biological perspective, because this avidity constant would not vary as a function of antigen capacity on the biosensor surface.

In summary, when both binding molecules are multivalent, the binding parameters determined by Biacore (from sensorgrams that truly are described by a simple 1:1 interaction model and are not complex) become ambiguous since two possible binding constants, site and stoichiometric (avidity), may be measured depending on the structure of the complex formed on the biosensor surface. In cases where sensorgrams with antigen–antibody complexes cannot be collected that fit a simple 1 : 1 interaction model (after eliminating all other human-induced causes of kinetic complexity, *vide infra*), it is advisable to change biophysical methodology and turn to an equilibrium method like KinExA (to be discussed in the next section).

Regardless of whether the K_D measured is a site or avidity K_D, most researchers in the therapeutic antibody field find this measurement crucial because it allows selection of the monoclonal antibody candidates with the required affinities for further development.

Myszka (1999a) has written an excellent tutorial on how to rigorously design a Biacore experiment and how to optimally process biosensor data to ensure the highest quality kinetic results. Figure 7.7 gives selected highlights from Myszka's paper (Myszka 1999a) on improving biosensor analysis. To ensure the highest quality biosensor results it is important to perform microfluidic washing steps before and after each antigen injection. It is also of utmost importance that the Biacore is scrupulously cleaned at regular intervals and before beginning each experiment (Myszka 1999a). Antibody should always be immobilized at a surface capacity that yields a maximum resonance unit response (R_{max}) that is no greater than 50–100 RUs to avoid mass transport artifacts. High antigen flow rates are also preferable to minimize any mass transport artifacts that may be present during the experiment. A wide concentration range (at least 10- to 100-fold above and below the K_D of the interaction is ideal) of antigen should be injected randomly in duplicate or triplicate. It is worth noting that pragmatically antigen concentrations that are 10- to 100-fold above and below the K_D are often impossi-

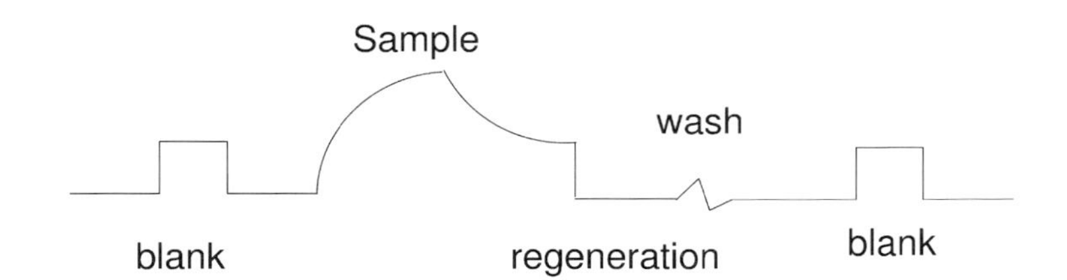

- Always inject sample buffer at the start and end of a cycle in order to wash out the microfludics system.
- Ligand surface density should give an R_{max} less than 50 – 100 RU to avoid or minimize mass transport artifacts, steric hindrance, or aggregation.
- If the ligand is an antibody, this should always be immobilized or captured on the biosensor surface to avoid avidity effects.
- Always use a high flow rate like 100 μl/min to minimize any mass transport considerations, and to deliver a more consistent analyte plug across the biosensor chip.
- Always replicate in duplicate, at least, and preferably, in triplicate all sample injections over a wide concentration range. Also, randomize the order of injection of all samples.
- Double reference all sensorgrams to correct for non-specific binding, bulk refractive index changes, and systematic instrument noise.
- Globally fit nonlinearly all kinetic data utilizing any specialized biosensor software.

Fig. 7.7 Suggested steps for rigorous biosensor methodology (adapted from Myszka 1999a).

ble to achieve owing to nonspecific binding interactions at high concentration and loss of signal at very low antigen concentration, especially for antigen–antibody complexes possessing picomolar affinities. In these cases, the antigen concentration range should be as wide as experimentally possible. In addition, if a small amount of decay is seen in the dissociation phase of the sensorgrams for a tight antigen–antibody complex (K_D <100 pmol L^{-1}), indicating the complex has a slow k_{off} (<5 × 10^{-4} s^{-1}), several additional relatively high antigen concentrations and buffer blanks should be injected and the dissociation phase followed for ~4 h or longer (Drake et al. 2004). Using this "long k_{off}" dissociation methodology, it is possible to measure a k_{off} ~10^{-5} s^{-1}, and if longer dissociation phases are collected (4–24 h), a k_{off} ~10^{-6} can be determined. It has been shown that this long k_{off} methodology results in reliable binding parameters for antigen–antibody complexes when compared with solution-phase equilibrium methods (Drake et al. 2004).

If the association phase of a biomolecular interaction is extremely slow or the K_D is extremely tight (being in the picomolar range, thus requiring very low analyte concentrations to be flowed in order to be near the K_D of the interaction), it is also possible to follow the association phase for hours, instead of the normal few minutes, when specialized Biacore procedures are invoked (Navratilova et al. 2005).

Sensorgram data should always be double-referenced. Double-referencing many times transforms a singly referenced sensorgram data set that is not capable of being fit into a data set that fits easily to a simple 1 : 1 interaction model. The importance of double-referencing sensorgram data cannot be emphasized enough, especially when low surface capacities are used. Lastly, all processed sensorgram data should be globally fit nonlinearly using any readily available specialty biosensor software program. It should also be noted that if complex sensorgrams are observed, it is not scientifically justified to fit the data to a complex interaction model (referred to as "model surfing") without a plethora of additional biosensor and extra-biosensor kinetic and equilibrium data that supports the complex binding mechanism. Many artifacts can cause complex sensorgram data such as: impure monoclonal antibody, impure antigen, heterogeneity introduced by immobilization chemistry, badly processed data (poorly y- or x-aligned sensorgrams or non-double-referenced sensorgrams), nonspecific binding to the actual immobilized molecule or to the carboxylmethyldextran surface matrix that was not properly corrected for in data processing, or even a dirty or contaminated Biacore instrument.

7.5.2
KinExA Technology

KinExA (kinetic exclusion assay) technology differs from Biacore in that it is a solution-based method and requires a secondary fluorescent reporter molecule. The KinExA instrument is a flow spectrofluorimeter in which equilibrated solutions of an antigen–antibody complex are flowed through a bead pack with immobilized antigen (Blake et al. 1999; Darling and Brault 2004). Detection of free monoclonal antibody from the equilibrated solution once flowed through the resin is accomplished with a secondary fluorescently labeled species-specific polyclonal antibody. The percentage of free monoclonal antibody versus the total antigen titrated into each equilibrated solution is fit to a standard 1:1 equilibrium model to yield a K_D for the given interaction. KinExA is ideally suited for measuring antigen–antibody affinities since it requires a secondary fluorescent reporter molecule of which many exist for mAbs. The sensitivity of fluorescence detection is also desirable for measuring interactions with very tight equilibrium dissociation constants.

The KinExA instrument is also capable of directly measuring $k_{on,}$ by following the decrease of free antibody as a function of time as an antigen–antibody complex approaches equilibrium, referred to as the "direct" method. Another alternative procedure for measuring the k_{on} with KinExA is to determine free antibody concentration as a function of added antigen known as the "injection method." The k_{off} is not measured directly in KinExA but rather is calculated from the product of $K_D \times k_{on}$. The timescale in which k_{on} is measured by KinExA (during a "direct" injection method) versus Biacore is drastically different. The longest injection time usually possible on a Biacore is about 2.5–10 min whereas with KinExA, kinetic data points are usually only collected every 5–15 min for a total time of hours until equilibrium is reached. For more specifics on these equilibrium and kinetic methods, please refer to the articles by Blake et al. and Darling and Brault (Blake et al. 1999; Darling and Brault 2004).

The shape of KinExA equilibrium titration curves is not dependent on the absolute antibody-binding site concentration but rather on the ratio of antibody-binding site concentration to the K_D of the interaction. If the antibody-binding site concentration is much higher than the K_D, then the curve is insensitive to the K_D and changes shape as a function of the active antibody-binding site concentration and is called an antibody-controlled curve. If the antibody-binding site concentration is near or slightly above the K_D, then the resulting curve is sensitive to the K_D and has little active antibody-binding site concentration information present in it and is referred to as a K_D-controlled curve. Both types of equilibrium titration curves can be fit simultaneously in a dual curve analysis (or n-curve analysis with "n" multiple curves) to yield reliable values of the active antibody-binding site concentration and the K_D. The most rigorous values for K_D and active antibody-binding site concentration result from an n-curve analysis since, obviously, it is more difficult to fit n curves globally than it is to fit each curve locally. In addition, there is more data with which to calculate the binding parameters from an n-curve analysis. Multiple (n) curve analysis also provides extra assurance that the equilibrium titration data truly do, indeed, fit a simple 1:1 equilibrium model.

With both types of titration curves being fit simultaneously, it is easier to see deviations from the theoretical model than in a single curve analysis. However, if the active antibody-binding site concentration is chosen correctly (discussed below) for a K_D-controlled titration, then a single curve analysis, in general, results in an acceptable value for K_D that is near identical to the K_D determined from an n-curve analysis. This single curve titration approach can be useful for obtaining replicate measurements quickly for the K_D of an antigen–antibody interaction so accurate 95% confidence limits can be calculated. It is important to realize that the 95% confidence intervals calculated for the fitted parameters by the KinExA software reflect the precision of the parameters derived from the data fitting, and do not represent the 95% confidence intervals for replicate measurements.

Several major considerations need to be realized to design and perform high-quality KinExA equilibrium and kinetic experiments. First, antigen–antibody mixtures must be given ample time to equilibrate. For antigen–antibody complexes with picomolar K_D values (usually corresponding to complexes with extremely slow k_{off}) this means days to weeks are needed to allow solutions to equilibrate (Drake et al. 2004). Secondly, solution volumes in KinExA are much larger than those in Biacore experiments so "old-fashioned" analytical volumetric technique is required for the highest quality equilibrium and kinetic results from KinExA. Thirdly, an active antibody-binding site concentration should be used that is no more than 3-fold above the K_D of the interaction for collection of a K_D-controlled titration. With equilibrium titration data of exceptional quality, it is possible to have the active antibody-binding site concentration be set at 10-fold above the K_D and still determine a satisfactory K_D. Fourth, for antibody-controlled curves it is best to work at antibody-binding site concentrations greater than or equal to 10-fold above the K_D. Fifth, for "direct" kinetic experiments, select antigen and antibody-binding site concentrations so that ~80% of the antibody-binding sites are bound at equilibrium. Also, in a "direct" kinetic experiment choose antigen and antibody-binding site concentrations that will allow enough exponential decay of % free antibody-binding site concentration as a function of time for accurate determination of k_{on}. Sixth, for an "inject" kinetic experiment, use a range of initial antigen concentrations that gives 20–100% antibody complexation with a given constant antibody-binding site concentration for the particular mixing time that has been chosen for the KinExA instrument.

One point important to consider is that even with KinExA technology, as mentioned earlier for Biacore, the same ambiguity could potentially exist in the interpretation of an equilibrium constant for a multivalent antigen–bivalent antibody complex. However, the advantage of KinExA over Biacore in this particular situation is that binding data are collected after equilibrium has been reached, thus avoiding following any complex kinetic steps that can be difficult to model and fit in Biacore. KinExA experiments usually result in a titration curve that fits a 1:1 equilibrium model satisfactorily, allowing determination of a K_D for both monovalent and multivalent antigen–antibody complexes. Once again, as to what type of complex the determined K_D refers to in the case of a multivalent antigen–bivalent antibody complex is not readily known without other structural biological studies. KinExA technology has also been used to measure on-cell affinities directly, as described in the next section (Xie et al. 2005).

Certainly, it is not totally correct to imply that all rate constants and equilibrium constants determined by biosensors for antigen–antibody complexes will match exactly with solution-phase measurements, especially for interactions having picomolar K_D values as many therapeutic antibodies do. However, if optimized and advanced biosensor techniques are followed, reliable binding parameters can be determined for very tight interactions. Agreement between biosensor data and solution-phase methodology like KinExA is often found satisfactory when rigorous experimental protocols are utilized for both methodologies (Day et al. 2002; Drake et al. 2004). Biacore and KinExA complement each other nicely as biophysical methods for the development of therapeutic antibodies. This is illustrated by the higher throughput capability of Biacore and its ability to measure picomolar or less tight affinities, and the ability of KinExA to measure picomolar to sub-picomolar affinities (Rathanaswami et al. 2005) more easily in certain circumstances than Biacore. It is also important to bear in mind when characterizing antigen–antibody interactions that analytical and biophysical techniques determine results only as reliable as their experimental design and data analysis methods.

7.5.3 Cell-based K_D Titrations

Obviously, biophysical techniques like Biacore and KinExA routinely require purified antigens in order to obtain high-quality kinetic and thermodynamic data. However, antigens like purified membrane receptors with transmembrane domains may not be amenable to easy purification, while a purified extracellular domain of a transmembrane receptor may still have the significant problem of losing native conformation and function. Fortunately, cell-based *in vitro* affinity determination methods can overcome the above-mentioned difficulties. One cell-based affinity method that has appeared recently uses KinExA technology to titrate an increasing concentration of cells expressing the antigen of interest into a constant concentration of antibody. By doing a dual curve titration and implementing an "antigen unknown" fitting process, the K_D of the interaction of the antigen–antibody interaction can be determined (Xie et al. 2005).

Another technique that is quite useful for determining K_D values for antibody binding to cell surface antigen is fluorescence-activated cell sorting (FACS) (Cardarelli et al. 2002). The FACS methodology usually consists of titrating monoclonal antibody into a constant amount of cells and allowing the binding reaction to come to equilibrium. Subsequently, the bound monoclonal antibody is labeled with a fluorescently-labeled secondary polyclonal antibody and the fluorescence signal, which is proportional to bound monoclonal antibody, is followed as a function of total monoclonal antibody concentration added. Analysis of the data is accomplished most often by a Scatchard plot.

Recently, Drake and Klakamp (2007) have developed a new four-parameter nonlinear equation and methodology based on the multiple, independent binding site equation that fits cell-based binding data much more rigorously than currently existing methods. This new equation and methodology allows a titration curve, whose shape is sensitive to either the active cell-based antigen concentration or the K_D of the interaction, to be fit accurately no matter what condition is controlling the shape of the titration curve.

With routine FACS-based Scatchard analysis, the K_D determined is not necessarily always accurate. The accuracy depends on the experimental conditions the titration is run under since this type of analysis always assumes total monoclonal antibody concentration titrated into the cells is equal to free monoclonal antibody concentration at equilibrium. Often this is true, however, for very tight interactions ($<100\,\text{pmol}\,\text{L}^{-1}$) this is sometimes not the case (Drake and Klakamp 2007).

Cell-based methodologies should only be utilized when the antigen is incapable of being highly purified for traditional biophysical characterization, or when it is known based on strong scientific arguments that the purified antigen is not as functional or potent as in its native state. One drawback of all cell-based methodologies is that the K_D determined could be a site-binding constant or avidity constant as discussed above for Biacore (to the first approximation the cell surface can be thought of as being analogous to the biosensor surface for this avidity discussion). For example, if a cell line expressing more antigen on its surface is used versus a cell line expressing less antigen, two different K_D values might be measured for the same antibody. What is the correct K_D and which cell line mimics most closely the real *in vivo* situation? Obviously, there are no easy answers to these questions. Hence, it is almost always more advantageous to measure antigen–antibody interactions with purified reagents in a well-defined system, unless as stated above, it is impossible to obtain pure or functional antigen preparations.

KinExA cell-based and FACS cell-based titrations, when used properly, add indispensable tools to the arsenal of biophysical methodologies available for studying the binding of mAbs to cell surface antigens.

7.6 Conclusions

Antibodies of high affinity are usually required to effectively neutralize or modulate antigens. Thus, affinity maturation has evolved into a critical element of the humoral response to antigen exposure. When dosing therapeutic antibodies, affinity is an important characteristic affecting the potency and dose requirement for a mAb directed against a given antigen target. The affinity and potency requirement for a therapeutic antibody should be considered on a case-by-case basis: in all cases, an affinity threshold exists beyond which further improvements in affinity will not produce further improvements in potency. Conversely, for a given target, selection of an antibody with inadequate affinity will result in clinical or commercial failure of the therapeutic product. In the selection process for a lead therapeutic antibody, the appropriate application of analytical methods for biophysical characterization of antibody affinity is essential. Rational application of basic, kinetic principles governing antigen–antibody interactions can greatly facilitate the development of therapeutic antibodies.

Acknowledgments

The authors would like to acknowledge Raffaella Faggione, PhD and Olivia Lecomte-Raeber, PhD for generation of the data shown in Fig. 7.2a,b.

References

Blake, R.C., Pavlov, A.R., Blake, D.A. (1999) Automated kinetic exclusion assays to quantify protein binding interactions in homogeneous solution. *Anal Biochem* 272: 123–134.

Boder, E.T., Midelfort, K.S., Wittrup, K.D. (2000) Directed evolution of antibody fragments with monovalent femtomolar antigen-binding affinity. *Proc Natl Acad Sci USA* 97: 10701–10705.

Canziani, G.A., Klakamp, S., Myszka, D.G. (2004) Kinetic screening of antibodies from crude hybridoma samples using Biacore. *Anal Biochem* 325: 301–307.

Cardarelli, P.M., Quinn, M., Buckman, D., Fang, Y., Colcher, D., King, D.J., Bebbington, C., Yarranton, G. (2002) Binding to CD20 by anti-B1 antibody or F(ab')(2) is sufficient for induction of apoptosis in B-cell lines. *Cancer Immunol Immunother* 51: 15–24.

Darling, R.J., Brault, P.A. (2004) Kinetic exclusion assay technology: characterization of molecular interactions. *Assay Drug Dev Technol* 2: 647–657.

Day, Y.S., Baird, C.L., Rich, R.L., Myszka, D.G. (2002) Direct comparison of binding equilibrium, thermodynamic, and rate constants determined by surface- and solution-based biophysical methods. *Protein Sci* 11: 1017–1025.

Defrance, T., Casamayor-Palleja, M., Krammer, P.H. (2002) The life and death of a B cell. *Adv Cancer Res* 86: 195–225.

Drake, A.W., Klakamp, S.L. (2007) A rigorous multiple independent binding site model for determining cell-based equilibrium dissociation constants. J. Immunol Methods, in press.

Drake, A.W., Myszka, D.G., Klakamp, S.L. (2004) Characterizing high-affinity antigen/antibody complexes by kinetic- and equilibrium-based methods. *Anal Biochem* 328: 35–43.

Foote, J., Eisen, H.N. (1995) Kinetic and affinity limits on antibodies produced during immune responses. *Proc Natl Acad Sci USA* 92: 1254–1256.

Foote, J., Eisen, H.N. (2000) Breaking the affinity ceiling for antibodies and T cell receptors. *Proc Natl Acad Sci USA* 97: 10679–10681.

Fulton, R.J., McDade, R.L., Smith, P.L., Kienker, L.J., Kettman, J.R., Jr. (1997) Advanced multiplexed analysis with the FlowMetrix system. *Clin Chem* 43: 1749–1756.

Hanes, J., Schaffitzel, C., Knappik, A., Pluckthun, A. (2000) Picomolar affinity antibodies from a fully synthetic naive library selected and evolved by ribosome display. *Nat Biotechnol* 18: 1287–1292.

Hoet, R.M., Cohen, E.H., Kent, R.B., Rookey, K., Schoonbroodt, S., Hogan, S., Rem, L., Frans, N., Daukandt, M., Pieters, H., van Hegelsom, R., Neer, N.C., Nastri, H.G., Rondon, I.J., Leeds, J.A., Hufton, S. E., Huang, L., Kashin, I., Devlin, M., Kuang, G., Steukers, M., Viswanathan, M., Nixon, A.E., Sexton, D.J., Hoogenboom, H.R., Ladner, R.C. (2005) Generation of high-affinity human antibodies by combining donor-derived and synthetic complementarity-determining-region diversity. *Nat Biotechnol* 23: 344–348.

Hoogenboom, H.R. (2005) Selecting and screening recombinant antibody libraries. *Nat Biotechnol* 23: 1105–1116.

Jia, X.C., Raya, R., Zhang, L., Foord, O., Walker, W.L., Gallo, M.L., Haak-Frendscho, M., Green, L.L., Davis, C.G. (2004) A novel method of multiplexed competitive antibody binning for the characterization of monoclonal antibodies. *J Immunol Methods* 288: 91–98.

Karlsson, R., Falt, A. (1997) Experimental design for kinetic analysis of protein-protein interactions with surface plasmon resonance biosensors. *J Immunol Methods* 200: 121–133.

Mendez, M.J., Green, L.L., Corvalan, J.R., Jia, X.C., Maynard-Currie, C.E., Yang, X. D., Gallo, M.L., Louie, D.M., Lee, D.V., Erickson, K.L., Luna, J., Roy, C.M., Abderrahim, H., Kirschenbaum, F., Noguchi, M., Smith, D.H., Fukushima, A., Hales, J.F., Klapholz, S., Finer, M. H., Davis, C.G., Zsebo, K.M., Jakobovits, A. (1997) Functional transplant of megabase human immunoglobulin loci recapitulates human antibody response in mice. *Nat Genet* 15: 146–156.

Miraglia, S., Swartzman, E.E., Mellentin-Michelotti, J., Evangelista, L., Smith, C., Gunawan, I., Lohman, K., Goldberg, E.M., Manian, B., Yuan, P.M. (1999) Homogeneous cell- and bead-based assays for high throughput screening using fluorometric microvolume assay technology. *J BiomolScreen* 4: 193–204.

Morton, T.A., Myszka, D.G. (1998) Kinetic analysis of macromolecular interactions using surface plasmon resonance biosensors. *Methods Enzymol* 295: 268–294.

Myszka, D.G. (1999a) Improving biosensor analysis. *J Mol Recognit* 12: 279–284.

Myszka, D.G. (1999b) Survey of the 1998 optical biosensor literature. *J Mol Recognit* 12: 390–408.

Navratilova, I., Eisenstien, E., Myszka, D.G. (2005) Measuring long association phases using Biacore. *Anal Biochem* 344: 295–297.

Rathanaswami, P., Roalstad, S., Roskos, L., Su, Q.J., Lackie, S., Babcook, J. (2005) Demonstration of an *in vivo* generated sub-picomolar affinity fully human monoclonal antibody to interleukin-8. *Biochem Biophys Res Commun* 334: 1004–1013.

Rich, R.L., Myszka, D.G. (2000) Survey of the 1999 surface plasmon resonance biosensor literature. *J Mol Recognit* 13: 388–407.

Rich, R.L., Myszka, D.G. (2005) Survey of the year 2004 commercial optical biosensor literature. *J Mol Recognit* 18: 431–478.

Roskos, L.K., Davis G.C., Schwab, G.M. (2004) The clinical pharmacology of therapeutic monoclonal antibodies. *Drug Dev Res* 61: 108–120.

Schier, R., Marks, J.D., Wolf, E.J., Apell, G., Wong, C., McCartney, J.E., Bookman, M.A., Huston, J.S., Houston, L.L., Weiner, L.M. (1995) *In vitro* and *in vivo* characterization of a human anti-c-erbB-2 single-chain Fv isolated from a filamentous phage antibody library. *Immunotechnology* 1: 73–81.

Sundberg, E.J., Mariuzza, R.A. (2002) Molecular recognition in antibody-antigen complexes. *Adv Protein Chem* 61: 119–160.

Tabrizi, M.A., Tseng, C.M., Roskos, L.K. (2006) Elimination mechanisms of therapeutic monoclonal antibodies. *Drug DiscovToday* 11: 81–88.

Van Regenmortel, M.H. (2003) Improving the quality of BIACORE-based affinity measurements. *Dev Biol (Basel)* 112: 141–151.

Wabl, M., Cascalho, M., Steinberg, C. (1999) Hypermutation in antibody affinity maturation. *Curr Opin Immunol* 11: 186–189.

Wu, X., Feng, J., Komori, A., Kim, E.C., Zan, H., Casali, P. (2003) Immunoglobulin somatic hypermutation: double-strand DNA breaks, AID and error-prone DNA repair. *J Clin Immunol* 23: 235–246.

Xie, L., Mark, J.R., Glass, T.R., Navoa, R., Wang, Y., Grace, M.J. (2005) Measurement of the functional affinity constant of a monoclonal antibody for cell surface receptors using kinetic exclusion fluorescence immunoassay. *J Immunol Methods* 304: 1–14.

Zahnd, C., Spinelli, S., Luginbuhl, B., Amstutz, P., Cambillau, C., Pluckthun, A. (2004) Directed *in vitro* evolution and crystallographic analysis of a peptide-binding single chain antibody fragment (scFv) with low picomolar affinity. *J Biol Chem* 279: 18870–18877.

8
Molecular Engineering III: Fc Engineering

Matthias Peipp, Thomas Beyer, Michael Dechant, and Thomas Valerius

8.1
Mechanisms of Action of Monoclonal Antibodies

8.1.1
Introduction

Monoclonal antibodies have gained increasing importance as therapeutic reagents in clinical medicine over the last decade (Reichert et al. 2005), but many patients obtain only suboptimal responses to this expensive treatment. Since the market for antibodies is expected to become significantly more competitive over the next few years, there is an urgent need for optimized antibody constructs with excellent safety and high efficacy. For many years, efforts to improve antibodies concentrated on reducing their potential immunogenicity (Riechmann et al. 1988), leading to humanized or even fully human antibodies. Another approach aims to optimize antibodies by improving their effector functions. These effector functions are conceptionally divided into direct effects, which are mediated by the variable antigen binding region of the antibody, and indirect effects mediated by the constant Fc region. Although the fine specificity of monoclonal antibodies against the same target antigen (e.g. CD20) may control the immune effector mechanisms of individual antibodies (Cragg and Glennie 2004), efforts to improve effector functions concentrate on modulating the Fc region.

Sound knowledge of the relevant mechanisms of action of a particular antibody may have significant relevance for its clinical development, since this information may suggest ways to further enhance its efficacy (e.g. by antibody engineering). Thus, following a short discussion of candidate mechanisms, we describe modifications of the Fc region designed to improve specific effector functions triggered by therapeutic antibodies.

Handbook of Therapeutic Antibodies. Edited by Stefan Dübel

ISBN 978-3-527-31453-9

8.1.2
Preclinical Evidence

For the majority of antibodies, a complex mixture of different mechanisms of action is supposed to cooperate *in vivo* (Glennie and van de Winkel 2003) (Fig. 8.1). *In vitro* assays have proved particularly helpful in demonstrating direct effects on tumor cells, such as apoptosis induction, inhibition or stimulation of cell signaling, blockade of growth factors and cell proliferation. However, Fc-mediated, indirect effects of antibodies, such as complement-dependent cytotoxicity (CDC) and antibody-dependent cellular cytotoxicity (ADCC), can also easily be investigated *in vitro*. In addition, the relative contribution of different effector cell populations, such as mononuclear cells (MNC) and polymorphonuclear cells (PMN), can be assessed under these assay conditions. The relevance of these findings for clinical situations requires further study.

Several lines of evidence support the role of Fc-mediated effector functions for the efficacy of antibodies in animal models. Probably the most direct evidence for individual effector mechanisms is derived from genetically modified mice. For example, several therapeutic antibodies (rituximab, Herceptin (trastuzumab), Campath-1H, anti-CD2, anti-CD25) lost most of their efficacy in FcRγ chain knockout mice, which lack activating Fcγ receptors, while antibody efficacy was enhanced in mice lacking the inhibitory mouse FcγRIIb receptor (Clynes et al. 2000; Nimmerjahn and Ravetch 2006). The contribution of individual Fcγ receptors and the influence of different murine IgG isotypes was elegantly investigated

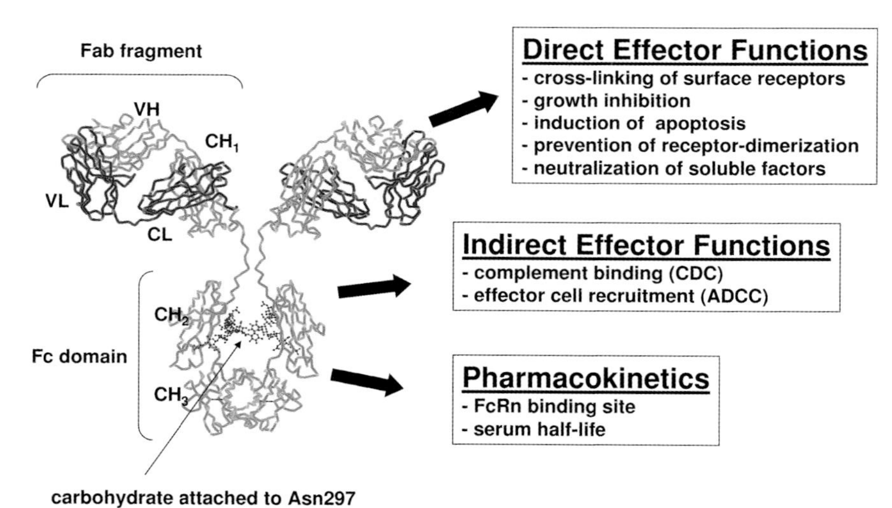

Fig. 8.1 Structure and function of IgG. Antibodies mediate different effector functions via distinct domains of the antibody molecule. Antibody model: computer-generated model structure of human IgG1 (Clark 1997). Dark gray: heavy chains; light gray: light chains.

using matched sets of antibody isotypes in mice deficient for select Fcγ receptors. (Nimmerjahn and Ravetch 2005). The role of the complement system for rituximab efficacy was assessed (e.g. in $C1q^{-/-}$ mice). Here, the response of a human CD20-transfected, complement-sensitive cell line to rituximab therapy was significantly diminished in $C1q^{-/-}$ mice compared to wildtype mice (Di Gaetano et al. 2003). However, when syngeneic B cell depletion by anti-mouse CD20 antibodies was investigated, antibody efficacy was diminished in $FcR\gamma^{-/-}$, $Fc\gamma RI^{-/-}$, or $Fc\gamma RIII^{-/-}$ mice, but not in $C3^{-/-}$, $C4^{-/-}$, or $C1q^{-/-}$ mice, indicating that under these conditions Fc receptor-mediated mechanisms predominate (Uchida et al. 2004). Induction of a tumor-directed immune response after antibody therapy has been reported in animal models (Clynes et al. 1998), but clinical examples of this probably ideal mechanism of action have been rare so far.

8.1.3 Clinical Evidence

Our knowledge about the clinically relevant mechanisms of action for monoclonal antibodies is rather limited. Early clinical studies with isotype switch variants of the rat CD52 antibody Campath-1H indicated that the rat IgG2b version of the antibody was more effective in depleting human B cells than its rat IgM and IgG2a counterparts. Interestingly, all three isotypes similarly activated human complement, but only the rat IgG2b version efficiently triggered ADCC (Dyer et al. 1989). Most of the more recent information is derived from studies with rituximab. Here, several lines of evidence point to an important role of Fc receptors. For example, clinical response to rituximab therapy correlated with the number of natural killer (NK) cells in the peripheral blood of lymphoma patients (Janakiraman et al. 1998). Furthermore, response to rituximab correlated with the expression of certain Fc receptor alloforms. Preclinical studies have demonstrated that donors expressing a histidine in position 131 of the Fcγ receptor IIa (FcγRIIa-131H) bound human IgG stronger and triggered higher levels of ADCC than donors with an arginine in this position (FcγRIIa-131R) (Parren et al. 1992). Corresponding results were observed for the FcγRIIIa receptor, where the FcγRIIIa-158V alloform was more active than the FcγRIIIa-158F alloform. In clinical trials, patients had significantly higher clinical and molecular response rates to rituximab if they expressed the FcγRIIIa-158V rather than the FcγRIIIa-158F alloform (Cartron et al. 2002; Treon et al. 2005). Another study reported a correlation between both the FcγRIIa and FcγRIIIa allotypes and response to rituximab (Weng and Levy 2003).

In addition to supporting the role of Fc receptor-mediated mechanisms of action, these studies also suggest particular effector cell populations to be clinically relevant. While the FcγRIIIa receptor is almost exclusively expressed by human NK cells and tissue macrophages, FcγRIIa is widely expressed by cells of the myeloid lineage, including monocytes and PMNs (Table 8.1). Interestingly, the same FcγRIIa and FcγRIIIa polymorphisms were not correlated with response to rituximab or Campath-1H in patients with chronic lymphocytic leukemia

Table 8.1 Human Fcγ receptors.

	Receptor function	*IgG affinity (K_a) (L mol^{-1})*	*Expression*		*Polymorphisms*	*Ligand specificity (human IgG)*
			Constitutive	*Inducible*		
FcγRI (CD64)	Activating	High (10^8–10^9)	Monocytes, macrophages	Neutrophils, eosinophils (?)		3 > 1 > 4 >>> 2
FcγRIIa (CD32a)	Activating	Low ($<10^7$)	Monocytes, macrophages, neutrophils, basophils, eosinophils, Langerhans cells, B cells, platelets, endothelial cells		131R	3 > 1 >>> 2,4
					131H	3 > 1 = 2 >>> 4
FcγRIIb (CD32b)	Inhibitory	Low ($<10^7$)	Monocytes, macrophages, neutrophils, basophils, eosinophils, Langerhans cells, B cells, platelets, endothelial cells		187I/T	3 ≥ 1> 4 >> 2
FcγRIIIa (CD16a)	Activating	Medium (3×10^7)	NK cells, macrophages, subtypes of monocytes and T cells	Monocytes	158V/F affecting IgG1	1 = 3 >>> 2,4
FcγRIIIb (CD16b)	(GPI-linked)	Low ($<10^7$)	Neutrophils	Eosinophils (?)	NA1/NA2	1 = 3 >>> 2,4

(CLL), suggesting that in CLL patients systems other than Fc receptor-mediated effector mechanisms may predominate (Farag et al. 2004; Lin et al. 2005).

Another potential Fc-mediated mechanism is CDC. While evidence for systemic complement activation after rituximab therapy has been observed in several studies (Kennedy et al. 2004), its contribution to the killing of lymphoma cells *in vivo* is more controversial. While some investigators reported a correlation between the *in vitro* sensitivity to CDC and clinical responses to rituximab (Manches et al. 2003), this was not observed in other studies (Weng and Levy 2001). Since systemic complement activation after rituximab application may play a key role in the side effects of this treatment (van der Kolk et al. 2001), potential benefits of modified antibody constructs with increased capacity to activate complement need to be balanced against the risk of more severe toxicity. Interestingly, a novel human CD20 antibody, which proved particularly effective in activating human complement *in vitro*, did not demonstrate unexpected toxicities in early clinical trials (Coiffier et al. 2005).

Today, the clinically relevant mechanisms of action for therapeutic antibodies against solid tumors are even more elusive. Here, available information suggests a stronger impact of tumor cell-related factors rather than antibody-mediated functions (Hirsch and Bunn 2005). Data on the influence of Fcγ receptor polymorphisms for the response to solid tumor antibodies have, to our knowledge, not been reported, but are eagerly expected.

8.2 Modifying Effector Functions

8.2.1 Antibody Isotype

Among the human antibody isotypes (IgG1–4, IgA1 and 2, IgM, IgD, and IgE), IgG1 is therapeutically the most widely used isotype today. In this section, we will discuss the rationale for using this isotype, and discuss potential alternatives, the selection of which will depend on the required effector functions and on pharmacokinetic characteristics.

8.2.1.1 IgG Antibodies

Murine IgG subclasses (IgG1, IgG2a, IgG2b, IgG3) display substantial differences in their ability to mediate effector functions. Recently, subclass-dependent differences in the affinities for specific activating or inhibitory murine Fcγ receptors were described. The ratio of activation to inhibition (A/I ratio) differed by several orders of magnitude between the murine IgG subclasses, and predicted the *in vivo* activity of antibodies in two different models (Nimmerjahn and Ravetch 2005). However, these models cannot be directly transferred to the human situation, because mouse and human IgG subclasses do not directly correspond, and the human FcγR system is more complex than the murine one (Woof 2005).

In the human system, four IgG subclasses (IgG1, IgG2, IgG3, and IgG4) have been distinguished serologically (Jefferis 1986) and genetically (Rabbitts et al. 1981). Serum half-lives of 21 days have been reported for IgG1, IgG2, and IgG4 antibodies, whereas IgG3 antibodies have a half-life of only 7 days (Waldmann and Strober 1969) (Table 8.2). Furthermore, human IgG subclasses also display differential specificity and affinity to activating and inhibiting FcγR (see Table 8.1), indicating that the selection of the appropriate isotype is an important consideration for clinical applications. Further variation in the human γ heavy chain genes and subsequent heterogeneity in the IgG proteins has been described in what is called the Gm allotype system. Today, at least 18 different Gm allotypes have been described (van Loghem 1986; de Lange 1989). The functional relevance of these allotypes is widely unclear, but at least two IgG3 allotypes were reported to differ in their capacity to activate complement (Bruggemann et al. 1987; Bindon et al. 1988).

Concerning their structure, the main differences between the IgG subclasses reside in the amino acid composition of the C_H2 domain in the lower hinge region. This region also contains the binding sites for complement and Fcγ receptors, and determines the flexibility of the molecule (Burton 1985) (Table 8.2).

The initial event in the activation of the classical complement pathway is binding of C1q. The capacity of the four human IgG subclasses to bind C1q was determined using matched sets of chimeric antibodies with identical variable regions (Bruggemann et al. 1987). Surprisingly, despite higher C1q binding capacity of IgG3, the final complement killing activity of IgG1 was higher (Bruggemann et al. 1987; Dechant et al. 2002). Furthermore, IgG1 proved to be the most effective subclass at high antigen concentrations, whereas IgG3 was superior at lower concentrations (Lucisano Valim and Lachmann 1991; Michaelsen et al. 1991). Interestingly, IgG2 was demonstrated to mediate effective complement lysis at very high antigen densities by additional activation of the alternative pathway, which is regarded to be especially important for opsonization and killing of bacteria (Lucisano Valim and Lachmann 1991). IgG4 was consistently inactive in complement activation.

Table 8.2 Common features of human IgG subclasses.

	IgG1	*IgG2*	*IgG3*	*IgG4*
Molecular weight (kDa)	146	146	170	146
% of serum IgG	70	20	6	4
Serum half-life (days)	21	21	7	21
Gm allotypes	4	1	13	–
Inter-heavy chain disulfide bonds	2	4	11	2
Hinge length (aa)	15	12	62	12
Protein A binding	+++	+++	(+)	+++
Protein G binding	+++	+++	+++	+++
CDC	+++	+	++	–
ADCC	+++	+/–	++	–

Antibodies interact with effector cells via binding of their Fc region to cellular Fcγ receptors (reviewed in van de Winkel and Capel 1993). This process results in either activation or inhibition of effector cells, depending on which Fcγ receptors are predominantly engaged. In humans, three activating FcγR have been characterized: FcγRI, FcγRIIa and FcγRIIIa, which demonstrate distinct cellular expression patterns (Table 8.1). FcγRIIb, the only inhibitory cellular FcγR, has low affinity and contains an intracytoplasmatic immunoreceptor tyrosine-based inhibitory motif (ITIM). Human FcγRs have been demonstrated to have differential IgG subclass specificity (Table 8.1). Importantly, FcγRI is the only high-affinity IgG receptor, while the other Fcγ receptors have low or intermediate affinity. FcγRIIa is the only FcγR that binds human IgG2. This binding affinity for IgG2 is critically affected by a genetic polymorphism of the FCGRIIA gene, resulting in two distinct allotypes: FcγRIIa-131H has significantly higher affinity for human IgG2 than the FcγRIIa-131R allotype. NK cell expressed FcγRIIIa binds IgG1 and IgG3, while no binding of IgG2 and IgG4 has been found. For FcγRIIIa an important biallelic polymorphism has been well characterized: the FcγRIIIa-158V allele binds human IgG1 significantly stronger than the FcγRIIIa-158F allele. Neutrophil-expressed FcγRIIIb is a GPI-linked molecule, which displays the functionally relevant NA1/NA2 polymorphism. The inhibitory FcγRIIb receptor has similar affinity for human IgG1 and IgG3, lower affinity for IgG4 and does not bind IgG2 (reviewed in van Sorge et al. 2003). Functionally, ADCC induction by matched sets of chimeric antibodies was determined to be in the following order: IgG1 > IgG3 >> IgG2 and IgG4 (Bruggemann et al. 1987; Dechant et al. 2002). Analyses of different effector cell populations revealed that IgG1 antibodies triggered NK cells very effectively via FcγRIIIa, whereas neutrophils were only poorly activated via FcγRIIa.

In conclusion, human IgG1 is the preferred IgG subclass, if activating effector mechanisms are required, because IgG1 effectively triggers complement and NK cells. On the other hand, human IgG4 or IgG2 appear to be candidate isotypes if interactions with the host immune system are undesired. However, IgG4 antibodies have a tendency to form half-molecules, which may even be exchanged between two IgG4 molecules (reviewed in Aalberse and Schuurman 2002). This instability of human IgG4 could be corrected by a single S228P mutation in the hinge region (Angal et al. 1993). For therapeutic purposes, IgG4 molecules were further engineered by a E235S mutation, resulting in complete incapacity to interact with cellular Fcγ receptors (Reddy et al. 2000). Human IgG2 also has low complement-activating capacity and poorly interacts with most cellular Fc receptors. However, this later interaction is strongly influenced by the FcγRIIa-131R/H polymorphism, suggesting that human IgG2 should also be modified for therapeutic applications (see below).

8.2.1.2 IgA Antibodies

IgA represents the most abundantly produced antibody isotype *in vivo* (Kerr 1990), is critically involved in the host defense at mucosal surfaces (Monteiro and van de Winkel 2003), and activates human neutrophils more effectively than IgG antibodies (Dechant and Valerius 2001). Neutrophils are the most numerous phagocytic cell population *in vivo*, and constitute the first line of defense against bacteria and fungi. Furthermore, neutrophils can kill a broad spectrum of tumor cells and have been shown to be critically involved in tumor rejection in animal models (Di Carlo et al. 2001). As outlined in the following paragraphs, human IgA has additional properties, which make IgA an attractive antibody isotype for immunotherapy.

Two subclasses – IgA1 and IgA2 – are distinguished, with IgA2 having a shorter hinge region, and an increased resistance against enzymatic degradation by bacterial proteases. Both isotypes contain *N*- and *O*-linked carbohydrates (Yoo and Morrison 2005). After covalent binding to plasma cell-produced joining (J) chain, IgA antibodies form natural dimers. Binding of these dimers to the polymeric immunoglobulin receptor (pIgR) leads to the directed transcellular secretion of IgA onto mucosal surfaces. At the luminal surface, secretory IgA (sIgA) is released, which consists of the IgA dimer, the J chain, and the proteolytically cleaved extracellular part of the pIgR (Fig. 8.2). Thus, monomeric, dimeric and secretory IgA are distinguished, with a predominance of monomeric IgA in plasma, and sIgA at mucosal surfaces.

The pharmacokinetic properties of IgA are fundamentally different from those of IgG. In contrast to IgG, IgA does not bind to FcRn, is therefore not protected from degradation, and its serum half-life (approx. 5 days) is significantly shorter than that of IgG. On the other hand, IgA, but not IgG, is actively transported to mucosal surfaces of the gut, the airways, and the urogenital tract. This offers the potential advantage that intravenously applied IgA could target pathogens or common tumors, such as lung or colon cancers, from the luminal surface, which is often enriched in neutrophilic effector cells.

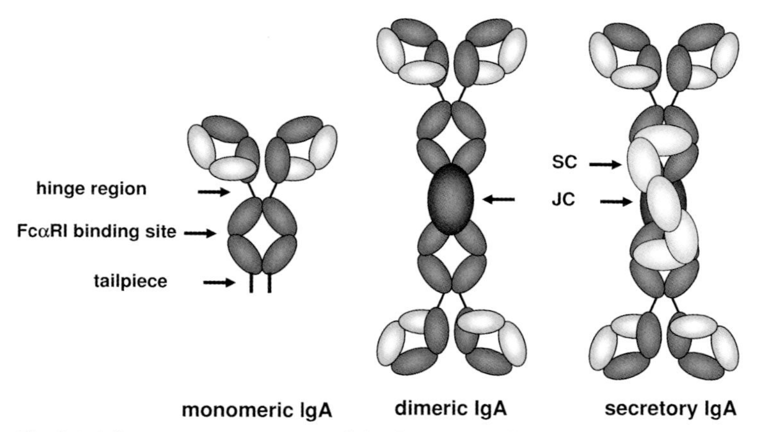

Fig. 8.2 Schematic representation of the three IgA isoforms. Dark gray: heavy chain; light gray: light chain. JC, J chain; SC, secretory component.

Functions of IgA include direct neutralization of pathogens on mucosal surfaces, intracellular neutralization of viruses during transepithelial transport, as well as activation of immune effector cells, which is triggered by the myeloid IgA receptor (FcαR; CD89) (Woof and Mestecky 2005). In contrast to IgG antibodies, both human IgA isotypes bind to their receptor at the C_H2/C_H3 interface, and activate human complement poorly. Interestingly, tumor cell killing by neutrophils was significantly enhanced in the presence of human IgA compared with human IgG antibodies (Valerius et al. 1997; Huls et al. 1999; Dechant et al. 2002). IgA-mediated tumor cell killing was further enhanced when blood or isolated effector cells from myeloid growth factor (G-CSF or GM-CSF)-treated patients was analyzed, suggesting that the combination of IgA antibodies and myeloid growth factors may act synergistically. However, the inhibitory effects of IgA antibodies on phagocyte function have also been described (Pasquier et al. 2005).

Until recently, animal models to investigate the function of human IgA *in vivo* were limited, since mice do not express a human FcαRI homolog. Transgenic mice, which express the human receptor under its physiological regulatory elements, served to reassess the role of IgA *in vivo* (van Egmond et al. 2000). In the future, the therapeutic potential of human IgA in infectious diseases (van Spriel et al. 1999; Vidarsson et al. 2001; Pleass et al. 2003) and for tumor immunotherapy may be investigated in these animals. So far, no clinical experience with the systemic application of human IgA antibodies has been reported.

8.2.2 Altered Fc Receptor Binding

8.2.2.1 Introduction

While fundamental structural requirements for the effector functions of IgG antibodies were apparent from early studies, such as glycosylation at Asn297 for C1q binding and interaction with Fcγ receptors (Nose and Wigzell 1983; Tao and Morrison 1989; Jefferis and Lund 2002), more refined insights were obtained from crystal structures. Thus, analyses of the co-crystal structure of the IgG Fc with FcγRIII revealed that direct interaction sites were mainly located in the protein moiety of IgG, while only minor contacts between sugar residues of the Fc and FcγRIII were observed (Sondermann et al. 2000). Since the activation of leukocyte Fc receptors was dependent on Fc glycosylation, it was supposed that the oligosaccharide moiety excerted its influence indirectly, probably through modulation of the Fc conformation (Jefferis et al. 1995; Wright and Morrison 1997). Co-crystal structures from a series of Fc glycosylation variants with FcγRIII indicated that the sugars act both to increase the distance and to decrease the mobility of the receptor-interacting segments of the C_H2 domains (Krapp et al. 2003). From these observations, two approaches to modulate Fc functions became evident: altering the glycosylation profile (glyco-engineering), or mutation of selected amino acids in the IgG Fc portion (protein engineering).

8.2.2.2 Glyco-Engineered Antibodies

All human antibody isotypes contain carbohydrates at conserved positions in the constant regions of their heavy chains, with each isotype possessing a distinct array of *N*-linked carbohydrate structures, which variably affects protein assembly, secretion or functional activity. The structure of the attached *N*-linked carbohydrate varies considerably, depending upon the degree of posttranslational processing, and can include high-mannose, multiply branched as well as biantennary complex oligosaccharides (Wright and Morrison 1997). IgG, the most abundant antibody isotype in human serum, has a single *N*-linked biantennary complex type structure attached to Asn297, which shows considerable heterogeneity (Jefferis et al. 1995). A "core" heptasacharide can be defined with variable addition of outer arm sugar residues (Jefferis 2005) (Fig. 8.3). The attached oligosaccharide is approximately as large as the C_H2 domain itself, and is buried between the two C_H2 domains, forming extensive contacts with amino acid residues within C_H2 (Rademacher et al. 1988; Wright and Morrison 1997).

Glycosylation is a cotranslational/posttranslational modification that results in the attachment of a glucosylated high mannose oligosaccharide (GlcNac-2Man9Glu3). This sugar is then trimmed to a GlcNac2Man9Glu structure, which is bound by chaperones that aid and monitor folding fidelity (Jefferis 2005). The resulting glycoprotein, which is targeted for secretion, then transits the Golgi apperatus, where the oligosaccharide is initially trimmed back by glycosidases to a GlcNac2Man5 structure, before being processed by the successive action of glycosyltransferases to generate the complex type biantennary structure (Jefferis 2005) (Fig. 8.3).

Several factors can influence glycosylation. Therefore, the selection of appropriate expression cell lines for the production of therapeutic antibodies is critical to obtain a favorable glycosylation profile. Early studies established that CHO, NS0, and Sp2/0 cells are able to produce chimeric and humanized IgG antibodies with major glycoforms identical to glycoforms present in polyclonal human IgG (reviewed in Jefferis 2005). Campath-1H, a humanized IgG1 antibody directed against CD52, has been expressed in several mammalian cell lines, including YB2/0 rat myeloma, NS0 mouse myeloma and CHO cells. Results from these studies demonstrated that major differences in antibody glycosylation occurred between these cell lines. YB2/0-expressed Campath-1H contained an antibody fraction with a bisecting GlcNAc on the core oligosaccharide, as well as a fraction of nonfucosylated antibody. In ADCC assays, the YB2/0-expressed antibody reached similar activity as Campath-1H expressed in CHO and NS0 cells, but required significantly lower antibody concentrations (Lifely et al. 1995). Several reports suggested that a bisecting GlcNAc, lack of fucose, and the content of galactose in the antibody oligosaccharide all affect the levels of ADCC (Kumpel et al. 1995; Umana et al. 1999; Shields et al. 2002). In a comparative analysis, the relative contribution of these sugar residues to enhance ADCC was analyzed. In this study, galactose had no effect on ADCC, and a high content of bisecting GlcNAc had a relatively weak effect on enhancing ADCC. More importantly, nonfucosylated oligosaccharide was demonstrated to have a prominent effect in

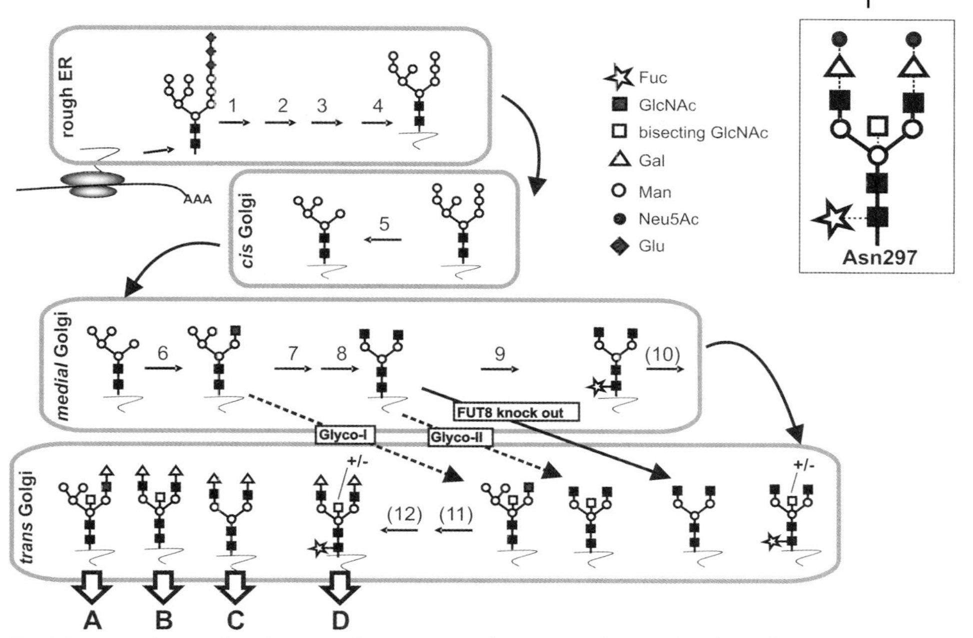

Fig. 8.3 Biosynthesis of *N*-glycans and bisected *N*-glycans and its manipulation to achieve reduced fucose content. Glyco-I approach: overexpression of engineered GnT-III with localization domain exchanges, leading to a high proportion of bisected, non-fucosylated, hybrid type glycans. Glyco-II approach: co-expression of GnT-III and Man-II, directing the pathway to bisected, complex type oligosaccharides (Ferrara et al. 2006). FUT8 knockout approach: no α1,6-FucT activity, leading to complex type nonfucosylated oligosaccharides. A dashed line indicates a certain balance between molecules modified in the modified and non-modified pathway. A: nonfucosylated, bisected, hybrid type; B: nonfucosylated, bisected, complex type; C: nonfucosylated, complex type; D: bisected or nonbisected complex type. 1: oligosaccharyltransferase; 2: α-glucosidase I; 3: α-glucosidase II; 4: α-1,2-mannosidase; 5: α-mannosidase I; 6: β1,2-*N*-acetylglucosaminyltransferase I; 7: Golgi α-mannosidase II (Man-II); 8: β1,2-*N*-acetylglucosaminyltransferase II; 9: core α1,6-fucosyltransferase (α1,6-FucT); 10: β1,4-*N*-acetylglucosaminyltransferase III (GnT-III); 11: β1,4-galactosyltransferase; 12: sialyltransferase. ER, endoplasmatic reticulum; Fuc, fucose; GlcNAc, *N*-acetylglucosamine; Gal, galactose; Man, mannose; Neu5Ac, *N*-acetylneuramic acid; Glu, glucose. Box: Biantennary oligosaccharide of the IgG Fc portion. Full lines define the core oligosaccharide structure; dashed lines show additional sugar residues that may be attached.

enhancing ADCC compared with a bisecting GlcNAc-containing oligosaccharide in the IgG1 molecule (Shinkawa et al. 2003).

Following these observations, two strategies have been followed to alter the glycosylation profile of therapeutic antibodies in order to specifically improve their capacity to trigger ADCC. Both strategies aimed at reducing core fucosylation to enhance FcR binding, which could be achieved by direct or indirect mechanisms. In the indirect approach, overexpression of β1,4-*N*-acetylglucosaminyltransferase III (GnT-III) led to the addition of a bisecting GlcNAc residue, which had an important influence on multiple subsequent enzymatic glycosylation reactions in the Golgi complex of the cell. Once an oligosaccharide is bisected by the action of GnT-III, it cannot serve as a suitable substrate for several glycosylation enzymes, especially Golgi α-mannosidase II (Man-II) and, importantly, α1,6-fucosyltransferase (α1,6-FucT) (Schachter 2000). Several groups investigated the influence of a bisecting GlcNAc by overexpression of GnT-III in cell lines used for the production of therapeutic antibodies. The glycosylation pattern of an anti-neuroblastoma antibody (chCE7) was engineered in CHO cells by tetracycline-regulated expression of the GnT-III enzyme. The results demonstrated that the 15- to 20-fold enhancement in ADCC activity correlated with the level of bisected, nonfucosylated oligosaccharides, and that there was an optimal range of GnT-III overexpression for maximal *in vitro* ADCC (Umana et al. 1999). Similar results were reported for the CD20-directed antibody rituximab produced in CHO cells overexpressing GnT-III (Davies et al. 2001), and for a CD19-directed chimeric antibody produced in HEK-293 cells (Barbin et al. 2006).

Recently, a Lewis Y-specific antibody with 10-fold enhanced ADCC activity has been reported by overexpressing GnT-III (Schuster et al. 2005). In this report, two approaches were followed. In the first approach (Glyco-I), the GnT-III was expressed along the Golgi apparatus by fusing the α-mannosidase-II localization domain to the catalytic domain of GnT-III. The authors observed increased levels of bisected, non-fucosylated hybrid type oligosaccharides, and these forms seemed to be related to a moderate reduction of complement activation. In the second approach (Glyco-II), GnT-III with its autologous localization domain, directing the enzyme to the *trans*-Golgi cisternae together with α-mannosidase-II, the CDC activity could be recovered and even slightly enhanced without affecting ADCC activity (Fig. 8.3). The authors stated that, depending on the clinical requirements, fine-tuning of complement activation might be possible (Schuster et al. 2005).

Applying the same technology, two glycovariants of Campath-1H were produced: Cam-Glyco-I, enriched in nonfucosylated, hybrid-type Fc carbohydrates, and Cam-Glyco-II, carrying nonfucosylated oligosaccharides both of hybrid and complex type. Both glycovariants demonstrated markedly enhanced ADCC in comparison with regular Campath-1H. Cam-Glyco-I mediated lower CDC than Cam-Glyco-II. The *in vivo* activity of the two glycoforms was assessed in cynomolgus monkeys. The pharmacokinetics of both glycoforms were not significantly different from wildtype Campath-1H and both variants appeared to deplete lymphocytes from both blood and lymph nodes better than wildtype Campath-1H. (Dyer et al. 2005).

The impact of fucose on ADCC was initially analyzed using cell lines with low α1,6-FucT activity (CHO-Lec13 and rat YB2/0). This resulted in antibody preparations with varying fucose content, and up to 50-fold enhanced ADCC by CD20-directed antibodies. C1q binding was not influenced by the lack of fucose, suggesting no impact on CDC activity (Shields et al. 2002). Aiming for fully defucosylated antibody preparations, a direct approach has also been described. Thus, α1,6-FucT was switched off by gene knockout of the *FUT8* gene, or by introducing small interfering RNA into the CHO producer cell line (Mori et al. 2004; Yamane-Ohnuki et al. 2004). α1,6-FucT is the key enzyme that catalyzes the transfer of fucose from GDP-fucose to the GlcNAc residue in an α1,6 linkage in the medial Golgi cisternae. Knockout of both *FUT8* alleles in CHO cells resulted in a complete loss of α1,6-FucT activity and enabled the production of fully defucosylated CD20 antibodies (Fig. 8.3). *FUT8*$^{-/-}$-produced CD20 antibodies were not altered with respect to antigen binding. However, they exhibited more than 100-fold higher ADCC activity compared with fully fucosylated antibodies, and were significantly more effective than their YB2/0-expressed counterparts (44% defucosylated). Importantly, elimination of α1,6-fucosylation activity by *FUT8* knockout did not alter other *N*-linked glycosylation patterns (Yamane-Ohnuki et al. 2004).

The increased ADCC activity of under-fucosylated IgG1 (and IgG2, IgG3, and IgG4) may be explained by enhanced binding to FcγRIIIa on NK cells, whereas binding to FcγRI and FcγRIIa/b was less influenced by the lack of fucose (Shields et al. 2002; Niwa et al. 2005). Both polymorphic forms of FcγRIIIa exhibited significantly improved binding to IgG1 that lacked fucose. Binding of dimeric nonfucosylated forms of Hu4D5 (a humanized anti p185HER2 antibody) to FcγRIIIa-158F or FcγRIIIa-158V revealed about 42-fold or 19-fold improvements, respectively (Shields et al. 2002). The structural basis for this affinity enhancement is not fully understood. Recent data suggest that the glycosylation of FcγRIIIa plays an important role and that the high affinity between glyco-engineered antibodies and FcγRIIIa is mediated by productive interactions formed between the receptor carbohydrate attached at N162 and regions of the Fc part that are only accessible when it is nonfucosylated (Ferrara et al. 2005). As FcγRIIIa and FcγRIIIb are the only human Fcγ receptors that are glycosylated at this position, the proposed interactions between the FcγRIIIa-attached carbohydrate and the Fc portion might explain the observed selective affinity increase of glyco-engineered antibodies for only this receptor (Ferrara et al. 2005). The effects of the FcγRIIIa gene polymorphism on ADCC mediated by fucosylated versus nonfucosylated antibodies were analyzed using different glycoforms of rituximab. Nonfucosylated forms were more potent in inducing ADCC than fucosylated rituximab, and this difference was independent of the FcγRIIIa polymorphism at position 158. Importantly, the use of nonfucosylated rituximab reduced the difference in ADCC activity between low-affinity and high-affinity Fc receptors. In contrast to V-carriers showing a 10-fold greater activity than F-carriers for fucosylated rituximab, there was no significant difference observed when nonfucosylated Rituximab was used (Niwa et al. 2004). Therefore, it is speculated that the use of low-fucose antibodies might improve the therapeutic effects of CD20-directed therapy for all patients independent of the FcγRIIIa phenotype.

8.2.2.3 **Protein-Engineered Antibodies**

In contrast to glyco-engineering, which so far has only been employed to enhance antibody binding to activating Fcγ receptors, engineering the protein backbone of antibodies was used to generate antibodies with either enhanced or diminished binding to individual Fcγ receptors. Decreased Fc receptor binding appears particularly attractive for immunotoxins, but also naked antibodies have been developed which should not bind to leukocyte Fc receptors.

Protein-engineered antibodies with diminished interactions with Fcγ receptors For antibodies which should not bind to cellular Fc receptors, $F(ab)_2$ fragments would be a logical step forward, but $F(ab)_2$ fragments are expensive to produce and have a short plasma half-life due to their lack of binding to FcRn. Therefore, whole IgG antibodies with reduced binding to Fcγ receptors have been engineered. Considering the different binding affinities of various human IgG isotypes to cellular Fc receptors (Table 8.1), human IgG4 or human IgG2 backbones appear as logical starting platforms for non-Fc receptor binding variants. For example, unmodified human IgG4 was selected to target toxins such as calicheamicin to either CD33 (Gemtuzumab)- or CD22 (CMC-544)-expressing tumor cells (Damle and Frost 2003). However, variants of human IgG1 with diminished Fc receptor binding have also been generated (see below).

Antibodies against CD3 are prototypic examples where Fc receptor engagement triggers significant clinical toxicity, probably without contributing to therapeutic efficacy (Chatenoud 2005). So far, OKT-3 (muromonab, mIgG2a) is the only approved CD3 antibody that is associated with severe clinical toxicity due to activation of resting T cells. This T-cell activation was demonstrated to be triggered by antibody-mediated crosslinking of T cells with Fcγ receptor-bearing bystander cells such as monocytes/macrophages. T-cell receptor crosslinking results in T-cell mitogenesis and in massive cytokine release, which limits the clinical applicability of mitogenic CD3 antibodies.

However, CD3-directed antibodies have significant potential as immunosuppressive agents for the prevention and treatment of transplant rejections, in the treatment of severe T cell-mediated autoimmune diseases and may even be employed to induce antigen-specific tolerance (Chatenoud 2003). Therefore, several approaches have been followed to generate less mitogenic CD3 antibodies by reducing Fcγ receptor binding. For example, a rat anti-human CD3 antibody (YTH 12-5) was humanized and expressed as human IgG (including all four subclasses), IgA, or IgE. All isotypes could elicit cytokine release *in vitro* (Bolt et al. 1993). Mutating amino acid 297 (N297A) in the humanized IgG1 version prevented antibody glycosylation, resulting in a CD3 antibody with impaired binding to all FcγRs and with significantly reduced complement-activating capacity. This antibody proved nonmitogenic *in vitro*, and demonstrated low toxicity and signs for immunosuppressive activity in a phase I clinical study (Friend et al. 1999). In a randomized phase II trial in patients with new-onset type 1 diabetes, a 6-day course of an aglycosylated chimeric CD3 antibody (human IgG1, ChAglyCD3) was effective in preserving residual beta cell function. However, all antibody-treated patients experienced infusional side effects, and the majority reported symptoms from transient Epstein–Barr virus (EBV) reactivation (Keymeulen et al. 2005).

hOKT3γ1(Ala-Ala) is a humanized IgG1 version of OKT3, in which the amino acids in positions 234 and 235 have been mutated to alanine. Thereby, hOKT3γ1(Ala-Ala) was reported to lose complement-activating capacity, Fcγ receptor binding, and mitogenicity. A subsequent phase I study with hOKT3γ1(Ala-Ala) demonstrated efficacy similar to that of conventional OKT3 in the treatment of renal allograft rejection with markedly fewer side effects (Woodle et al. 1999). hOKT3γ1(Ala-Ala) was also tested in patients with psoriatic arthritis (Utset et al. 2002) or type I diabetes (Herold et al. 2002). In both patient populations, no significant cytokine release was observed, infusion-related toxicity was low and, importantly, these phase II trials suggested clinical efficacy.

Another approach used human IgG2 as a template for the introduction of mutations, because human IgG2 interacts only with FcγRIIa, and CD3-directed human IgG2 antibodies required 10- to 100-fold higher antibody concentrations to induce T-cell proliferation. Binding to FcγRII receptors was further reduced by two engineered mutations in the constant regions of HuM291 (V234A and G237A) (Cole et al. 1997). As expected, this construct was nonmitogenic *in vitro*. As HuM291 dissociated quickly from cell surface CD3 molecules, only minimal internalization, but sustained signaling by the T-cell receptor was observed. Thereby, HuM291 effectively triggered apoptosis in activated human T cells (Carpenter et al. 2002). In clinical phase I studies in renal allograft or allogeneic bone marrow transplantation patients, the majority of patients did not demonstrate measurable cytokine levels after antibody application, infusion-related toxicity was low, and immunosuppressive activity was observed (Norman et al. 2000; Carpenter et al. 2002).

Protein-engineered antibodies with improved Fcγ receptor binding Clinical observations and animal studies indicate that efficient recruitment of immune effector cells and triggering of cellular effector functions such as ADCC are major mechanisms of action for some therapeutic antibodies (see above). The therapeutic efficacy of these antibodies may be improved by increasing binding of their Fc domains to activating FcγRs. For this strategy, the balance between activating and inhibitory receptors is an important consideration (Nimmerjahn and Ravetch 2005), and optimal effector functions may result from Fc parts with enhanced affinity for activating Fcγ receptors relative to the inhibitory FcγRIIb isoform (Clynes et al. 2000; Nimmerjahn and Ravetch 2005). Several different strategies have been followed to identify critical amino acids that could be altered to modify FcR binding and enhance immune effector mechanisms.

In a comprehensive study, all solvent-exposed residues in the Fc part of human IgG1 were individually changed to alanine, and binding to the different FcγRs was analyzed (Shields et al. 2001). These studies identified several groups of mutants that discriminated between binding to FcγRI, FcγRIIb, and FcγRIIIa. Interestingly, several Fc variants, in which two or more amino acids were simultaneously altered to alanine, exhibited additive binding characteristics. For example, the triple Fc mutant S298A/E333A/K334A demonstrated improved binding to FcγRIIIa and diminished binding to FcγRIIb (Shields et al. 2001)

(Fig. 8.4). Several variants exhibited significant improvements in ADCC with effector cells from either FcγRIIIa-V/V or FcγRIIIa-F/F homozygous donors. Using this triple mutant with FcγRIIIa-F/F donor cells, maximal ADCC levels were increased by more than 100%, and 1–2 logs improvement in potency were observed, as reflected by shifts in the EC_{50} values to lower concentrations (Shields et al. 2001; Lazar et al. 2006).

In a second approach using a yeast display system (Stavenhagen et al. 2004), mutated human IgG1 Fc regions were screened for altered binding affinity to different Fcγ receptors. For this purpose, a mutant Fc library was generated by error-prone PCR, and the mutated sequences were fused to the Aga2p cell wall protein, which allowed display on the yeast cell wall. This library was screened by soluble tetrameric FcγR complexes (FcγRIIIa or FcγRIIb) and cell sorting. Different screening strategies were applied, combining positive selection and depletion cycles. A variety of Fc mutants with higher affinity for FcγRIIIa and enhanced ADCC activity were isolated (see Table 8.3 and Fig. 8.4 for a selection of mutants). Interestingly, several variants with enhanced ADCC activity were identified, in

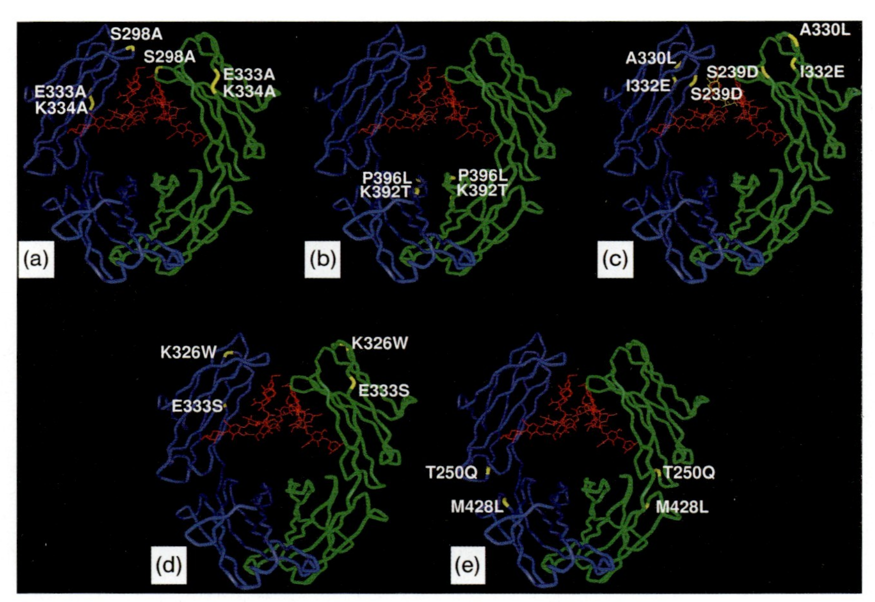

Fig. 8.4 Engineered Fc portions with altered effector functions. (a–c) Enhanced ADCC; (d) altered complement activation; (e) prolonged half-life. (a) S298A/E333A/K334A variant (Shields et al. 2001). (b) K392T/P396L variant (Stavenhagen et al. 2004). (c) S239D/I332E/A330L variant (Lazar et al. 2006). (d) K326W/E333S variant (Idusogie et al. 2001). (e) T250Q/M428L variant (Hinton et al. 2006). Altered amino acid positions resulting in modified effector functions are indicated. Human IgG1 Fc portion: crystal structures adapted from Krapp et al. (2003). Blue and green: C_H2-C_H3 domains; red: carbohydrate; yellow: modifed amino acid position.

Table 8.3 Characteristics of selected engineered Fc variants.

Variant	*FcγRIIIa binding*	*FcγRIIb binding*	*IIIa/IIb ratio*	*ADCC induction*	*Fold reduction in EC_{50} value*	*Complement activation*	*Reference*
Wildtype	↑	↑	1	↑	–	+	
S298A-E333A-K334A	↑↑	↓	10	↑↑↑	10–100	No data available	Shields et al. 2001
S239D-I332E	↑↑↑	↑↑	4	↑↑↑	100–1000	+	Lazar et al. 2006
S239D-I332E-A330L	↑↑↑↑	↑↑↑	9	↑↑↑	100–1000	–	Lazar et al. 2006
K288N-A330S-P396L	↑↑	↑↑	2	↑↑↑		No data available	Stavenhagen et al. 2004
K392T-P396L	↑↑	↑↑	2	↑↑↑	10–100	No data available	Stavenhagen et al. 2004

IIIa/IIb: Fold FcγRIIIa binding/FcγRIIb binding; EC_{50}: effective concentration 50%; ↑ enhanced activity/binding compared with wildtype; ↓ reduced activity/binding compared with wildtype.

which the mutated amino acids were located outside of the Fc receptor binding region in the C_H3 domain of the antibody (see Fig. 8.4). These data suggest that some mutants increase Fc receptor binding indirectly (e.g. by altering the structural properties of the Fc domain or by influencing the glycosylation of the Fc region) (Stavenhagen et al. 2004).

Lazar and colleagues used computational design algorithms and high-throughput screening to engineer Fc variants with optimized FcγR affinity and specificity (Lazar et al. 2006). When structural information was available (Fc/FcγRIII complex), affinity was directly optimized by designing substitutions that provided more favorable interactions at the Fc/FcγR interface. When structural information was incomplete or lacking (as for the Fc/FcγRIIb complex), calculations provided a set of variants enriched for stability and solubility. A number of engineered Fc variants demonstrated significant enhancements in binding affinity to both human FcγRIIIa-158V and FcγRIIIa-158F alleles along with an improved FcγRIIIa/IIb ratio (Table 8.3; Fig. 8.4). In the context of Campath-1H, both variants S239D/I332E and S239D/I332E/A330L displayed more than two orders of magnitude enhanced ADCC potency compared with wildtype Campath-1H, and were 10-fold more active than the S298A/E333A/K334A triple mutant described by Shields and colleagues. Interestingly, in the rituximab background variant S239D/I332E elicited CDC comparable to wildtype rituximab, while the addition of the A330L mutation ablated CDC. Thus, the set of S239D/I332E and S239D/I332E/A330L variants provides the option for enhancing ADCC with or without triggering CDC as additional mechanism of action. To demonstrate the superior cytotoxic potential of the Fc variants *in vivo*, B-cell depletion using CD20-directed antibodies was analyzed in cynomolgus monkeys. The approximate dose required for 50% B-cell depletion by wildtype antibody was approximately 10 μg kg^{-1} per day. For the S239D/I332E variant, a dose of 0.2 μg kg^{-1} per day was sufficient to achieve 50% depletion – an apparent 50-fold increase in potency (Lazar et al. 2006).

In conclusion, protein-engineered Fc variants demonstrated substantial improvements in Fc receptor-mediated effector functions *in vitro* and in primate models. Whether these types of second-generation antibodies will improve clinical outcomes needs to be evaluated in future clinical trials.

8.2.3
Altered Complement Activation

Since the contribution of complement to the therapeutic efficacy of antibodies is controversial, and since complement activation may contribute to the side effects of antibodies in patients, approaches to either enhance or to diminish complement activation have been proposed. Activation of the classical complement pathway requires binding of the C1q serum protein to the Fc portion of IgG. This C1q binding is completely abolished in Asn297 mutated, aglycosylated IgG, but these mutants also lack Fc receptor binding (see above). Therefore, more complement-specific approaches were required.

Initial studies with a mouse IgG2b antibody mapped the core C1q-binding site to amino acids E318, K320, and K322 (Duncan and Winter 1988). More recent mutational analyses of human IgG1 revealed that the C1q-binding region of human IgG1 is centered around D270, K322, P329, and P331 in the C_H2 domain. Thus, two mutants – D270A and P329A – were particularly ineffective in binding C1q and activating human complement, but still retained some complement-activating capacity at higher complement concentrations. Interestingly, the P329A mutant also demonstrated significantly impaired ADCC activity compared to wildtype, while the ADCC activity of the D270A mutant was similar to the control antibody (Idusogie et al. 2000).

Further studies revealed that amino acids K326 and E333 – located at the edges of the C1q-binding region – profoundly influenced complement activation by IgG1 mutants (Idusogie et al. 2001). Thus, a K326W substitution provided the highest increase in C1q binding (3-fold) and complement-mediated killing (2-fold). In amino acid position 333, the E333S substitution resulted in the highest increase in C1q binding (2-fold) and CDC activity (1.6-fold). The K326W/E333S double mutant demonstrated additive increases in C1q binding (5-fold), but CDC activity was not further increased compared to the single mutants. Interestingly, the K326W and the K326W/E333S mutants were completely inactive in ADCC assays (Fig. 8.4). These observations make them very interesting tools to elucidate the effector mechanism of antibodies *in vivo*, as they were supposed to act only by complement, but not by ADCC.

8.3
Modifying the Pharmacokinetics of Antibodies

8.3.1
Introduction

Studies in rituximab-treated patients have indicated that clinical responses were correlated with favorable pharmacokinetics in patients (Berinstein et al. 1998). Therefore, improving the serum half-life of therapeutic antibodies is an attractive approach, which may reduce the amount of required antibodies, and may increase their convenience for patients by prolonging treatment intervals. For most human IgG antibodies, serum half-life is in the range of 3 weeks. Elegant studies have demonstrated that this prolonged half-life of IgG compared with other serum molecules or antibody isotypes is not merely a function of protein size, but requires interaction with a specific Fc receptor. This receptor (FcRn) is a heterodimer of β_2 microglobulin and an HLA class I-related α-chain, which – in humans – is mainly expressed by endothelial cells (Ghetie and Ward 2000). After internalization by fluid phase pinocytosis, IgG is routed to acidic endosomes, where binding to FcRn is believed to occur at low pH. This binding prevents lysosomal degradation of IgG, and triggers sorting and transport to the cell surface, where IgG is released from FcRn at near neutral pH. Thereby, FcRn actively protects IgG from lysosomal degradation, and controls serum half-life of endogenous and exogenous antibodies.

Recent studies have demonstrated that antibody consumption (e.g. by widely expressed and rapidly internalized antigens) may also dramatically influence their serum half-life (Lammerts van Bueren *et al.* 2006; Shih et al. 1994). The relative contribution of FcRn binding and target antigen-mediated antibody consumption for the pharmacokinetic profile of therapeutic antibodies is unknown. Whether antibody consumption is amenable to molecular engineering – without impeding antibody efficacy – has not been addressed.

8.3.2
Modifying Binding to FcRn

Analysis of the crystal structure of FcRn with Fc has mapped their interaction site to the interface between C_H2 and C_H3 in IgG molecules (Burmeister et al. 1994), while leukocyte Fc receptors bind to IgG in the lower hinge region. This region between C_H2 and C_H3 contains several histidine residues, which may account for the sharply pH-dependent interaction between FcRn and IgG. Considering the potential clinical relevance of this interaction, approaches to modify binding of IgG to FcRn were obvious. However, considering the biology of FcRn, mutations should probably not interfere with the pH dependency of binding.

Initial mutagenesis studies identified 10 human IgG1 mutants with higher affinity toward murine and human FcRn at pH 6.0. However, these mutants exhibited parallel increases in binding to murine FcRn at pH 7.4, which may explain why their serum half-lives in mice were not prolonged (Dall'Acqua et al. 2002). In another report, modeling the binding of human IgG2 to human FcRn guided the selection of amino acid positions 250, 314, and 428 of the human IgG2 heavy chain for further mutagenesis studies (Hinton et al. 2004). Random mutagenesis identified IgG variants at position 250 and 428 with increased binding to FcRn, while none of the 314 mutants resulted in increased FcRn binding. *In vitro*, the optimal mutants – T250Q, M428L, and T250Q/M428L – demonstrated approximately 4-, 8-, or 27-fold higher binding to FcRn at pH 6.0. In rhesus monkeys, the M428L and the T250Q/M428L mutants showed an approx. 2-fold increase in serum half-lives compared with the wildtype antibody. Furthermore, both amino acid substitutions proved to be transferrable to human IgG1 in a recent report, since amino acids 250 and 428 are conserved between the four human IgG isotypes. The serum half-life in rhesus monkeys was prolonged by 2.5-fold. It may be expected that these mutations also improve the pharmacokinetics of IgG isotypes in humans, provided that the data from rhesus monkeys can be transferred, and assuming that these mutations do not increase the immunogenicity in clinical studies (Hinton et al. 2006) (Fig. 8.4).

8.4
Summary and Conclusions

Despite considerable efforts to elucidate the mechanisms of action for therapeutic monoclonal antibodies, our current understanding of these mechanisms is still rather incomplete. Furthermore, the contribution of individual mechanisms may significantly vary between different target antigens, but also between antibodies targeting the same molecule. At present, it is not determined which preclinical assay will optimally predict clinical responses in patients, but with more antibodies coming along the line, our knowledge about relevant mechanisms will certainly increase. Meanwhile, molecular engineering will provide solid platforms for second-generation antibodies, which allow particular aspects of therapeutic antibodies to be selectively improved. Hopefully, these novel reagents will then translate into further improvements in antibody therapy.

References

Aalberse, R.C., Schuurman, J. (2002) IgG4 breaking the rules. *Immunology* 105: 9–19.

Angal, S., King, D.J., Bodmer, M.W., Turner, A., Lawson, A.D., Roberts, G., Pedley, B., Adair, J.R. (1993) A single amino acid substitution abolishes the heterogeneity of chimeric mouse/human (IgG4) antibody. *Mol Immunol* 30: 105–108.

Barbin, K., Stieglmaier, J., Saul, D., Stieglmaier, K., Stockmeyer, B., Pfeiffer, M., Lang, P., Fey, G.H. (2006) Influence of variable *N*-glycosylation on the cytolytic potential of chimeric CD19 antibodies. *J Immunother* 29: 122–133.

Berinstein, N.L., Grillo-Lopez, A.J., White, C.A., Bence-Bruckler, I., Maloney, D.,

Czuczman, M., Green, D., Rosenberg, J., McLaughlin, P., Shen, D. (1998) Association of serum rituximab (IDEC-C2B8) concentration and anti-tumor response in the treatment of recurrent low-grade or follicular non-Hodgkin's lymphoma. *Ann Oncol* 9: 995–1001.

Bindon, C.I., Hale, G., Bruggemann, M., Waldmann, H. (1988) Human monoclonal IgG isotypes differ in complement activating function at the level of C4 as well as C1q. *J Exp Med* 168: 127–142.

Bolt, S., Routledge, E., Lloyd, I., Chatenoud, L., Pope, H., Gorman, S.D., Clark, M., Waldmann, H. (1993) The generation of a humanized, non-mitogenic CD3 monoclonal antibody which retains in vitro immunosuppressive properties. *Eur J Immunol* 23: 403–411.

Bruggemann, M., Williams, G.T., Bindon, C. I., Clark, M.R., Walker, M.R., Jefferis, R., Waldmann, H., Neuberger, M.S. (1987) Comparison of the effector functions of human immunoglobulins using a matched set of chimeric antibodies. *J Exp Med* 166: 1351–1361.

Burmeister, W.P., Huber, A.H., Bjorkman, P.J. (1994) Crystal structure of the complex of rat neonatal Fc receptor with Fc. *Nature* 372: 379–383.

Burton, D.R. (1985) Immunoglobulin G: functional sites. *Mol Immunol* 22: 161–206.

Carpenter, P.A., Appelbaum, F.R., Corey, L., Deeg, H.J., Doney, K., Gooley, T., Krueger, J., Martin, P., Pavlovic, S., Sanders, J., Slattery, J., Levitt, D., Storb, R., Woolfrey, A., Anasetti, C. (2002) A humanized non-FcR-binding anti-CD3 antibody, visilizumab, for treatment of steroid-refractory acute graft-versus-host disease. *Blood* 99: 2712–2719.

Cartron, G., Dacheux, L., Salles, G., Solal-Celigny, P., Bardos, P., Colombat, P., Watier, H. (2002) Therapeutic activity of humanized anti-CD20 monoclonal antibody and polymorphism in IgG Fc receptor FcγRIIIa gene. *Blood* 99: 754–758.

Chatenoud, L. (2003) CD3-specific antibody-induced active tolerance: from bench to bedside. *Nat Rev Immunol* 3: 123–132.

Chatenoud, L. (2005) CD3-specific antibodies restore self-tolerance: mechanisms and clinical applications. *Curr Opin Immunol* 17: 632–637.

Clark, M.R. (1997) IgG effector mechanisms. *Chem Immunol* 65: 88–110.

Clynes, R., Takechi, Y., Moroi, Y., Houghton, A., Ravetch, J.V. (1998) Fc receptors are required in passive and active immunity to melanoma. *Proc Natl Acad Sci USA* 95: 652–656.

Clynes, R.A., Towers, T.L., Presta, L.G., Ravetch, J.V. (2000) Inhibitory Fc receptors modulate in vivo cytoxicity against tumor targets. *Nat Med* 6: 443–446.

Coiffier, B., Tilly, H., Pedersen, L.M., Plesner, T., Frederiksen, H., van Oers, M. H.J., Wooldridge, J., Kloczko, J., Holowiecki, J., Hellmann, A., Walewski, J. J., Flensburg, M.F., Petersen, J., Robak, T. (2005) HuMax CD20 fully human monoclonal antibody in chronic lymphocytic leukemia. Early results from an ongoing phase i/ii clinical trial. *Blood* 160: 448A.

Cole, M.S., Anasetti, C., Tso, J.Y. (1997) Human IgG2 variants of chimeric anti-CD3 are nonmitogenic to T cells. *J Immunol* 159: 3613–3621.

Cragg, M.S., Glennie, M.J. (2004) Antibody specificity controls in vivo effector mechanisms of anti-CD20 reagents. *Blood* 103: 2738–2743.

Dall'Acqua, W.F., Woods, R.M., Ward, E.S., Palaszynski, S.R., Patel, N.K., Brewah, Y. A., Wu, H., Kiener, P.A., Langermann, S. (2002) Increasing the affinity of a human IgG1 for the neonatal Fc receptor: biological consequences. *J Immunol* 169: 5171–5180.

Damle, N.K., Frost, P. (2003) Antibody-targeted chemotherapy with immunoconjugates of calicheamicin. *Curr Opin Pharmacol* 3: 386–390.

Davies, J., Jiang, L., Pan, L.Z., LaBarre, M.J., Anderson, D., Reff, M. (2001) Expression of GnTIII in a recombinant anti-CD20 CHO production cell line: Expression of antibodies with altered glycoforms leads to an increase in ADCC through higher affinity for FcγRIII. *Biotechnol Bioeng* 74: 288–294.

de Lange, G.G. (1989) Polymorphisms of human immunoglobulins: Gm, Am, Em and Km allotypes. *Exp Clin Immunogenet* 6: 7–17.

Dechant, M., Valerius, T. (2001) IgA antibodies for cancer therapy. *Crit Rev Oncol Hematol* 39: 69–77.

Dechant, M., Vidarsson, G., Stockmeyer, B., Repp, R., Glennie, M.J., Gramatzki, M., van De Winkel, J.G., Valerius, T. (2002) Chimeric IgA antibodies against HLA class II effectively trigger lymphoma cell killing. *Blood* 100: 4574–4580.

Di Carlo, E., Forni, G., Lollini, P., Colombo, M.P., Modesti, A., Musiani, P. (2001) The intriguing role of polymorphonuclear neutrophils in antitumor reactions. *Blood* 97: 339–345.

Di Gaetano, N., Cittera, E., Nota, R., Vecchi, A., Grieco, V., Scanziani, E., Botto, M., Introna, M., Golay, J. (2003) Complement activation determines the therapeutic activity of rituximab in vivo. *J Immunol* 171: 1581–1587.

Duncan, A.R., Winter, G. (1988) The binding site for C1q on IgG. *Nature* 332: 738–740.

Dyer, M.J., Hale, G., Hayhoe, F.G., Waldmann, H. (1989) Effects of CAMPATH-1 antibodies in vivo in patients with lymphoid malignancies: influence of antibody isotype. *Blood* 73: 1431–1439.

Dyer, M.J.S., Moser, S., Brünker, P., Bird, P., Almond, N., Püentener, U., Wheat, L.M.W., Bolam, E., Berrie, E., Grau, R., Buckby, E., Kennedy, B., Stebbings, R., Hale, G., Umana, P. (2005) Enhanced potency of glycoengineered anti-CD52 monoclonal antibodies (MAbs). *Blood* 160: 2958A.

Farag, S.S., Flinn, I.W., Modali, R., Lehman, T.A., Young, D., Byrd, J.C. (2004) Fcγ RIIIa and FcγRIIa polymorphisms do not predict response to rituximab in B-cell chronic lymphocytic leukemia. *Blood* 103: 1472–1474.

Ferrara, C., Stuart, F., Sondermann, P., Brunker, P., Umana, P. (2005) The carbohydrate at FcγRIIIa ASN162: An element required for high affinity binding to non-fucosylated IgG glycoforms. *J Biol Chem* 281: 5032–5036.

Ferrara, C., Brunker, P., Suter, T., Moser, S., Puntener, U., Umana, P. (2006) Modulation of therapeutic antibody effector functions by glycosylation engineering: influence of Golgi enzyme localization domain and co-expression of heterologous beta1:4-*N*-acetylglucosaminyltransferase III and Golgi alpha-mannosidase II. *Biotechnol Bioeng* 93: 851–861.

Friend, P.J., Hale, G., Chatenoud, L., Rebello, P., Bradley, J., Thiru, S., Phillips, J.M., Waldmann, H. (1999) Phase I study of an engineered aglycosylated humanized CD3 antibody in renal transplant rejection. *Transplantation* 68: 1632–1637.

Ghetie, V., Ward, E.S. (2000) Multiple roles for the major histocompatibility complex class I- related receptor FcRn. *Annu Rev Immunol* 18: 739–766.

Glennie, M.J., van de Winkel, J.G. (2003) Renaissance of cancer therapeutic antibodies. *Drug Discov Today* 8: 503–510.

Herold, K.C., Hagopian, W., Auger, J.A., Poumian-Ruiz, E., Taylor, L., Donaldson, D., Gitelman, S.E., Harlan, D.M., Xu, D., Zivin, R.A., Bluestone, J.A. (2002) Anti-CD3 monoclonal antibody in new-onset type 1 diabetes mellitus. *N Engl J Med* 346: 1692–1698.

Hinton, P.R., Johlfs, M.G., Xiong, J.M., Hanestad, K., Ong, K.C., Bullock, C., Keller, S., Tang, M.T., Tso, J.Y., Vasquez, M., Tsurushita, N. (2004) Engineered human IgG antibodies with longer serum half-lives in primates. *J Biol Chem* 279: 6213–6216.

Hinton, P.R., Xiong, J.M., Johlfs, M.G., Tang, M.T., Keller, S., Tsurushita, N. (2006) An engineered human IgG1 antibody with longer serum half-life. *J Immunol* 176: 346–356.

Hirsch, F.R., Bunn, P.A., Jr. (2005) Epidermal growth factor receptor inhibitors in lung cancer: smaller or larger molecules, selected or unselected populations? *J Clin Oncol* 23: 9044–9047.

Huls, G., Heijnen, I.A., Cuomo, E., van der Linden, J., Boel, E., van de Winkel, J.G., Logtenberg, T. (1999) Antitumor immune effector mechanisms recruited by phage display-derived fully human IgG1 and IgA1 monoclonal antibodies. *Cancer Res* 59: 5778–5784.

Idusogie, E.E., Presta, L.G., Gazzano-Santoro, H., Totpal, K., Wong, P.Y., Ultsch, M., Meng, Y.G., Mulkerrin, M.G. (2000) Mapping of the C1q binding site on rituxan, a chimeric antibody with a human IgG1 Fc. *J Immunol* 164: 4178–4184.

Idusogie, E.E., Wong, P.Y., Presta, L.G., Gazzano-Santoro, H., Totpal, K., Ultsch, M., Mulkerrin, M.G. (2001) Engineered antibodies with increased activity to recruit complement. *J Immunol* 166: 2571–2575.

Janakiraman, N., McLaughlin, P., White, C. A., Maloney, D.G., Shen, D., Grillo-Lopez, A.J. (1998) Rituximab: correlation between effector cells and clinical activity in NHL. *Blood* 92: 337A.

Jefferis, R. (1986) Polyclonal and monoclonal antibody reagents specific for IgG subclasses. *Monogr Allergy* 19: 71–85.

Jefferis, R. (2005) Glycosylation of recombinant antibody therapeutics. *Biotechnol Prog* 21: 11–16.

Jefferis, R., Lund, J. (2002) Interaction sites on human IgG-Fc for FcγR: current models. *Immunol Lett* 82: 57–65.

Jefferis, R., Lund, J., Goodall, M. (1995) Recognition sites on human IgG for Fcγ receptors: the role of glycosylation. *Immunol Lett* 44: 111–117.

Kennedy, A.D., Beum, P.V., Solga, M.D., DiLillo, D.J., Lindorfer, M.A., Hess, C.E., Densmore, J.J., Williams, M.E., Taylor, R.P. (2004) Rituximab infusion promotes rapid complement depletion and acute CD20 loss in chronic lymphocytic leukemia. *J Immunol* 172: 3280–3288.

Kerr, M.A. (1990) The structure and function of human IgA. *Biochem J* 271: 285–296.

Keymeulen, B., Vandemeulebroucke, E., Ziegler, A.G., Mathieu, C., Kaufman, L., Hale, G., Gorus, F., Goldman, M., Walter, M., Candon, S., Schandene, L., Crenier, L., De Block, C., Seigneurin, J.M., De Pauw, P., Pierard, D., Weets, I., Rebello, P., Bird, P., Berrie, E., Frewin, M., Waldmann, H., Bach, J.F., Pipeleers, D., Chatenoud, L. (2005) Insulin needs after CD3-antibody therapy in new-onset type 1 diabetes. *N Engl J Med* 352: 2598–2608.

Krapp, S., Mimura, Y., Jefferis, R., Huber, R., Sondermann, P. (2003) Structural analysis of human IgG-Fc glycoforms reveals a correlation between glycosylation and structural integrity. *J Mol Biol* 325: 979–989.

Kumpel, B.M., Wang, Y., Griffiths, H.L., Hadley, A.G., Rook, G.A. (1995) The biological activity of human monoclonal IgG anti-D is reduced by beta-galactosidase treatment. *Hum Antibodies Hybridomas* 6: 82–88.

Lammerts van Bueren, J.J., Bleeker, W.K., Bøgh, H.O., Houtkamp, M., Schuurman, J., van de Winkel, J.G.J., Parren, P.W.H.I. (2006) Effect of target dynamics on pharmacokinetics of a novel therapeutic antibody against the epidermal growth factor receptor: implications for the mechanisms of action. *Cancer Res* 66: 7630–7638.

Lazar, G.A., Dang, W., Karki, S., Vafa, O., Peng, J.S., Hyun, L., Chan, C., Chung, H. S., Eivazi, A., Yoder, S.C., Vielmetter, J., Carmichael, D.F., Hayes, R.J., Dahiyat, B.I. (2006) Engineered antibody Fc variants with enhanced effector function. *Proc Natl Acad Sci USA* 103: 4005–4010.

Lifely, M.R., Hale, C., Boyce, S., Keen, M.J., Phillips, J. (1995) Glycosylation and biological activity of CAMPATH-1H expressed in different cell lines and grown under different culture conditions. *Glycobiology* 5: 813–822.

Lin, T.S., Flinn, I.W., Modali, R., Lehman, T. A., Webb, J., Waymer, S., Moran, M.E., Lucas, M.S., Farag, S.S., Byrd, J.C. (2005) FCGR3A and FCGR2A polymorphisms may not correlate with response to alemtuzumab in chronic lymphocytic leukemia. *Blood* 105: 289–291.

Lucisano Valim, Y.M., Lachmann, P.J. (1991) The effect of antibody isotype and antigenic epitope density on the complement-fixing activity of immune complexes: a systematic study using chimaeric anti-NIP antibodies with human Fc regions. *Clin Exp Immunol* 84: 1–8.

Manches, O., Lui, G., Chaperot, L., Gressin, R., Molens, J.P., Jacob, M.C., Sotto, J.J., Leroux, D., Bensa, J.C., Plumas, J. (2003) In vitro mechanisms of action of rituximab on primary non-Hodgkin lymphomas. *Blood* 101: 949–954.

Michaelsen, T.E., Garred, P., Aase, A. (1991) Human IgG subclass pattern of inducing complement-mediated cytolysis depends on antigen concentration and to a lesser extent on epitope patchiness, antibody affinity and complement concentration. *Eur J Immunol* 21: 11–16.

Monteiro, R.C., van de Winkel, J.G. (2003) IgA Fc receptors. *Annu Rev Immunol* 21: 177–204.

Mori, K., Kuni-Kamochi, R., Yamane-Ohnuki, N., Wakitani, M., Yamano, K., Imai, H., Kanda, Y., Niwa, R., Iida, S., Uchida, K., Shitara, K., Satoh, M. (2004) Engineering Chinese hamster ovary cells to maximize effector function of produced antibodies using FUT8 siRNA. *Biotechnol Bioeng* 88: 901–908.

Nimmerjahn, F., Ravetch, J.V. (2005) Divergent immunoglobulin g subclass activity through selective Fc receptor binding. *Science* 310: 1510–1512.

Nimmerjahn, F., Ravetch, J.V. (2006) Fcγ receptors: old friends and new family members. *Immunity* 24: 19–28.

Niwa, R., Hatanaka, S., Shoji-Hosaka, E., Sakurada, M., Kobayashi, Y., Uehara, A., Yokoi, H., Nakamura, K., Shitara, K. (2004) Enhancement of the antibody-dependent cellular cytotoxicity of low-fucose IgG1 Is independent of FcγRIIIa functional polymorphism. *Clin Cancer Res* 10: 6248–6255.

Niwa, R., Natsume, A., Uehara, A., Wakitani, M., Iida, S., Uchida, K., Satoh, M., Shitara, K. (2005) IgG subclass-independent improvement of antibody-dependent cellular cytotoxicity by fucose removal from Asn297-linked oligosaccharides. *J Immunol Methods* 306: 151–160.

Norman, D.J., Vincenti, F., de Mattos, A.M., Barry, J.M., Levitt, D.J., Wedel, N.I., Maia, M., Light, S.E. (2000) Phase I trial of HuM291: a humanized anti-CD3 antibody, in patients receiving renal allografts from living donors. *Transplantation* 70: 1707–1712.

Nose, M., Wigzell, H. (1983) Biological significance of carbohydrate chains on monoclonal antibodies. *Proc Natl Acad Sci USA* 80: 6632–6636.

Parren, P.W., Warmerdam, P.A., Boeije, L.C., Arts, J., Westerdaal, N.A., Vlug, A., Capel, P.J., Aarden, L.A., van de Winkel, J.G. (1992) On the interaction of IgG subclasses with the low affinity FcγRIIa (CD32) on human monocytes, neutrophils, and platelets. Analysis of a functional polymorphism to human IgG2. *J Clin Invest* 90: 1537–1546.

Pasquier, B., Launay, P., Kanamaru, Y., Moura, I.C., Pfirsch, S., Ruffie, C., Henin, D., Benhamou, M., Pretolani, M., Blank, U., Monteiro, R.C. (2005) Identification of FcαRI as an inhibitory receptor that controls inflammation: dual role of FcRγ ITAM. *Immunity* 22: 31–42.

Pleass, R.J., Ogun, S.A., McGuinness, D.H., van de Winkel, J.G., Holder, A.A., Woof, J. M. (2003) Novel antimalarial antibodies highlight the importance of the antibody Fc region in mediating protection. *Blood* 102: 4424–4430.

Rabbitts, T.H., Forster, A., Milstein, C.P. (1981) Human immunoglobulin heavy chain genes: evolutionary comparisons of Cμ, Cδ and Cγ genes and associated switch sequences. *Nucleic Acids Res* 9: 4509–4524.

Rademacher, T.W., Parekh, R.B., Dwek, R. A., Isenberg, D., Rook, G., Axford, J.S., Roitt, I. (1988) The role of IgG glycoforms in the pathogenesis of rheumatoid arthritis. *Springer Semin Immunopathol* 10: 231–249.

Reddy, M.P., Kinney, C.A., Chaikin, M.A., Payne, A., Fishman-Lobell, J., Tsui, P., Dal Monte, P.R., Doyle, M.L., Brigham-Burke, M.R., Anderson, D., Reff, M., Newman, R., Hanna, N., Sweet, R.W., Truneh, A. (2000) Elimination of Fc receptor-dependent effector functions of a modified IgG4 monoclonal antibody to human CD4. *J Immunol* 164: 1925–1933.

Reichert, J.M., Rosensweig, C.J., Faden, L. B., Dewitz, M.C. (2005) Monoclonal antibody successes in the clinic. *Nat Biotechnol* 23: 1073–1078.

Riechmann, L., Clark, M., Waldmann, H., Winter, G. (1988) Reshaping human antibodies for therapy. *Nature* 332: 323–327.

Schachter, H. (2000) The joys of HexNAc. The synthesis and function of N- and O-glycan branches. *Glycoconj J* 17: 465–483.

Schuster, M., Umana, P., Ferrara, C., Brunker, P., Gerdes, C., Waxenecker, G., Wiederkum, S., Schwager, C., Loibner, H., Himmler, G., Mudde, G.C. (2005) Improved effector functions of a therapeutic monoclonal Lewis Y-specific antibody by glycoform engineering. *Cancer Res* 65: 7934–7941.

Shields, R.L., Namenuk, A.K., Hong, K., Meng, Y.G., Rae, J., Briggs, J., Xie, D., Lai, J., Stadlen, A., Li, B., Fox, J.A., Presta, L.G. (2001) High resolution mapping of the binding site on human IgG1 for FcγRI, FcγRII, FcγRIII, and FcRn and design of IgG1 variants with improved binding to the FcγR. *J Biol Chem* 276: 6591–6604.

Shields, R.L., Lai, J., Keck, R., O'Connell, L.Y., Hong, K., Meng, Y.G., Weikert, S.H., Presta, L.G. (2002) Lack of fucose on human IgG1 N-linked oligosaccharide improves binding to human FcγRIII and antibody-dependent cellular toxicity. *J Biol Chem* 277: 26733–26740.

Shih, L.B., Lu, H.H., Xuan, H., Goldenberg, D.M. (1994) Internalization and intracellular processing of an anti-B-cell lymphoma monoclonal antibody, LL2. *Int J Cancer* 56: 538–545.

Shinkawa, T., Nakamura, K., Yamane, N., Shoji-Hosaka, E., Kanda, Y., Sakurada, M., Uchida, K., Anazawa, H., Satoh, M., Yamasaki, M., Hanai, N., Shitara, K. (2003) The absence of fucose but not the presence of galactose or bisecting N-acetylglucosamine of human IgG1 complex-type oligosaccharides shows the critical role of enhancing antibody-dependent cellular cytotoxicity. *J Biol Chem* 278: 3466–3473.

Sondermann, P., Huber, R., Oosthuizen, V., Jacob, U. (2000) The 3.2-A crystal structure of the human IgG1 Fc fragment-FcγRIII complex. *Nature* 406: 267–273.

Stavenhagen, J., Vijh, S., Rankin, C., Gorlatov, S., Huang, L. (2004) Identification and engineering of antibodies with variant Fc regions and methods of using same US 2005/0064514 A1 MacroGenics, Inc.

Tao, M.H., Morrison, S.L. (1989) Studies of aglycosylated chimeric mouse-human IgG. Role of carbohydrate in the structure and effector functions mediated by the human IgG constant region. *J Immunol* 143: 2595–2601.

Treon, S.P., Hansen, M., Branagan, A.R., Verselis, S., Emmanouilides, C., Kimby, E., Frankel, S.R., Touroutoglou, N., Turnbull, B., Anderson, K.C., Maloney, D.G., Fox, E.A. (2005) Polymorphisms in FcγRIIIA (CD16) receptor expression are associated with clinical response to rituximab in Waldenstrom's macroglobulinemia. *J Clin Oncol* 23: 474–481.

Uchida, J., Hamaguchi, Y., Oliver, J.A., Ravetch, J.V., Poe, J.C., Haas, K.M., Tedder, T.F. (2004) The innate mononuclear phagocyte network depletes B lymphocytes through Fc receptor-dependent mechanisms during anti-CD20 antibody immunotherapy. *J Exp Med* 199: 1659–1669.

Umana, P., Jean-Mairet, J., Moudry, R., Amstutz, H., Bailey, J.E. (1999) Engineered glycoforms of an antineuroblastoma IgG1 with optimized antibody-dependent cellular cytotoxic activity. *Nat Biotechnol* 17: 176–180.

Utset, T.O., Auger, J.A., Peace, D., Zivin, R.A., Xu, D., Jolliffe, L., Alegre, M.L., Bluestone, J.A., Clark, M.R. (2002) Modified anti-CD3 therapy in psoriatic arthritis: a phase I/II clinical trial. *J Rheumatol* 29: 1907–1913.

Valerius, T., Stockmeyer, B., van Spriel, A. B., Graziano, R.F., van den Herik-Oudijk, I.E., Repp, R., Deo, Y.M., Lund, J., Kalden, J.R., Gramatzki, M., van de Winkel, J.G. (1997) FcαRI (CD89) as a novel trigger molecule for bispecific antibody therapy. *Blood* 90: 4485–4492.

van de Winkel, J.G., Capel, P.J. (1993) Human IgG Fc receptor heterogeneity: molecular aspects and clinical implications. *Immunol Today* 14: 215–221.

van der Kolk, L.E., Grillo-Lopez, A.J., Baars, J.W., Hack, C.E., van Oers, M.H. (2001) Complement activation plays a key role in the side-effects of rituximab treatment. *Br J Haematol* 115: 807–811.

van Egmond, M., van Garderen, E., van Spriel, A.B., Damen, C.A., van Amersfoort, E.S., van Zandbergen, G., van Hattum, J., Kuiper, J., van de Winkel, J.G. (2000) FcαRI-positive liver Kupffer cells: reappraisal of the function of immunoglobulin A in immunity. *Nat Med* 6: 680–685.

van Loghem, E. (1986) Allotypic markers. Monogr Allergy 19: 40–51.

van Sorge, N.M., van der Pol, W.L., van de Winkel, J.G. (2003) FcγR polymorphisms: Implications for function, disease susceptibility and immunotherapy. *Tissue Antigens* 61: 189–202.

van Spriel, A.B., van den Herik-Oudijk, I.E., van Sorge, N.M., Vile, H.A., van Strijp, J. A., van de Winkel, J.G. (1999) Effective phagocytosis and killing of *Candida albicans* via targeting FcγRI (CD64) or

FcαRI (CD89) on neutrophils. *J Infect Dis* 179: 661–669.

Vidarsson, G., van Der Pol, W.L., van Den Elsen, J.M., Vile, H., Jansen, M., Duijs, J., Morton, H.C., Boel, E., Daha, M.R., Corthesy, B., van De Winkel, J.G. (2001) Activity of human IgG and IgA subclasses in immune defense against *Neisseria meningitidis* serogroup B. *J Immunol* 166: 6250–6256.

Waldmann, T.A., Strober, W. (1969) Metabolism of immunoglobulins. *Prog Allergy* 13: 1–110.

Weng, W.K., Levy, R. (2001) Expression of complement inhibitors CD46: CD55: and CD59 on tumor cells does not predict clinical outcome after rituximab treatment in follicular non-Hodgkin lymphoma. *Blood* 98: 1352–1357.

Weng, W.K., Levy, R. (2003) Two immunoglobulin G fragment C receptor polymorphisms independently predict response to rituximab in patients with follicular lymphoma. *J Clin Oncol* 21: 3940–3947.

Woodle, E.S., Xu, D., Zivin, R.A., Auger, J., Charette, J., O'Laughlin, R., Peace, D., Jollife, L.K., Haverty, T., Bluestone, J.A., Thistlethwaite, J.R., Jr. (1999) Phase I trial of a humanized, Fc receptor nonbinding OKT3 antibody, huOKT3gamma1(Ala-Ala) in the treatment of acute renal allograft rejection. *Transplantation* 68: 608–616.

Woof, J.M. (2005) Immunology. Tipping the scales toward more effective antibodies. *Science* 310: 1442–1443.

Woof, J.M., Mestecky, J. (2005) Mucosal immunoglobulins. *Immunol Rev* 206: 64–82.

Wright, A., Morrison, S.L. (1997) Effect of glycosylation on antibody function: implications for genetic engineering. *Trends Biotechnol* 15: 26–32.

Yamane-Ohnuki, N., Kinoshita, S., Inoue-Urakubo, M., Kusunoki, M., Iida, S., Nakano, R., Wakitani, M., Niwa, R., Sakurada, M., Uchida, K., Shitara, K., Satoh, M. (2004) Establishment of FUT8 knockout Chinese hamster ovary cells: an ideal host cell line for producing completely defucosylated antibodies with enhanced antibody-dependent cellular cytotoxicity. *Biotechnol Bioeng* 87: 614–622.

Yoo, E.M., Morrison, S.L. (2005) IgA: an immune glycoprotein. *Clin Immunol* 116: 3–10.

Part II
The Way into the Clinic

9
Production and Downstream Processing

Klaus Bergemann, Christian Eckermann, Patrick Garidel, Stefanos Grammatikos, Alexander Jacobi, Hitto Kaufmann, Ralph Kempken, and Sandra Pisch-Heberle

9.1
Introduction

Biopharmaceuticals such as monoclonal antibodies (mAbs) are complex molecules that, today, can only be produced economically by using mammalian cells. As mAbs are highly susceptible to physical and chemical stress, supply for clinical studies and therapy requires the development of a suitable production process. This process should ensure high quality and stability of the product in accordance with the quality obligations of the regulatory authorities at an acceptable cost of goods. The development of a suitable production process for mAbs requires the concerted and coordinated activities of a number of disciplines such as molecular and cell biology, upstream and downstream processing, formulation development, filling operations, quality testing, and quality control.

Molecular and cell biology techniques are involved in cloning the antibody genes into an appropriate expression vector followed by the generation of a production cell line. The goal is to develop a production cell line possessing a high specific productivity which is capable of growing to high cell densities in serum-free and chemically defined media.

Upstream processing consists of all operations that are involved in the generation of the crude mAb. This is generally achieved by cultivation of the production cells, mostly by means of fermentation using bioreactors. Development of an optimal cell culture medium and process design are important elements for an efficient process capable of delivering the product in high quantities. After separation of the cells from the culture medium, which contains the crude product, the purification of the mAb from contaminants originating from the cells and the cell culture medium is addressed in the downstream processing steps. These consist mainly of chromatographic and filtration methods. High recovery of the protein, efficient removal of contaminants and the demonstration of an efficient inactivation and removal of potential viruses, while retaining the molecule's correct structure, are crucial at this stage of development.

Handbook of Therapeutic Antibodies. Edited by Stefan Dübel

ISBN 978-3-527-31453-9

The resulting drug substance undergoes final formulation prior to filling into the primary packaging container, which can be either a glass vial or a syringe. The development of the final formulation has to take the physicochemical and biological characteristics of the product, as well as the intended application route, into account in order to secure the defined product quality specifications during the shelf-life of the product.

The detailed description of the molecular characteristics of the product represents the basis for the definition of the quality parameters to ensure the safety and biological activity of the product. Quality parameters such as identity, purity, potency, strength, and stability are closely monitored on a lot-by-lot basis prior to release of the product for human use.

As the market for biopharmaceuticals continues to grow rapidly, key factors for successful development of mAbs are the so-called platform technologies. These technologies represent a company's know-how and have proven efficient and successful in previous development programs. Use of these technologies may significantly speed up development times.

Another important factor for a successful process development program is the coordination and optimal timing of the various activities and work packages of the different disciplines. This can be achieved by the implementation of an interdisciplinary project team with clearly defined responsibilities, roles and tasks for the team members and the project manager. This team is responsible for all different aspects of a product development program in different organizational units, which includes the monitoring of timelines and costs and the establishment of tools for further improvements.

9.2 Upstream Processing

The upstream process consists of all operations that are involved in the generation of the crude therapeutic antibody product, up to or even including the harvest as separation of the cells from the product-containing culture fluid.

Antibody products are generally generated by fermentation (cell cultivation) mostly in bioreactors, for all kinds of purposes, extending from product for research and development through product for clinical studies and up to the market supply for approved therapeutic antibody products.

9.2.1 Expression System

Currently, development of production processes for biopharmaceuticals is undergoing significant changes as the technology matures. Recombinant proteins can be produced in various expression systems such as bacterial, mammalian, insect, plant, and *in vitro* translation systems. A fundamental prerequisite for successful production of biologics from any of these expression systems is, of course, efficient transcription and translation. However, the choice of system is mainly driven by the overall yield of the production process and the biological activity and efficacy of the therapeutic entity.

Expression of recombinant proteins from *E. coli* cells is a well-established technology (Swartz 2001). *E. coli* cells can produce proteins in large quantities, fast growth rates enable short fermentation times, and the system has been successfully scaled up to supply the market demand for important drugs such as insulin (Lee 1996). However, *E. coli* cells are not capable of glycosylating proteins. Therefore, expression of glycoproteins such as mAbs requires other expression systems. The vast majority of such proteins are currently being expressed in mammalian cells, mostly Chinese hamster ovary (CHO) and mouse myeloma cells (NS0) cells. The two most prominent challenges for mammalian cell-based systems relate to (i) product titer in the cell culture fluid at the end of cultivation and (ii) development times from final drug candidate selection to an established process to produce clinical grade material. On both fronts recent developments indicate that the newest generations of mammalian cell culture production processes evolve rapidly to overcome these limitations (Wurm 2004). Many improvements, including novel or modified genetic elements to improve transcription rate, high-throughput screening concepts to obtain highly productive clones reliably, and host cell lines that grow to high densities in serum-free chemically defined media, have achieved specific productivities of values above 50 pg per cell and day (pg/c*d) for mAbs. Figure 9.1 shows the productivity profile for a CHO DG-44 high-producer cell line. Serial cultivation in spinner flasks (typical for industrial inoculum settings) displays a constant specific productivity of 55 pg per cell and day for 120 days in culture. In addition to high productivities, cell lines need to have stable phenotypic and genotypic product expression profiles to enable economically attractive production campaigns with many fed-batch runs initiated from one inoculum culture.

Furthermore, sophisticated design of clone-screening platforms, including platform media and well-characterized down-scale models, is crucial to ensure that high titers can be achieved with little or no time for process optimization. Such a high-expression concept can result in titers above 1 g L^{-1} without the need for optimizing the upstream process subsequent to clone screening. Due to the biochemical complexity of biologics and the influence of host cell, media, and other factors on the microheterogeneity of the product, comparability issues are of great importance. It is therefore essential to any fast-track cell line generation concept that it results in a production cell line that has the potential to form the basis for a high titer process as it is needed to generate material all the way to the market. A well-characterized media platform is crucial to minimize changes throughout further process optimization. Many recent approaches have demonstrated the great potential of host cell engineering, suggesting that future mammalian host cells will comprise optimized molecular pathways that control growth, apoptosis, transport, and metabolic fluxes (Kaufmann and Fussenegger 2003).

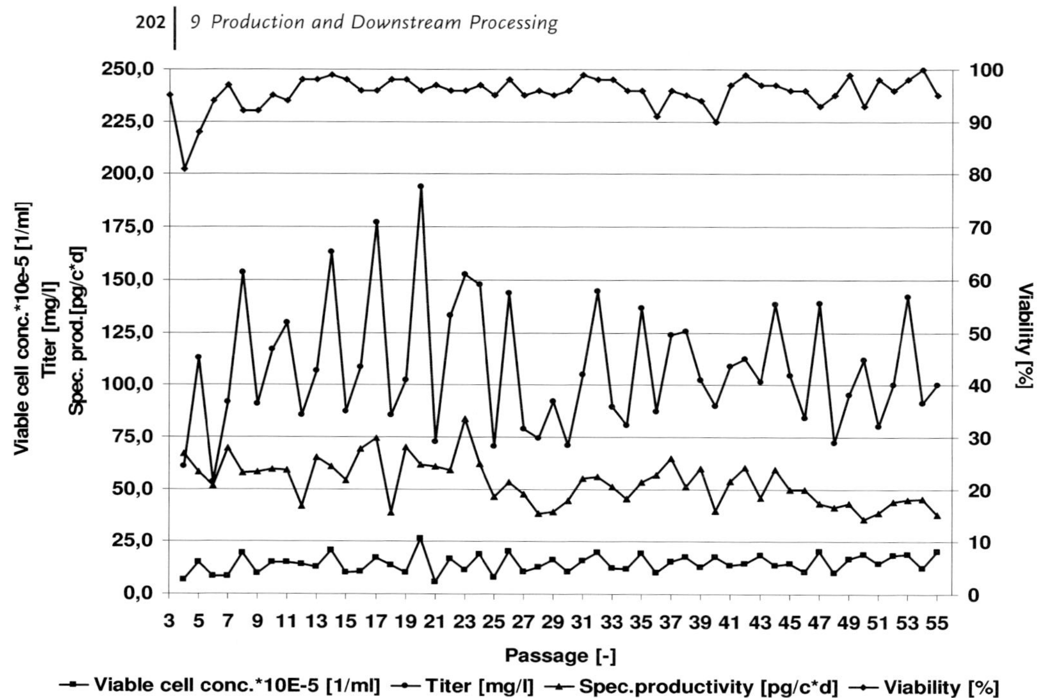

Fig. 9.1 BI HEX CHO (Boehringer Ingelheim's proprietary high-expression platform). Stable high expression of a monoclonal antibody in a clonal CHO cell line.

Although, for mAbs, the current expression platform of choice is mammalian cells, in the future, other hosts may represent interesting alternatives in the production of new biopharmaceuticals. Processing of yeast glycoproteins within the endoplasmic reticulum and the Golgi apparatus is different from the processing of their mammalian counterparts. This results in distinct nonmammalian glycopatterns of yeast proteins. In particular, high-mannose type *N*-glycosylation precludes the use of recombinant glycoproteins produced from yeasts for therapeutic use in humans as it leads to a short circulatory half-life *in vivo* and possibly altered activity, thereby compromising drug efficacy. Recently a lot of work has been focused on generating genetically engineered yeast strains that confer a synthetic *N*-glycosylation pathway designed to enable production of proteins with a more mammalian-like glycosylation pattern (Wildt and Gerngross 2005). Although these approaches are promising, many problems such as the lack of terminal sialylation need to be solved before yeast cells can be viewed as an alternative expression platform for producing glycoproteins for use in humans.

Another fundamental question that needs to be answered in the future concerns the robustness, growth performance and general phenotype of yeasts cells that have been genetically modified to substitute entire biological pathways such as glycosylation.

Another interesting field with regard to future expression systems is the evaluation of the production of biopharmaceuticals from plants. A variety of different plant-based systems may provide interesting options, including production within the leaves of transgenic plants, secretion from roots or leaves, or growth of plant cells in a bioreactor (Fischer et al. 2004). However, several fundamental problems need to be overcome before recombinant proteins produced from plants can be approved for therapeutic use in humans. Yields for antibodies or other proteins are commonly low, protein stability is often poor and in general downstream processing of plant-derived recombinant proteins is cumbersome. Recent studies point to directions this field may take in order to overcome these limitations. Strategies include targeting of heterologous proteins to the secretory pathway of plant cells, product expression in seeds and the use of strong inducible promoters (Fischer et al. 2004).

9.2.2 Cell Culture Media

A second key element for the upstream process is the development and use of suitable media for the fermentation. Media for microbial cells or yeast cells consist of a carbon and energy source, nutrient salts and trace elements, and sometimes a limited number of vitamins or amino acids. Media for mammalian cells are much more complex. About 30–80 ingredients are dissolved in purified (deionized) water and a proper balance of buffering and osmoregulating substances, trace elements, precursors for the primary and secondary cell metabolism (e.g. amino acids, vitamins, lipids) and nutrients is critical for a successful long-term cultivation of mammalian cells *in vitro*. Until recently it was industry standard to supplement the basal medium consisting of defined components with 5–20% serum (mostly from calf or cattle). Preparations of animal sera consist of a complex nondefined mixture of growth factors and other biomolecules that mediate cell survival and growth. The major disadvantages of using sera in cell culture processes are the residual risk of human pathogens present in animal sera, limited supply, high cost and high batch-to-batch variability (Allison et al. 2005).

During the last 25 years large efforts were devoted to the removal and substitution of serum. In the first generation of serum-free culture media, serum was replaced by defined substances still from animal sources added at defined concentrations such as bovine serum albumin (BSA), bovine insulin, bovine transferrin, lipoproteins, and cholesterol (Hewlett 1991).

Today's state-of-the-art culture media are free of bovine components and chemically defined. Here animal components are substituted by chemically defined mixtures or single substances such as recombinant insulin or IGF (insulin-like growth factor), lipid concentrates, iron salts or complexes, precursors, or stimulating substances. Media development over the past decades has led to a reduction in the protein content of cell culture media from about $20\,g\,L^{-1}$ to about $0.01\,g\,L^{-1}$ or even to zero. Currently commercial media suppliers offer a variety of basic media as powders or granulates or liquid formulations optimized for specific cell lines and the needs of serum-free cell culture processes (Kempken et al. 2006). Today's cell culture media provide an excellent platform for high-performance cell culture processes for the generation of therapeutic antibodies.

9.2.3
Cell Culture Process Design

The third key element is the development of an efficient, robust and reproducible cell culture process itself. Mammalian cells grow either anchorage-dependent or in suspension. *In vitro* cultivation systems for anchorage-dependent cells (Fig. 9.2a) range from T-flasks and roller bottles through fluidized bed bioreactors, where the cells are attached to microcarriers (i.e. massive or porous beads approx. 50–300 μm diameter) suspended in a bioreactor, up to fixed bed bioreactors, where the cells are attached in the capillaries of solid surfaces (e.g. in hollow-fiber membrane modules, Fig. 9.2b) or in ceramic modules. In all cases, culture medium and aeration is provided continuously or at intervals by exchange of medium and gas, or by perfusion (see below) of the bioreactors (Murakami et al. 1991).

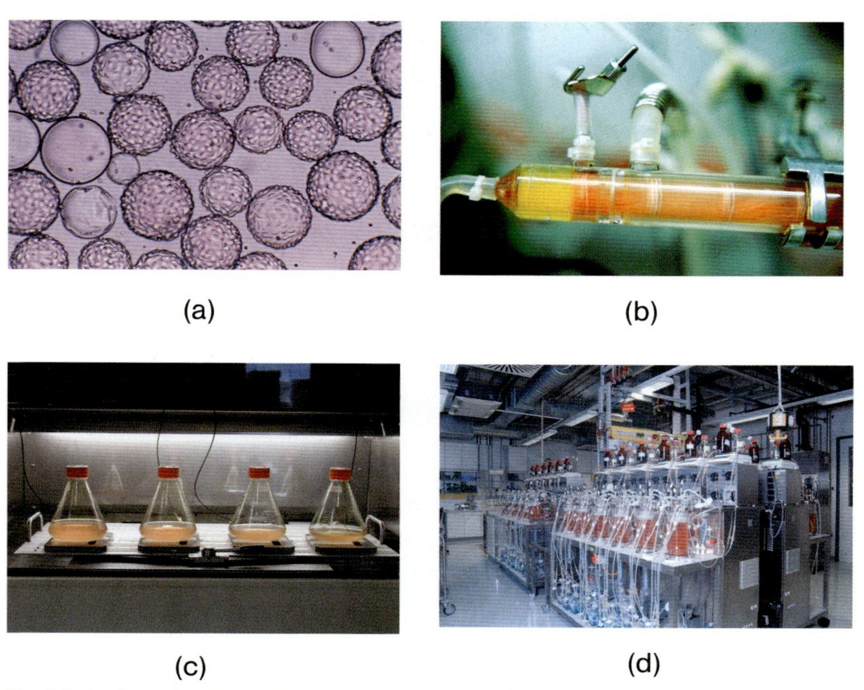

Fig. 9.2 *In vitro* cultivation systems: (a) microcarrier culture; (b) hollow fiber culture; (c) shake flask culture; (d) small-scale bioreactor culture.

Despite of the very high cell densities in fluidized and fixed bed cultivation systems, there are numerous disadvantages such as limited scalability, complex process technology, necrotic cells within multilayer cell populations, instability or degradation of multilayer cells in long-term cultivation, difficulties in cleaning for reuse of the systems and the need for numerous product harvests at short intervals (Preissmann et al. 1997).

Upstream processes with suspension cultures are the preferred format for industrial applications, such as the production of therapeutic antibodies in large quantities. For culturing mammalian cells in milliliter or low liter quantities, shake flasks (Fig. 9.2c) or spinner vessel systems are used. At a larger scale (5–20000 L) continuous stirred tank reactors (CSTR) are widely used (Fig. 9.2d). This type of fermenter is standardized, easy to scale-up, allows efficient cleaning-in-place (CIP) and sterilization-in-place (SIP) procedures, and can be flexibly used for a variety of different processes (Chu and Robinson 2001).

In addition to the fermentation equipment, the cell cultivation method has to be developed and optimized. The standard cultivation process is the "batch" process. A batch process is initiated by adding medium and a seed cell suspension of defined cell density to the bioreactor and subsequently cells are cultivated for a defined period of time under suitable conditions without further manipulations (Fig. 9.3).

A more sophisticated process format is the "fed-batch" process. Here, concentrated solutions of specific cell culture additives are added (fed) during the cultivation period (e.g. nutrients, growth factors, inductors or enhancers for product generation). Enhanced cell growth, higher cell densities, fewer nutrient limitations, and prolongation of culture viability are the key characteristics of a fed-batch process, generally leading to high product yields (Fig. 9.3).

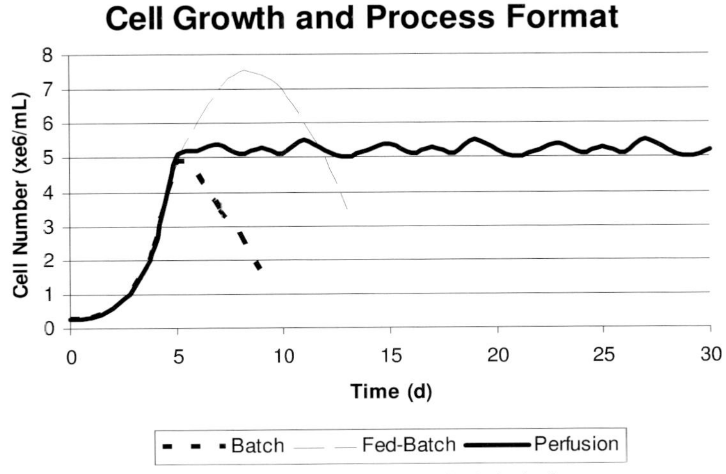

Fig. 9.3 Cell culture processes: batch process (thick dashed line), fed-batch process (thin dashed line), perfusion process (solid line).

In "continuous" processes such additions (feeds) are made continuously or periodically while an equal volume of culture fluid is removed simultaneously. Thereby a steady state can be established, leading to constant cultivation conditions and allowing continuous removal of cells, cell debris, and cytotoxic substances derived from apoptotic cells. Hence the run time of continuous fermentations can be very long, up to several months (Storhas 1994).

If a cell-free supernatant is removed from the bioreactor, this mode of continuous cultivation is called "perfusion" (Fig. 9.3). Very high cell densities can be obtained in perfusion cultures as well as high volumetric productivities (due to the large volume of perfused medium, while product concentrations themselves are rather low).

To date, the majority of manufacturing processes for therapeutic antibodies is based on the fed-batch format as they are characterized by short process development times, short generation time for individual product batches, robust and easily scalable mode of cultivation, fast and easy process validation and registration, and flexible production schedules in multipurpose manufacturing facilities (i.e. facilities in which several products are manufactured in campaigns concurrently, overlapping or sequentially).

9.2.4 Cell Culture Process Optimization

The key to high-performing upstream processes lies in a good combination of high-expression vector systems, robust cell lines capable of secreting high quantities of product per time, optimized media and additions (feeds), state-of-the art cell culture hardware, the appropriate cultivation method and optimized process design and operating conditions.

Technology platforms for all of these key elements can serve as a good basis for successful fast-track development of upstream processes for individual therapeutic antibodies. The developmental work leading to an optimized process is generally performed in small-scale systems. Crucial factors for optimization can be divided into cell parameters (such as seed density), physical parameters (such as temperature, pH, dissolved oxygen content, osmolality), biochemical parameters (such as concentrations of nutrients), process technology parameters (such as stirrer speed, characteristics of controllers), and mode of additions (such as times and quantities of feeds, compositions and concentrations of feeds).

Factorial design of experiments (DOE) is a valuable experimental and statistical tool for successful process optimization (Eriksson et al. 2000).

Recently, the described high-specific productivities were successfully translated into product titers of up to $6\,g\,L^{-1}$ for both CHO and NS0 cells. Figure 9.4 shows an example of how such high titers could be achieved through a process optimization program for a BI HEX CHO (Boehringer Ingelheim's proprietary high-expression platform) production cell. In this particular case, 11-day fed-batch cultivation led to approximately $4\,g\,L^{-1}$ of a monoclonal IgG1 subtype antibody.

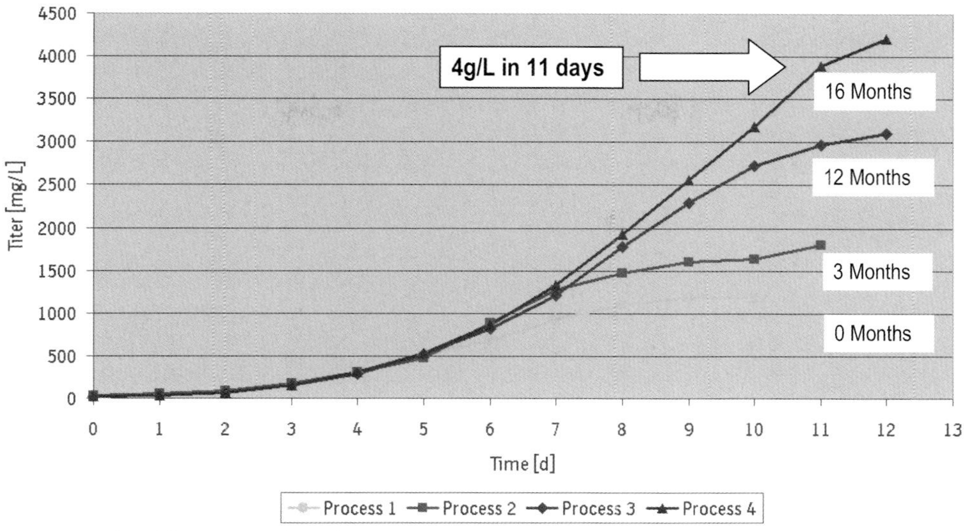

Fig. 9.4 High-yield processes through process optimization (0–16 months): $4\,g\,L^{-1}$ in 11-day fed-batch CHO BI HEX process (Boehringer Ingelheim's proprietary high expression platform).

9.2.5
Scale-up, Economy of Scale

Another challenge during process development is the scalability of results obtained in small-scale cultivations (e.g. 1 L scale) to the manufacturing scale (e.g. 10 000 L scale). Due to physical and technical reasons, not all characteristics and parameters from a 10 000-L bioreactor can be mimicked exactly in a 1-L bioreactor. But with technical expertise and experience and solid technical characterization of the bioreactor systems, results from the most relevant process parameters can be transformed from small scale to large scale and vice versa (Zlokarnik 2002). This means that the experiments at laboratory scale are meaningful for the large-scale systems and that the optimal small-scale process can be transferred successfully to the final manufacturing scale.

Nearly all therapeutic antibodies are applied in high doses. Treatments with antibodies also require multiple doses and/or long-term administrations, especially when targeted at tumor growth in cancerous diseases. Such proteins are therefore needed in large quantities, and consequently large manufacturing capacities are bound for the annual market supply of therapeutic antibodies. Cell culture facilities with large bioreactors (e.g. 10 000 L working volume) provide substantial cost benefits due to their economy of scale. Table 9.1 illustrates how over recent years the increase in titer moved production of antibodies from CHO cells towards the metric tonne-scale (yields in kilograms may be lower for proteins of small molecular weight). With the assumption that antibody product from a

Table 9.1 Producing antibodies from mammalian cells: towards the tonne scale.

Product	*mAb1*	*mAb2*	*mAb3*	*mAb4*
Titer ($mg\,L^{-1}$)	500	1 000	2 000	4 000
Yield (%)	60	70	70	70[a]
Batch (g)	3 750	8 750	17 500	35 000
Yield p.a. (kg)	375	875	1 750	3 500

Basis: 1 batch = 12 500 L. 100 batches/year.
a Assumption.

commercial $4\,g\,L^{-1}$ process could be purified with a yield of 70%, a production site that runs 100 successful batches of such a process annually at 12 500-L scale could supply the market with more than 3 tonnes of antibody each year (Table 9.1). However, it is important to note that the high titers of current state-of-the-art mammalian cell culture processes remain a challenge for establishing economical downstream (purification) processes at large scale.

9.2.6 Harvest

As antibodies expressed from mammalian cells contain signal peptides within their primary amino acid sequence they are secreted into the culture medium during the process. Therefore the product harvest at the end of the cultivation process occurs by simply separating cell culture fluid from the cells. In general, the cell separation method should be gentle to minimize cell disruption to avoid the release of proteases and other molecules that could affect the quality of the antibody product (Berthold and Kempken 1994).

In principle, three methods are used for harvest of products from mammalian cell cultures: filtration, centrifugation, and EBA (expanded bed adsorption).

Filtration can be done in a dead-end modus (static filtration) or in a tangential flow modus (dynamic filtration). If the cell suspension is filtered directly through a dead-end filtration system, a cascade of several filter steps is needed. The preferred system involves pre-filters with sieving and adsorption effects such as lenticular filters used to separate the cells, cell debris, and large particles. This is followed by clarification filters and optionally by sterilizing-grade filters to remove the particle load for the subsequent downstream processing. The latest developments of such systems have resulted in a combination of all these steps in a single filter cassette.

The separation of cells and large particles can be enhanced by tangential flow filtration (TFF). Here, the cell suspension is fed into hollow-fiber or flat plate microfiltration membrane modules and recirculated into the bioreactor (Fig. 9.5a).

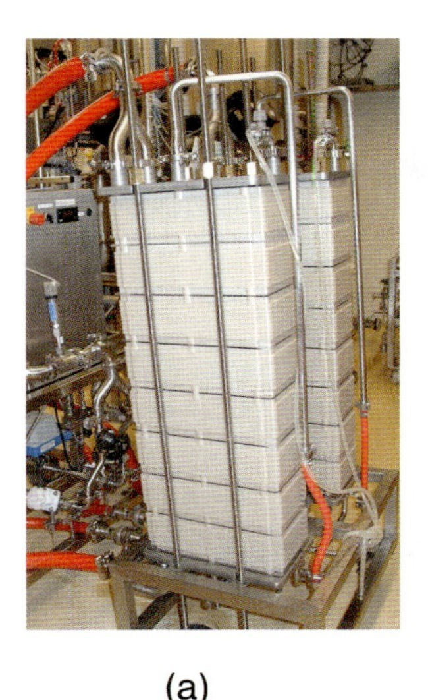

(a) (b)

Fig. 9.5 Harvest systems for mammalian cell cultures: (a) tangential flow filtration system, (b) disk stack centrifugation system.

Cells are pumped at high flow rates through the TFF system and concentrated approx. 10-fold, while the product-containing culture fluid can pass the membranes and is collected in a harvest hold tank (van Reis et al. 1991). To remove cell debris and small particles from the TFF filtrate, an additional dead-end clarification step is needed.

An attractive alternative to TFF membrane filtration is centrifugation. Standard disk stack centrifuges can be modified to allow gentle handling of the fragile mammalian cells and protein products (Tebbe et al. 1996). Centrifuges can be cleaned in place (CIP) and can be used for multiple products. The centrifugation step is followed by dead-end clarification which is similar to the clarification step behind a TFF system. The major advantage of centrifugation over TFF-based harvest regimes is the avoidance of problems such as membrane clogging and fouling (Fig. 9.5b).

EBA offers another alternative to filter-based separation. It is a chromatographic method which combines three unit operations (the separation of cells, the subsequent ultrafiltration/diafiltration of the cell culture fluid, the first chromatography step for product capture and purification) into one single step (Blank et al. 2001). The cell suspension is floated into the bottom of a chromatographic column which is operated as a fluidized bed (instead of a fixed bed as in regular chromatography processes). The product binds to the chromatographic matrix while the cells and particles pass the EBA column. The product can be eluted with high purity in the same way as in conventional fixed-bed chromatographic columns.

9.3
Downstream Processing

Most critical issues in downstream processing are the overall yield, sustained product quality, and feasibility in pilot- and large-scale production (Werner 2005). Therefore, a sophisticated purification strategy has to be developed in order to meet predefined criteria for purity, quality, efficacy, and safety of a therapeutic antibody as well as for process economy.

In general, the protein of interest represents only 1–40% of the total protein content of a mammalian production cell, the rest being a complex mixture of unwanted compounds including protein impurities, lipids, carbohydrates, nucleotides, fermentation ingredients (growth hormones, vitamins, trace elements), cell particulates, and, to a large extent, water. Moreover, adventitious agents such as virus and bacterial endotoxins may contaminate the target antibody solution and must be eliminated for therapeutic application by specific steps for removal and inactivation. Today, product purity and safety is achieved by a combination of various chromatography and dia- and ultrafiltration principles in conjunction with appropriate buffer systems and chemical or physical virus inactivation methods.

From a biochemical point of view any treatment during purification may exert stress on the overall three-dimensional structure of the protein, due to drastic changes in pH values, protein or salt concentrations, buffers or solvents, and by shear forces at liquid stream and surface interfaces. Such stress conditions may result in denaturation or aggregation of the antibody with losses in yield and efficacy. It is therefore necessary to monitor product quality and functionality during downstream processing by appropriate and fast analytical tools.

With regard to economy a limited number of robust process steps are focused on and the design of the process must enable up-scaling to pilot and production scale, that means processing of 400–2000 L and 12 000 L fermentation volumes, concomitant with suitable process turnover times. Finally, a crucial aspect of process development is that product quality and the production process have to comply with regulatory requirements.

The following sections describe and discuss challenges and solutions for the purification of efficacious, safe therapeutic antibodies with emphasis on requirements for up-scaling, economy, and regulatory acceptance using platform technologies. Future trends, such as affinity ligands, will be mentioned at the bottom of the article.

9.3.1
Platform Technologies for Downstream Processing of Monoclonal Antibodies

Therapeutic antibodies are frequently used at high doses and, therefore, pressure for development of robust and economic purification processes is an increasing demand. Currently, downstream processing of mAbs focuses mainly on the application of platform technologies such as represented in Fig. 9.6. Four main sections can be distinguished: primary recovery, viral clearance, purification and

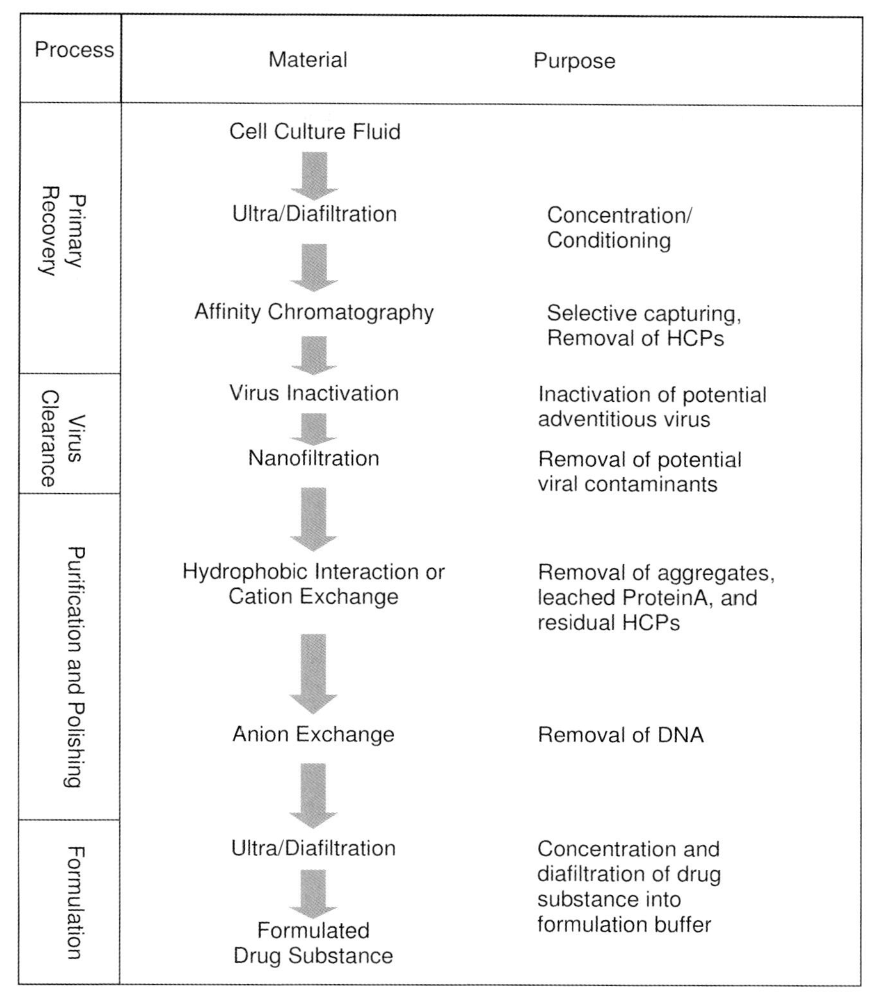

Fig. 9.6 Scheme of a standard platform purification process for therapeutic monoclonal antibodies.

polishing, and formulation of the drug substance. Neither the order in which the individual steps are carried out nor the applied principles used have to be applied as shown; each mAb requires a unique treatment for optimum results.

9.3.2 Primary Recovery

9.3.2.1 Ultra-/Diafiltration (UF/DF)

For the purposes of concentration and conditioning crossflow units are commonly used. Large fermentation volumes are readily reduced by ultrafiltration (UF), and during diafiltration (DF) ideal and constant starting conditions for the following capture step can be adjusted with regard to pH, conductivity, or buffer strength. Membranes can be operated quickly and with no significant loss of product. Time is an issue in this regard since high turnover rates of proteases can degrade significant amounts of product during the comparably slow loading procedure to the capture column.

9.3.2.2 Affinity Chromatography

Affinity chromatography is a very powerful capture technology. For purification of therapeutic antibodies, chromatography resins with Protein A or variants thereof are widely spread. Protein A has strong affinity to the Fc part of antibodies with an affinity constant K_D of 70 nmol L^{-1} (Li et al. 1998). Product purities of more than 95% can be achieved in one step (Follman and Fahrner 2004). The highly specific interaction takes place in a broad pH range and is nearly independent of the conductivity of the loading buffer. However, Protein A affinity chromatography is the most expensive operation unit in downstream processing with up to 50% of total costs (Fig. 9.7).

9.3.3 Virus Clearance

Virus clearance methods positioned downstream of the bioreactor are a regulatory requirement for product release and generally comprise two steps: virus inactivation and virus removal. Viral contaminants can enter production from a variety of sources. Rodent cell lines, such as Chinese hamster ovary (CHO) or mouse (NS0) cell lines, are routinely used in the production of mAbs, raising the possibility of contamination with human and rodent viruses, for example parainfluenza or reovirus. According to International Conference on Harmonisation of

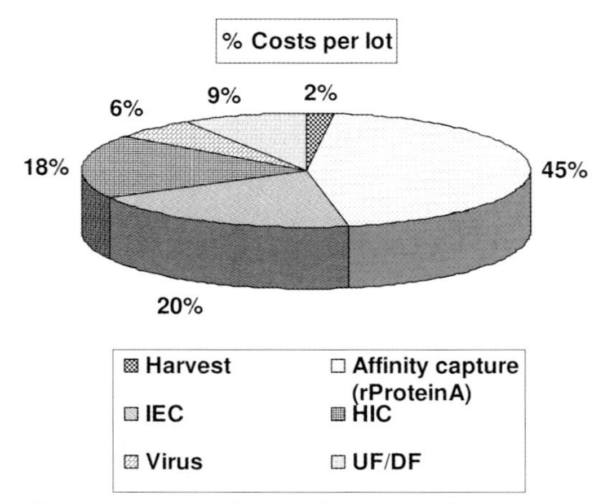

Fig. 9.7 Major cost drivers of raw materials used in downstream processing are chromatography resins.

Technical Requirements for Registration of Pharmaceuticals for Human Use (ICH) guideline Q5A (http://www.emea.eu.int/index/indexh1.htm), processes must be validated to remove or inactivate four to six orders of magnitude more virus than is estimated to be present in the starting material. Additionally it is recommended that a purification scheme harbors at least two chromatography steps, both having the capability of reducing potential virus load.

Effective virus inactivation can be achieved by:

- Chemical inactivation (extreme pH; pH < 3.9 or pH > 13)
- Heat (e.g. by microwave)
- Photochemical inactivation (UV irradiation)
- Solvent-detergent inactivation.

Methods for virus removal are:

- Filtration (nanofiltration)
- Chromatographic separation.

Effective virus inactivation and removal has to be proven experimentally using a validated scale-down model of the actual purification process. Virus solutions are spiked to the samples and analyzed before and after each process step. A selection of the following viruses is commonly used: murine leukemia virus (MuLV), pseudorabies virus (PRV), reovirus 3 or parvovirus (MVM, PPV, BPV).

9.3.4
Purification and Polishing

In many cases the antibody's purity and monomer content exceed 95% after affinity chromatography. However, additional purification steps and final formulation are necessary to ensure high and defined product quality as well as product safety to suit administration into humans. Polishing of the target antibody should make use of orthogonal chromatographic principles to remove any residual impurities or adventitious agents interfering with product quality. Defined low levels of residual HCP (host cell proteins), leached rProteinA and DNA must be achieved by appropriate downstream processing steps in order to be accepted by regulatory authorities.

9.3.4.1 Hydrophobic Interaction Chromatography

Hydrophobic interaction chromatography (HIC) provides a powerful tool for depletion of protein aggregates, leached rProteinA and HCP. Gel filtration (size-exclusion chromatography (SEC)), often described in textbooks, is not supposed to be a feasible alternative for production scales of more than 1–2 g of the desired protein (Table 9.2). HIC is able to cope with over 30 g antibody per L resin with yields beyond 85%, high flow rates, 10-fold lower costs, moderate initial product concentrations, and no dilution of product. Even higher product throughput (processing time) is feasible with HIC if used in negative mode (i.e. selective binding of unwanted constituents while the antibody appears in the flowthrough).

Table 9.2 Comparison of process and economic aspects of hydrophobic interaction chromatography (HIC) and gel filtration or size-exclusion chromatography (SEC) for production of 100 g monoclonal antibody.

	HIC	*SEC*
Column size (L)	4	40
Column load	25 $g L^{-1}$	25 g per cycle
Initial product concentration ($g L^{-1}$)	10	25
Column dimension (cm)	14 × 26	20 × 120
No of cycles	1	4
Resin cost (euros)	7200	80000
Flow rate		
$cm h^{-1}$	150	30
$ml min^{-1}$	400	167
Yield (%)	85	80
Buffer volume[a] (L)	18	160
Process time[a] (h)	0.75	17
Final product concentration ($g L^{-1}$)	12	5

a Only bind/elution, no equilibration, wash, regeneration/sanitization.

One drawback of HIC is that high salt concentrations are necessary for protein binding, bearing the risk of aggregation.

9.3.4.2 **Ion Exchange Chromatography**

Ion exchange chromatography (IEC) is a very efficient method for numerous aspects of downstream processing. Today's IEC resins have high binding capacities of 50 mg antibody per mL resin or more and they can be run in either bind/elute (positive mode) or flowthrough mode (negative mode). Scale-up of column dimensions at equal resin performance is feasible and costs are moderate compared with affinity or hydrophobic resins. Sanitization can be achieved by strong acids and base solutions (e.g. 1 $mol L^{-1}$ HAc and 1 $mol L^{-1}$ NaOH). A feature of IEC is that samples have to be applied at low conductivity, often necessitating a UF/DF step in advance.

Cation exchange chromatography Since therapeutic antibodies often have basic pI values of 8–9.5, cation exchange (CIEX) chromatography is applied in positive mode purifications at physiological or mild acidic buffer conditions. CIEX chromatography is capable of depletion of HCPs, leached rProteinA, aggregates, and fermentation ingredients, giving rise to nonproduct-related precipitation. Due to its capabilities, CIEX chromatography is also attractive as a primary capture step; however, product enrichment and depletion of HCPs is normally less efficient than by affinity chromatography and depends strongly on the type of protein applied.

Anion exchange chromatography and removal of DNA Although suited for purification, binding of antibodies to anion exchange (AIEX) resins often requires pH values above 9–10 and therefore bears an elevated risk for deamidation of Asn residues (Creighton 1996).

Among the impurities to be eliminated during the downstream purification process, one component of major interest for safety is residual host cell DNA. It is necessary to guarantee that this impurity is reduced to a level of less than 100 pg/dose in the product administered to a patient. Due to its high content of phosphate, DNA is highly negatively charged at physiological pH and thus well suited to be removed quantitatively by binding to AIEX ligands.

Today's purification processes often make use of membranes functionalized with AIEX ligands on their surface. Membranes such as Sartobind Q are designed as single-use entities, making regeneration procedures superfluous. Pricing is moderate and further advantages are higher yields and savings in total labor time, buffer, and buffer tank capacities, which are usually needed for sanitization or for validation of a chromatography column. Furthermore, the risk of cross-contamination is excluded. The binding capacity of AIEX membranes is easily sufficient for quantitative removal of DNA, since capture and/or midstream purification steps normally leave only low DNA concentrations of 50–500 pg mg^{-1} protein (Walter 1998). The filter area is dictated by requirements for flux rather than DNA-binding capacity. Finally, AIEX membranes are amenable to scale-up from lab to production scale.

In some cases AIEX chromatography will be preferred as a polishing step if removal of DNA and residual interfering impurities can advantageously be combined.

Validation of DNA removal Regulatory authorities, like the US Food and Drug Administration (FDA) or the European Agency for the Evaluation of Medicinal Products (EMEA), demand to monitor DNA depletion at various steps of the downstream process. For this purpose, radiolabeled DNA is spiked to protein samples and depletion factors are evaluated in a validated scale-down purification model. In addition, binding capacity of the main DNA removal step in the downstream purification scheme is determined to ensure that the residual cellular DNA content in the final product is reduced to a defined level of no more than 100 pg per dose. A major drawback of AIEX chromatography compared with single-use membranes is that repeated cycles have to be run in order to ensure safe reuse of a chromatography column.

9.3.5
Final Formulation

As a final step within downstream processing, formulation is conducted by an appropriate UF/DF system. Membranes with low protein binding, such as polyethersulfone or regenerated cellulose, are applied. Membranes are not resistant to extensive sterilization or sanitization with hot steam or NaOH. Therefore, sterility of the formulated bulk is taken special care of by determining the levels endotoxins and bioburden.

9.3.6
Integrated Downstream Process Development

With regard to an ambitious product development timeline an integrated development strategy (Fig. 9.8) is mandatory, which means that overlapping rather than sequential development activities are realized in upstream, downstream, and pharma development units. Support and early feedback to upstream colleagues is important with regard to product and process quality. Clone selection finally giving rise to the valuable production clone is also assisted by the downstream department. Low milligram quantities of the target protein being obtained from small culture samples are indispensable for the assessment of biochemical and biophysical protein properties and initial studies on purification and filtration steps. A process is deemed feasible if a series of process steps jointly result in reasonable yield, product quality, and a scaleable format. Development during the feasibility study encompasses purification of milligram amounts up to 1 g of target protein.

Once a feasible downstream process has been developed, up-scaling is performed to challenge robustness, which means the reproducibility of the process in larger scale at constant product quality, as well as economic aspects, for example processing times and buffer volumes. A fine-tuning process is applied to purify several grams to 100 g quantities of therapeutic protein in order to support toxicology and preclinical studies or, under Good Manufacturing Practise (GMP) regulations, clinical phase I and II studies. Production scale normally aims for delivery of hundreds of grams to several kilograms in the case of therapeutic mAbs. Facility constraints, such as space or turnover times, become an important issue and often intensive development effort is encountered to fulfill the goals for economy, quality, and regulatory demands.

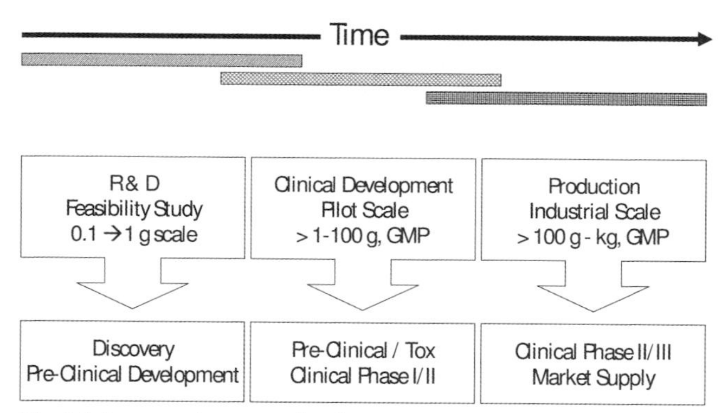

Fig. 9.8 Integrated process development strategy.

9.3.7
Future Perspectives

Time constraints and increasing demands on product quality require rapid and comprehensive methods for the development of downstream processes. The competitive market environment requires short- and long-term solutions to cope with these challenges. Currently, state-of-the-art techniques have almost been improved to their theoretical optima and possible alternatives have not yet been developed to a degree of maturity applicable to biopharmaceutical production scale.

Monolithic resins (Zmak et al. 2003) could overcome the diffusion limitations of traditional bead resins, but the cost and size of monoliths are currently out of scope for production scale (Branovic et al 2003). Simulated moving bed (SMB) is a powerful approach with respect to large-scale applications and low buffer consumption (Nicoud 1998). The drawbacks of this technology are the high number of valves and pumps involved, bearing an enhanced risk of malfunction and germ contamination during the production process. The power of continuous operation of SMB renders the definition of a product lot difficult.

Novel affinity ligands on resins or innovative tags on mAbs could overcome the main disadvantages of Protein A resins: high cost and moderate stability. The cost structure may change with competitive scaffold proteins or small chemical or peptidomimetic ligands having comparable properties to Protein A. The advantages of small ligands are that they are cheap and easy to synthesize in large quantities, they have high chemical stability, resistance against proteases and no leaching. Leaching is of relevance because the removal of these process-related impurities has to be demonstrated by biochemical assays prior to release for human use.

Protein tags, such as His-, Strep-, Intein- or GST-tags – to mention just a few – are well characterized and widespread in molecular biology, in research and diagnostic applications. Some of them, as short as six amino acids, can be inserted within a polypeptide chain to undergo chemically induced self-cleavage. However, these modifications raise concerns with regard to immunogenicity which has to be addressed during preclinical and clinical development.

Ambitious development timelines with increasing demands for throughput and product qualities require automated parallel screening of resins and chromatography conditions with implemented fast process analytics. Miniaturized robot platforms provide an excellent tool to address these needs.

9.4
Formulation Development

9.4.1
Challenges during Early Formulation Development Phase of Biopharmaceuticals

It is a prerequisite that a marketable protein formulation must be safe to administer, inducing no or minimal local irritation, meets the specific clinical and delivery requirements, and that its physical, chemical, as well as biological stability is given during the recommended product shelf-life (LeHir 2001; Langguth et al. 2004). Therefore, the development of an optimal protein formulation requires a sound understanding of the protein's physicochemical and biophysical properties, evaluated by sensitive analytical tools.

During early process development, challenges with regard to protein stability are encountered with special emphasis on protein downstream processing, where protein degradation as well as physical instabilities (e.g. aggregation) have to be avoided. With this in mind this section focuses on protein stability in a liquid formulation and analytical characterization tools in the early development phase. Depending on the chosen formulation and environmental conditions for a certain process step during the manufacturing of biopharmaceuticals, stability of the protein may be affected leading to a loss of structure by conformational change, resulting in the formation of, for example, aggregates. The appearance of aggregates is commonly observed in highly concentrated protein formulations due to the higher probability of protein–protein interactions.

Most proteins adopt a secondary, tertiary, or even a quaternary structure which is essential for biological activity. Various forces are involved in determining protein-folding pathways and thus overall protein stability. The forces involved in the formation of higher protein structure include hydrophobic and electrostatic interactions, the formation of hydrogen bonding networks, covalent bonds, and van der Waals interactions. Any event (e.g. shear stress during sterile filtration) that upsets the sensitive balance between these forces and interactions can lead to conformational changes, for example the exposure of hydrophobic areas inducing reduced solubility and thus an increased tendency to form aggregates.

9.4.2
Strategies and Analytical Tools for Rapid and Economic Formulation Development

In the early development phase of an optimal protein formulation various test protocols are used to get a deeper understanding of the chemical and physical stability/instability of the investigated protein formulation. An overview of physicochemical test methods is shown in Table 9.3. A key parameter to be investigated is induced protein denaturation as a function of the temperature applied. Other stability-indicating studies include shaking, shear, and/or freeze/thaw stress studies in order to obtain information about potential degradation pathways.

Sensitivity to oxidation and light also needs to be considered. Important data are derived from accelerated stress stability studies at elevated temperatures, for example, storage for 4–12 weeks at 25 and/or 40 °C (Waterman and Adami 2005). To facilitate increased intramolecular protein interactions various excipients are screened using the methods summarized in Table 9.3 (class II) with the aim of developing formulations inducing highest protein stability.

Table 9.3 Analytical tools and physicochemical methods used for the development of an optimal protein (e.g. antibody) formulation.

Class I	*Class II*	*Class III*
Appearance, color, clarity	Infrared spectroscopy	X-ray
Rheology	Raman spectroscopy	Nuclear magnetic resonance
Surface tension	Circular dichroism	Atomic force microscopy
Analytical centrifugation	Fluorescence spectroscopy (IF, EF)	Electron microscopy
Turbidimetry	Scattering techniques (DLS, SLS, RALS, light, neutron)	Mass spectrometry (MALDI-TOF, MS-MS)
Optical microscopy	Electrophoresis (SDS-PAGE, IEF, CE)	Chemical analysis (peptide mapping, sequencing, AA analysis)
	Chromatography (RP-HPLC, HP-SEC, IEC)	
	Calorimetry	
	Surface plasmon resonance	

RP-HPLC, reverse-phase high-performance liquid chromatography; HP-SEC, high-performance size-exclusion chromatography; IEC, ion-exchange chromatography; IF, intrinsic fluorescence; EF, extrinsic fluorescence; MALDI-TOF, matrix-assisted laser desorption/ionization time of flight; MS-MS, tandem mass spectrometry; DLS, dynamic light scattering; SLS, static light scattering; RALS, right angle light scattering; SDS-PAGE, sodium dodecyl sulfate polyacrylamide gel electrophoresis; IEF, isoelectric focusing; CE, capillary electrophoresis; AA, amino acid.

Such studies are aimed at developing a protein formulation suitable for toxicological studies and for early clinical studies. Further development programs have to be performed for the development of the final formulation appropriate for pivotal clinical studies and market supply due to the fact that the prerequisites for the formulation, namely dosage form, primary packaging, delivery device and sometimes even the application route may change during the clinical development. However, in order to reduce cost in the early development phase, physicochemical and biophysical test methods are used for rapid formulation screening and thus to identify formulation conditions and potential excipients which may stabilize the protein in solution. The analysis of protein properties in formulations requires often specific tailor-made analytical methods able to detect small differences against a more or less complex formulation background, and consequently provide information on the protein structure of various formulations.

Table 9.3 summarizes the most important analytical tools used for the development of an optimal protein formulation. These techniques can be separated into classes I, II, and III. The analytical tools in class I give information on protein macroscopic bulk properties (i.e. on the properties of protein populations, such as surface tension or viscosity) (Liu et al. 2005), whereas the techniques in class III (e.g. nuclear magnetic resonance, X-ray techniques) are more appropriate for the elucidation of specific structural alterations and detailed information at atomic level resolution. Methods in class II are appropriate for the characterization of protein structures at the molecular/microscopic as well macroscopic level. For rapid formulation screening, class I and II methods are most appropriate (Rouessac and Rouessac 2004, Kellner et al. 2004).

The specific analytical techniques used for formulation screening depend on the protein behavior as well as on the dosage form and protein concentration in the formulation. The challenge is to use techniques that have the desired sensitivity, allowing the detection of even small changes of the protein, and also to ensure that the techniques do not affect the composition of the sample being investigated. As an example, depending on the analytical method used, the sample cannot be analyzed at its initial protein concentration. As a consequence, the protein solution has to be diluted, and this may directly interfere with, for example, the formation of particles in the formulation. Using fluorescence spectroscopic techniques for the analysis of protein formulations the protein concentration is usually between 0.05 and 0.2 mg mL^{-1}. This allows the investigation of formulation of extremely diluted samples. However, if stability information are requested for protein formulation with protein concentrations between 5 to at least 100 mg mL^{-1}, other techniques such as infrared spectroscopy can be used (Garidel and Schott 2006).

In early formulation development the avoidance of protein aggregation is a challenge, because it may result in significant product loss during downstream processing. In addition, protein aggregates present in the drug product may compromise the safety of the product (e.g. by the generation of immunological responses) (Chi et al. 2003). Therefore, various analytical tools summarized in Table 9.4 have been developed to investigate soluble as well as insoluble aggregates in solutions (Sine 2003; Tatford et al. 2004; Fraunhofer and Winter 2004). Spectroscopic techniques (Hollas 2003) such as circular dichroism and infrared spectroscopy are used to determine the microscopic causes for protein aggregation (Manning 2005).

It is important to get a deeper understanding of the protein's degradation pathway in order to be able to develop a strategy for its stabilization in solution. One example to illustrate this approach is shown in Fig. 9.9. In this example, an increase of beta-sheet secondary protein structure is observed in relation to protein aggregation.

There are indications that the aggregation mechanism of certain proteins involves two steps: first a transition from random coiled and helical secondary protein structures to beta-sheet structures and in a second step, aggregation of the beta-sheet structures. In order to avoid such an aggregation, environmental conditions including the presence of certain excipients have to be found, enabling the stabilization of the helical secondary protein structure to reduce and even avoid the transition to beta-sheet structures. However, it has also been shown that proteins (IgG1) may aggregate without obvious changes in the secondary structure (Schüle et al. 2004, 2005). Therefore, various test methods are used together to understand the aggregation mechanism. Furthermore, these exam-

Table 9.4 Analytical tools for the characterization of the formation and presence of particles/aggregates in protein solutions.

Method	*Individual technique*	*Remarks*
Visual	"Eye"	Visual particle
Particle counting	Electrical impedance	Particle number per volume unit, size, quantification, micrometer scale
Microscopy	LM	Shape and size determination, quantification, micrometer scale
	EM	Shape and size determination of insoluble aggregates, nanometer scale
	AFM	Shape and size determination of insoluble aggregates, morphology, roughness, nanometer scale
Light scattering	SLS	Size and shape
	DLS	Size and relative distribution of soluble aggregates
	LS/turbidimetry	Size estimation and relative distribution
	RALS	Aggregation process, soluble/insoluble particles
Rheology	DSR	Gelation characteristics, shear module, protein–protein interaction
Chromatography	HP-SEC	Size estimation and quantification, soluble, small particles, quantification
	RP-HPLC	Aggregate iso-forms, quantification
	FFF, AFFF	Size determination and quantification, small and large particles
Electrophoresis	SDS-PAGE	Size estimation and nature of aggregate formation, quantification
	Native PAGE	Aggregation process and mechanistic formation
Calorimetry	DSC	Thermally induced protein unfolding and aggregation
Centrifugation	AC	Size, mass, and shape estimation
Spectroscopy	FTIR	Aggregation process and mechanistic, change in protein secondary structure
	CD	Aggregation process, change in protein secondary and tertiary structure
	Fluorescence	Aggregation process
	UV-VIS	Detection of soluble and insoluble aggregates
	NMR	Aggregation formation

LM, light microscopy; EM, electron microscopy; AFM, atomic force microscopy; LS, light scattering; DSR, dynamic shear rheometry; FFF, field flow fractionation; AFFF, asymmetric field flow fractionation; DSC, differential scanning calorimetry; AC, analytical centrifugation; FTIR, Fourier transform infrared; CD, circular dichroism; UV-VIS, ultraviolet-visible; NMR, nuclear magnetic resonance.
For an overview and further details see Kellner et al. (2003).

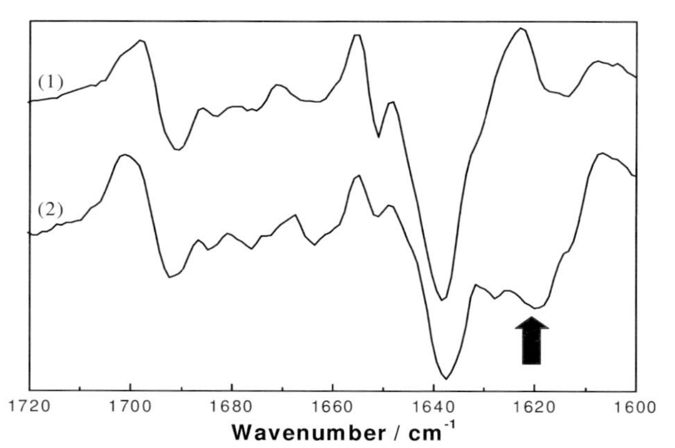

Fig. 9.9 Second derivative infrared spectroscopy of a liquid antibody (IgG1) formulation at a concentration of $10\,mg\,mL^{-1}$. (1) Infrared spectrum of the initial protein conformation. (2) Infrared spectrum of the protein formulation after a 3-day shaking stress at room temperature. The appearance of the formation of intermolecular antiparallel beta-sheet structure which is characterized by a low-frequency band around $1620\,cm^{-1}$ (marked by an arrow) is indicative for the formation of protein particles and thus protein aggregates (Garidel 2004; Manning 2005).

ples show that the formation of protein aggregates may be induced by different pathways.

As mentioned above, one of the most common mechanisms leading to loss of bioactivity of a protein is the time-dependent formation of aggregates and loss of solubility (Arakawa and Timasheff 1985; Schein 1990). Protein solubility in a certain formulation is often unknown. Various methods are described in the literature to determine protein solubility but most of them have strong drawbacks and can only be employed for certain cases. It is a challenge to obtain highly concentrated protein solutions without inducing protein degradation. One approach is to slowly concentrate a given protein solution (e.g. by ultrafiltration) using membranes with low protein binding capacity. Using this technique it is possible to saturate the solution with the concomitant appearance of a solid phase which, in most cases, represents protein crystals. In this case, the determined protein solubility is that of the crystalline form of the protein which could differ substantially from that of the amorphous solid (Arakawa and Timasheff 1985).

Another approach for the determination of protein solubility uses an inert, extraneous excipient like polyethylene glycol to induce protein precipitation (Middaugh et al. 1979). Based on thermodynamic considerations, the protein solubility is determined by plotting the logarithm of protein solubility versus the precipitation excipient concentration, which is often linear. Solubility is derived by extrapolation to zero precipitation excipient. However, this method should be used with care and is best employed in a comparative manner. It should also be considered that the derived protein solubility data refer to the particular solid state of the protein obtained by excipient precipitation.

9.4.3
Stabilization of Liquid Protein Formulations by Excipients

Besides its concentration, other factors that have a large impact on the stability of a protein in solution are formulation conditions such as pH, ionic strength, presence or absence of solutes, and storage conditions. The best choice of pH for a protein formulation depends on the chemical structure and the amino acid sequence of the protein. The most common chemical degradation pathways in proteins are deamidation, isomerization, cyclic imide formation, cleavage, oxidation, pyroglutamate formation, beta-elimination, and crosslinking. The kinetics and activation energies of a number of degradations are pH dependent. The hydroxinium ion is directly involved in specific acid-catalyzed reactions (e.g. Asp-Pro cleavage, direct Asp-Gly hydrolysis, or succinimide formation at Asn-X residues). Deamidation of Asn is more pronounced at alkaline pH, because this reaction is base catalyzed (Ahern and Manning 1992). The degree of protonation of a protein also influences its overall polarity (e.g. changes in ionic and/or dipole–dipole interactions) and, thus, protein–protein interactions and solubility (pH dependent).

The choice of pH and formulation components also plays a critical role in the maintenance of glass container (primary packaging) integrity, and leaching of extractables from rubber stoppers. An alkaline pH could promote dissolution of silica from glass, for example, resulting in pH changes in the formulation during long-term storage.

Excipients for the prevention of protein degradation and aggregation of therapeutic antibodies seem to exert their effects via a number of different pathways. These include:

- Binding of the excipient to the protein and stabilization of the native conformation. Examples of such direct protein–excipient interactions have been observed in polysulfates and cyclodextrins.
- Preferential exclusion from the protein surface which allows a preferential protein hydration. According to this mechanism the protecting role of excipients such as polyethylene glycol, amino acids, polymers, polyols, sugars, or other substances with multiple hydroxyl groups, has been described.
- Prevention of protein–protein or protein–surface interactions or increasing solubility of the aggregates. Nonionic surfactants are believed to function according to this mechanism (Cleland et al. 1993).

Many of the abovementioned excipients (Ahern and Manning 1992) are able to increase protein stability and/or inhibit irreversible aggregation and thus maintain protein integrity. These effects can be explained by an increased difference in free energy between the native and the denatured states, stabilizing the native state by making it even more energetically favorable than any unfolded state. Stabilizing excipients acting as cosolvent increase the protein stability by inducing a preferential hydration of the protein. This effect reduces the frequency of protein–protein interactions and limits their ability to initiate nucleation and so protein aggregation. If a protein is unstable in liquid formulation, the formulation can be frozen in order to minimize degradation kinetics or to develop a solid, dry formulation (e.g. freeze-dried, spray-dried formulation). However, other degradation pathways and stresses have then to be taken into account and addressed specifically.

In summary, the successful development of an optimal protein formulation depends on an intimate understanding of the protein's physicochemical and biological characteristics, including chemical and physical stability (Frokjaer and Otzen 2005). Because each protein molecule (and even mAbs) acts as a single individual with a large number of often underestimated specific properties, only general strategies for the development are available. The formulations of most biopharmaceuticals are therefore developed case by case, taking the individual demands for clinical studies and marketing as well as the application route into account.

9.5 Protein Characterization and Quality Control Testing

9.5.1 Protein Characterization

Initial protein characterization typically starts during preclinical development and is further intensified during clinical development phases. Characterization activities for investigational new drug applications generally are limited in their extent but should address the influence of product heterogeneity on biological activity. For example, the role of glycosylation concerning the efficacy of an mAb is product-dependent, thus the required depth of carbohydrate analysis will be dictated by the product. For marketing applications, all possible aspects of protein heterogeneity (protein variants) and possible influences of the manufacturing process on the physicochemical and biological properties in relation to the biological activity should be elucidated. The key to characterization is founded in the appropriate combination of the different methods that analyze the sample from substantially orthogonal and independent directions (Harris et al. 2001).

9.5.1.1 Protein Variants

Characterization of the protein variants comprises identification of the modification, including the respective site(s) and the mechanism of accumulation. Thus a variant may be categorized as a product-related substance or a product-related impurity, which results in an assessment of whether routine monitoring of the variant has to be performed and limits established. Variants with decreased biological activity would be classified as product-related impurities. Variants comparable to the unaltered protein in its potency would be regarded as a protein-related variants or product characteristics unless efficacy and safety of the protein is impaired (ICH S6 guideline; CPMP/3097/02 2003). Generation of stress samples to create variant-enriched material is a common approach to identify sites most susceptible to degradation. The most common product variants observed in antibody production are described below.

Aggregates and oligomers Aggregation is a key issue underlying multiple deleterious effects, including loss of efficacy and immunogenicity. Size exclusion chromatography (HP-SEC) is most suitable for routine quantification of soluble aggregates and sodium dodecyl sulfate polyacrylamide gel electrophoresis (SDS-PAGE) detects covalent non-SDS dissociable aggregates. Orthogonal methods, such as sedimentation velocity analytical ultracentrifugation (AUC) (Laue and Stafford 1999), field flow fractionation (FFF, shortcomings, e.g. robustness; Schimpf et al. 2000), and light scattering (Liu and Chu 2002) are commonly used to ensure that HP-SEC is providing a complete characterization of aggregates, especially detecting very large species that might not enter the column and non-covalent, reversible, or weakly associated aggregates that might break up during analysis by HP-SEC. SDS-PAGE, and HP-SEC are commonly used for lot release, whereas AUC, FFF, and LS are carried out for characterization and cross-validation purposes, only. A more detailed characterization of aggregates the individual HP-SEC species (obtained by collection of the respective peaks from multiple injections) or AUC fractions could be analyzed by reduced and nonreduced SDS-PAGE in combination with liquid chromatography/mass spectrometry (LC-MS) techniques, N-terminal sequence analysis, denaturing HP-SEC, capillary electrophoresis (CE), Fourier transform infrared spectroscopy (FTIR), and circular dichroism (CD) spectroscopy (Andya et al. 2003).

Protein degradants/fragments Impurities such as degradants or fragments induced by physical stress (exposure to accelerated temperatures, low pH conditions and shear stress as a function of the processing conditions), for example, should also be investigated. For manifestation of internal cleavages, the protein might have to be converted into its reduced state since disulfide bonds may hold the fragments in place. Degradants/fragments can be analyzed by SDS-PAGE and further characterization of the respective bands by a combination of LC-MS and N-terminal protein sequence analysis (NTS).

HP-SEC under native and denaturizing conditions is suitable to determine the overall number of fragments present and to isolate them for further investigation. Resolution of fragments can be achieved by SDS-PAGE analysis and characterization can be performed by enzymatic cleavage of the respective bands in the polyacrylamide matrix or by blotting the bands from the gels onto PVDF (polyvinylidine difluoride) membranes, for example, and subsequent analysis by MS techniques (Eckerskorn et al. 1992; Patterson and Aebersold 1995).

Deamidation, oxidation, disulfide pairing, N- and C-terminal variants and other less common modifications Enzymatic cleavage (e.g. by trypsin) in combination by mass spectrometry (Nguyen et al. 1995; Perkins et al. 2000) is sensitive to primary and secondary protein structure and can detect the following product modifications:

- Perturbations due to the formation of disulfide mispairings and scrambling (additional peptides with specific molecular weight: Gorman et al. 2002) and formation of trisulfides (+32 Da: Andersson et al. 1996).
- Deamidation, induced by high pH, usually occurs at the sequences of Asn (Gln) followed by glycine (−1 Da: Stephenson and Clark 1989; Zhang et al. 2002) and results in the formation of a 3 : 1 ratio of isoaspartate/aspartate. Formation of succinimide variants (−18 Da) has been detected in proteins (Teshima 1991).
- Methionine oxidation (+16 Da).
- Cyclization of N-terminal glutamate and glutamine residues to pyroglutamate (−17 Da and −18 Da, respectively).
- Enzymatic cleavage of C-terminal lysines (−128 Da) by mammalian carboxypeptidases (Harris 1995) and possible subsequent enzymatic processing of the new C-terminal glycine (Prigge 1997) to result in amidation of the amino acid penultimate to the glycine residue (−186 Da).
- Posttranslational phosphorylation at serine, threonine, and tyrosine residues (+80 Da): PEGylation introduced at free amino groups to increase bioavailability (making MALDI-TOF without enzymatic digestion also feasible), carbamylation introduced during processing of the protein at high concentrations of urea and elevated temperatures (+43 Da), adducts as a result of protein processing (refolding buffer additives, e.g. cysteine), and drug conjugates with molecules such as maytansine (Wang et al. 2005).

Ion-exchange chromatography (IEC) is an especially simple and powerful technique to resolve and quantitate charge variants introduced, for example, by deamidation (cation-exchange chromatography is most suitable) or enzymatic C-terminal processing. Alternative methods are capillary electrophoresis-based methods (cIEF and CZE) (Good 2004) and isoelectric focusing (IEF) (Gianazza 1995).

Oxidation may be resolved by HIC and is also detected by IEC as a blurred profile or a shift of retention times. Structural integrity with respect to disulfide pairing can be analyzed by, for example, HIC, reverse-phase high-performance liquid chromatography (RP-HPLC), and differential scanning calorimetry (DSC). Integrity of the N-terminus can be assessed by N-terminal sequencing.

Carbohydrate heterogeneity To assess glycosylation heterogeneity a combination of several chromatographic and mass spectrometric techniques is commonly used (Sheeley et al. 1997; Mechref and Novotny 2002). The chromatographic techniques include enzymatic or chemical release of the *N*-linked oligosaccharides followed by fluorescence derivatization and a subsequent chromatographic or capillary electrophoretic separation. The thus released oligosaccharides can be characterized by comparison to known standards, the structural composition might be confirmed by exoglycosidase digestion studies and molecular masses can be determined by matrix-assisted laser desorption/ionization time-of-flight mass spectrometry (MALDI-TOF-MS).

To identify the site occupancy of the oligosaccharides, and determine their composition electrospray ionization time-of-flight mass spectrometry (ESI-TOF-MS) is performed on the enzymatically digested protein. For this purpose, several proteases are available typically used enzymes are trypsin, lys-C, AspN, Papain (to generate Fab and Fc fragments) or a combination thereof.

9.5.1.2 Overall Structural Confirmation (Higher Order Structure)

The secondary, tertiary, or quaternary structure of a protein can be impaired by any of the aforementioned chemical changes, or under the influence of other factors (e.g. temperature and pH). In depth overall structural characterization is usually performed prior to filing for a marketing authorization application. For this purpose the following methods can be applied: CD spectroscopy (Perczel et al. 1991), DSC (Remmele et al. 2000), FTIR and Raman spectroscopy (Van de Weert et al. 2001) and nuclear magnetic resonance spectroscopy (NMR) (Fersht and Daggett 2002).

9.5.1.3 Relationship Between Physicochemical/Structural Properties and Biological Activity

Protein variants that are likely to affect potency can occur from, for example, disulfide scrambling, misfolding, oxidation of methionines, and mainly unspecific deamidation, aggregation, and clippings. Aggregates or fragments resolved by HP-SEC should also be analyzed for their *in vitro* potency to determine if the respective species is biologically active. The glycosylation pattern of mAbs is known to affect their biological performance (Wright and Morrison 1998; Jefferis 2001; Shields et al. 2002). Differences in the sialylation of a glycoprotein may cause attenuated pharmacokinetics and pharmacodynamics (Stockert 1995) and α-galactose motives are potentially immunogenic (it should be taken into account that there is a corresponding natural antibody that constitutes 1% of circulating IgG in humans; Galili 1992).

With the start of phase I clinical trials, a functional potency assay should be available, with "functional" meaning relevant to the mechanism of action of the therapeutic antibody. Common cell-based potency assays include proliferation, inhibition of proliferation, and apoptosis.

9.5.2
Quality Control Testing

At the start of clinical development, regulatory test methods (specified analytical methods with defined acceptance criteria for the product) should have been established. The development and selection of regulatory test methods and criteria to characterize the drug product and establish standards of conformance for product release are guided by the principles and practices described in ICH guideline Q6B or are considered to be product specific. The set of testing methods and specifications for release of drug substance and drug product should be designed to monitor the overall quality, safety, identity, strength, purity, and potency as well as general characteristic properties. Development and justification of specification criteria is based on the overall manufacturing experience, with consideration of results from preclinical and clinical production lots and data from the validation of analytical methods. Release testing methods usually are a subset of the tests used for characterization of the molecule. A typical set of lot release and stability testing methods is given in Table 9.5.

In addition to lot release testing, process monitoring and in-process testing of process-related impurities has to be performed and is a key to product quality. Impurity levels allowed depend upon, for example, the dose administered, the schedule of administration, duration (chronic versus acute) etc. since these may comprise a risk potential with respect to immunogenicity, toxicity, genotoxicity, and transmission of transmissible spongiform encephalopathy (TSE) (see ICH Q3 and S6 guidelines). Potential process-related impurities are media components (e.g. transferrin, insulin), cell components (host cell proteins and DNA), chemical additives (e.g. antibiotics, methotrexate, antifoam agents), and leachables (e.g. protein A, heavy metals).

Protein impurities (e.g. protein A, insulin, host cell protein) are most often measured using immunoassays as well as SDS-PAGE or immunoblot assays, whereas lower molecular weight impurities (e.g. methotrexate, hydrocortisone) are usually analyzed by HPLC. For measuring host cell proteins it is a common strategy in many companies to develop a multiproduct ELISA for all products derived from a particular cell type, or to use a commercially available "generic" ELISA. Internationally accepted specifications for impurities are not established, except for DNA, for which specification is available (World Health Organization 1997).

Several approaches are possible for the control of an impurity: a specification may be established for routine release testing on the drug substance/drug product stage or a limit may be set for in-process samples for batch-to-batch testing without implementing a release specification. Most preferably, when approaching an application for marketing authorization, a manufacturer would demonstrate consistent removal of existing impurity components by comprehensive process

Table 9.5 Lot release and stability testing methods (excluding safety) and acceptance criteria for a typical monoclonal antibody for clinical phase I.

Test	*Method*	*Acceptance criterion*	*DS release*	*DP release*	*DP stability*	*Comments*
Characteristics	Degree of coloration	NMT Y6	x	x	x	Pharmacopeial method, limit depends on formulation
	Clarity and degree of opalescence	NMT Reference III	x	x	x	Pharmacopeial method, limit depends on concentration
	pH	value ± 0.2	x	x	x	Pharmacopeial method
	Osmolality	value ± 30 $mOsm\,kg^{-1}$	x	x	x	Pharmacopeial method
	Extractable Volume/ volume in container	NLT (sum of) nominal volume (according Ph. Eur./ USP)		x		Pharmacopeial method
	Appearance	according Ph. Eur./USP		x	x	May be difficult to evaluate
	Solubility	according Ph. Eur./USP		x	x	
	UV-scan	Value ± 10%	x	x	x	
Identity and heterogeneity (structural integrity)	IEF or IEC (CEC)	Pattern qualitatively comparable to standard material (+ for IEC: report result for acidic species)	x	x	x	IEC more suitable for evaluation (quantitatively) of deamidation
	Peptide map	Profile qualitatively comparable to standard material	(x)			

Table 9.5 *Continued*

Test	*Method*	*Acceptance criterion*	*DS release*	*DP release*	*DP stability*	*Comments*
	Oligosaccharide map	Profile qualitatively comparable to standard material	(x)			Often not required as specified assay depending on (biological functionality of the antibody) but may be performed for internal information to monitor cell culture processing
Purity	SDS-PAGE (Coomassie)	Reduced: Sum of heavy and light chains NLT 90% Non reduced: main band NLT 85% Reduced and non reduced: number, intensity and molecular weight of main bands comparable to reference standard	x	x	x	
	HP-SEC	IgG monomer: NLT 95%	x	x	x	
	Particulate matter/ subvisible particles	≥10 μm: NMT 6000 ≥25 μm: NMT 600		x	(x)	Pharmacopeial method, may not have to be performed, in case an on-line filter is used for application
Potency	Cell-based bioassay	60–140% of standard material	x	x	x	If justified, a binding ELISA can be used instead

NMT, not more than; NLT, not less than; DS, drug substance; DP, drug product; Y6, yellow 6; IEF, isoelectric focusing; IEC, ion exchange chromatography; CEC, cation exchange chromatography; SDS-PAGE, sodium dodecyl sulfate polyacrylamide gel electrophoresis; HP-SEC, high-performance size-exclusion chromatography; ELISA, enzyme-linked immunosorbent assay; Ph Eur, European Pharmacopoeia; USP, United States Pharmacopeia.

validation and characterization (e.g. by performing additional small-scale spiking studies and thus avoiding to set limits for routine in process or release testing or reduce the extent of in-process testing).

Improvement, adjustment, and establishment of analytical methods and related acceptance criteria for in-process and product release testing appropriate to the development stage of the product and in response to process changes during development is generally required.

Prior to release of GMP material for phase I clinical trials, an adequate validation of regulatory tests should have been completed. Typically, the scope of a phase I validation is limited compared with the full ICH validation performed prior to submission of a market application (ICH Q2A).

9.5.3 Stability Testing

The biochemical, physicochemical, and biological methods of the stability testing program should be designed to control the degradation profile and to monitor the potency of the product for the intended shelf-life. Typically, analytical methods employed for stability testing are a subset of the ones developed for release testing (Table 9.5). Since methods designed to test for identity, process-related impurities, and contaminants are not stability indicating, they do not need to be included in the stability program (ICH Q5C and ICH Q1A). If a different specification for release and expiration is used, this should be justified by sufficient data to demonstrate that the clinical efficacy is not affected. Stability studies may be economized by applying sample selection criteria based on ICH Q1D.

Changes in the manufacturing process and formulation of the product, either during clinical development or post marketing authorization are likely to initiate an appropriate stability program. With submission of an investigational new drug application, typically a minimum of 3 months real time stability data for the intended storage temperature as well as data from accelerated stability studies should be available to ensure the defined quality of the respective clinical lots. Ideally, preceding stability data stemming from developmental lots (toxicology studies) should be available.

The drug product should be stable at least for the duration of the planned clinical study at the intended storage temperature. Storage time of the product may therefore not be allowed to exceed real-time stability data unless supportive stability data justify an extension. With filing of an marketing authorization application, stability data stemming from three drug product lots, possibly from different drug substance lots, data stemming from a photo-stability study of at least one batch (ICH Q1B), and data demonstrating stability of relevant hold steps such as intermediates or drug substance should be available (for more details please refer to USP 1049 and Q5C, Q1E, Q1A).

9.6
Overall Development Strategy and Outlook

A growing demand during early product development is to shorten the timeline for the supply of material for toxicological studies in animals and clinical studies in humans. Usually the production process needs to be developed before material can be supplied, which can be very time consuming. Thus, supply of material appropriate for toxicological studies in animals and first clinical studies in humans is often on the critical path in early stages of development. For later clinical studies, however, resupply can be managed timely on the basis of the existing process.

State-of-the-art process development has to comply with the growing demand of shortening the timeline and, at the same time, the importance of a "do it right the first time" paradigm to prevent unnecessary costs and delays at later development stages. The design of such a (state-of-the-art) development concept is presented in Fig. 9.10. In this strategy, material for toxicological studies produced from CHO cells can be supplied as little as 15 months after the cloning of product-encoding genetic sequences into high-expression vectors. At the same time it ensures that the production cell generated during this program will have the high

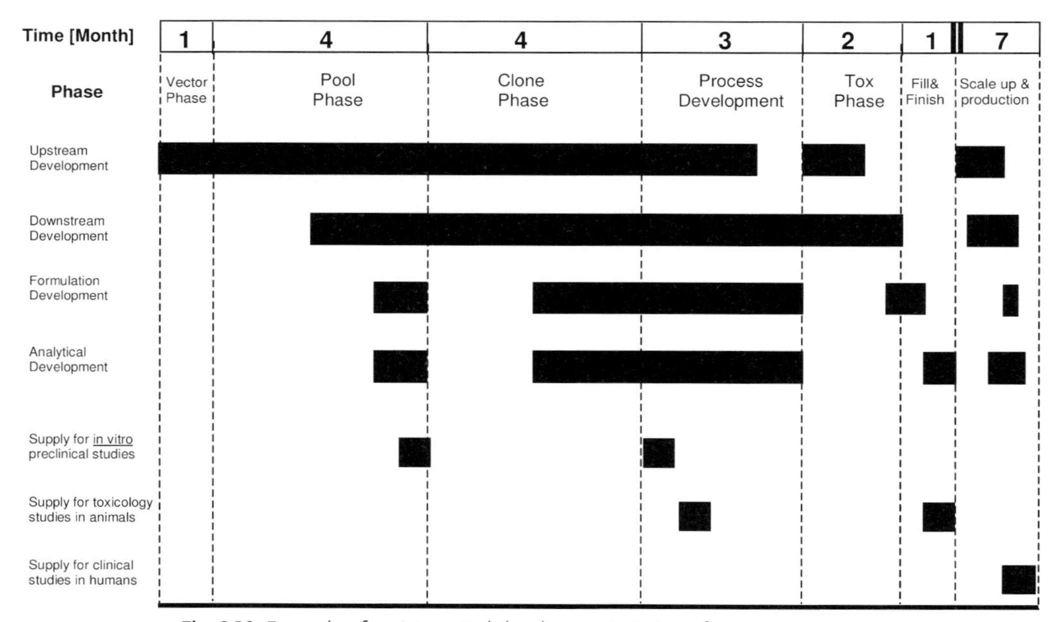

Fig. 9.10 Example of an integrated development strategy for supply of monoclonal antibody material appropriate for toxicological studies within 15 months and clinical material within 22 months.

expression potential needed to avoid a change in the production cell line at later stages in development.

Using the newly designed basic production process, sufficient quantities of material for clinical studies can be supplied after an additional 7 months, a total of 22 months after the start of development (i.e. cloning of the gene into the expression vector) (Fig. 9.10). Once the product has proven successful in early clinical trials (e.g. has shown a favorable safety profile) and proof-of-concept in humans has been demonstrated, the basic production process can be further developed. The final commercial process has to be capable of supplying sufficient material appropriate for pivotal studies and market at acceptable cost of goods. The optimal timing for development of the commercial process is in parallel with the ongoing clinical studies in order to secure product supply for clinical studies and market without any delay or gap.

The key to success for any such strategy is the well-coordinated early parallel development of the upstream, downstream, and formulation activities as well as of analytical methods for production of toxicological and phase I clinical material in order to generate a solid basis for development of the commercial process later on. The driving force for this development program is an interdisciplinary project team with defined roles, tasks, and responsibilities for the project manager and the team members. One of the project team's first tasks is the generation of the detailed project plan containing timeline, work packages, and the required capacities, milestones, and deliverables. This plan, which has to be approved by the senior management, is the basis for the execution of the project. Regular presentations of the project status to the senior management allows close monitoring of any deviations to the plan and ensures the quick approval of any major changes to scope, timing, and cost. Using these tools it is possible to run several development projects in parallel and to perform a portfolio management based on the resources available and on the success of the development candidates.

9.7 Outlook

One of the main obstacles to the development of successful therapeutic antibodies (recombinant proteins) has been the relatively low productivity of the production systems, which sometimes leads to limited availability, high production costs, and hence high costs for therapy. The current state-of-the-art production processes for mAbs in mammalian cells have shown titers up to 6 g/L, and at least 10-fold improvement compared with routine processes of the late 1990s. These achievements, as well as further improvements that can be expected in the near future, present an optimistic perspective for patients worldwide for an improved availability of mAbs for treatment of diseases with a highly unmet medical need at acceptable costs.

Acknowledgments

The authors like to thank Drs Dorothee Ambrosius, Stefan Bassarab, Helmut Hoffmann, Michael Schlüter, and Franziska Thomas for critical reading of this article and valuable comments and Ms Karin Jedrysiak for preparing the manuscript.

References

Ahern, T.J., Manning, M.C. (1992) *Stability of Protein Pharmaceutical. Part A, Chemical and Physical Pathways of Protein Degradation*. New York: Plenum Press.

Allison, D.W., Aboytes, K.A., Fong, D.K., Leugers, S.L., Johnson, T.K., Loke, H.N., Donahue, L.M. (2005) Development and optimization of cell culture media. *BioProcess Int* (January 2005), 38–45.

Andersson, C., Edlund, P.O., Gellerfors, P., Hansson, Y., Holmberg, E., Hult, C., Johansson, S., Kordel, J., Lundin, R., Mendel-Hartvig, I.B., Noren, B., Wehler, T., Widmalm, G., Ohman, J. (1996) Isolation and characterization of trisulfide variant of recombinant human growth hormone formed during expression in *Escherichia coli*. *Int J Pept Protein Res* 47: 311–321.

Andya, J.D., Hsu, C.C., Shire, S.J. (2003) Mechanisms of aggregate formation and carbohydrate excipient stabilization of lyophilized humanized monoclonal antibody formulations. *AAPS PharmaSci* 5: 1–11.

Arakawa, T., Timasheff, S.N. (1985) Theory of protein solubility. *Methods Enzymol* 114: 49–77.

Berthold, W., Kempken, R. (1994) Interaction of cell culture with downstream purification: a case study. *Cytotechnology* 15: 229–242.

Blank, G.S., Zapata, G., Fahrner, R., Milton, M., Yedinak, C., Knudsen, H., Schmelzer, C. (2001) Expanded bed adsorption in the purification of monoclonal antibodies: a comparison of process alternatives. *Bioseparation* 10: 65–71.

Branovic, K., Buchacher, A., Barut, M., Strancar, A., Josic, D. (2003) *J Chromatogr B* 790: 175–182.

Chi, E.Y., Krishnan, S., Randolph, T.W., Carpenter, J.F. (2003) Physical stability of proteins in aqueous solution: mechanism and driving forces in nonnative protein aggregation. *Pharm Res* 20: 1325–1336.

Chu, L., Robinson, D.K. (2001) Industrial choices for protein production by large-scale cell culture. *Curr Opin Biotechnol* 12: 180–187.

Cleland, J.L., Powell, M.F., Shire, S.J. (1993) The development of stable protein formulations: a close look at protein aggregation, deamidation, and oxidation. *Crit Rev Ther Drug Carrier Syst* 10: 307–377.

CPMP/3097/02 (2003) *Note for Guidance on Comparability of Medicinal Products containing Biotechnology-deriverd Proteins as Drug Substance – Non Clinical and Clinical Issues*. http://www.emea.eu.int/index/indexh1.htm

Creighton, T.E. (ed.) (1996) *Proteins*, 2nd edn. New York: Freeman & Company.

Eckerskorn, C., Strupat, K., Karas, M., Hillenkamp, F., Lottspeich, F. (1992) Mass spectrometric analysis of blotted proteins after gel- electrophoretic separation by matrix-assisted laser desorption ionization mass spectrometry. *Electrophoresis* 13: 664–665.

Eriksson, L., Johansson, E., Kettaneh-Wold, N., Wikström, C., Wold, S. (2000) *Design of Experiments: Principles and Applications*. Umeå: Umetrics Academy Book.

Fersht, A.R., Daggett, V. (2002) Protein folding and unfolding at atomic resolution. *Cell* 108: 573–582.

Fischer, R., Stoger, E., Schillberg, S., Christon, P., Twyman, R.M. (2004) Plant-based production of biopharmaceuticals. *Curr Opin Plant Biol* 7: 152–158.

Follman, D., Fahrner, R.L. (2004) *J Chromatogr A* 1024: 79–85.

Fraunhofer, W., Winter, G. (2004) The use of asymmetrical flow field-flow fractionation in pharmaceutics and biopharmaceutics. *Eur J Pharm Biopharm* 58: 369–383.

Frokjaer, S., Otzen, D.E. (2005) Protein drug stability: a formulation challenge. *Nature Rev* 4: 298–306.

Galili, U. (1992) The natural anti-Gal antibody: evolution and function. *Immunology Series* 55: 355–373.

Garidel, P. (2004) What can we learn from infrared spectroscopy for the development of protein formulations? In: *Formulation and Drug Delivery Strategies for Biopharmaceuticals*. Munich: IBC Life Sciences, pp. 1–20.

Garidel, P., Schott, H. (2006) Fourier transform mid-infrared spectroscopy for the analysis and screening of liquid protein formulations in pharmaceutical development. *Int BioProcess* 4: 40–46, 48–55.

Gianazza, E. (1995) Isoelectric focusing as a tool for the investigation of post-translational processing and chemical modifications of proteins. *J Chromatogr A* 705: 67–87.

Good, D.L. (2004) Capillary electrophoresis of proteins in a quality control environment. In: *Capillary Electrophoresis of Proteins and Peptides, Methods in Molecular Biology*, Vol. 276. New Jersey: Humana Press.

Gorman, J.J., Wallis, T.P., Pitt, J.J. (2002) Protein disulfide bond determination by mass spectromety. *Mass Spectrom Rev* 21: 183–216.

Harris, R.J. (1995) Processing of C-terminal lysine and arginine residues of proteins isolated from mammalian cell culture. *J Chromatogr A* 705: 129–134.

Harris, R.J., Kabakoff, B., Macchi, F.D., Shen, F.J., Kwong, M., Andya, J.D., Shire, S.J., Bjork, N., Totpal, K., Chen, A.B. (2001) Identification of multiple sources of charge heterogeneity in a recombinant antibody. *J Chromatogr B* 752: 233–245.

Hewlett G. (1991) Strategies for optimizing serum-free media. *Cytotechnology* 5: 3–14.

Hollas, J.M. (2003) *Spectroscopie*. Paris: Edition Dunod.

Jefferis, R. (2001) Glycosylation of human IgG antibodies: relevance to therapeutic applications. *Biopharm* 14: 19–26.

Kaufmann, H., Fussenegger, M. (2003) Metabolic engineering of mammalian cells for higher protein yield. In: Makrides, S.C. (ed.) *New Advances in Biochemistry, Gene Transfer and Expression in Mammalian Cells*. Amsterdam: Elsevier Science, pp. 457–469.

Kellner, R., Mermet, J.M., Otto, M., Valcárcel, Widmer, H.M. (2004) *Analytical Chemistry*. Weinheim: Wiley-VCH.

Kempken, R., Howaldt, M., Walz, F. (2006) Kultur von Tierzellen. In: Chmiel, H. (ed.) *Bioprozeßtechnik*. München: Elsevier GmbH.

Langguth, P., Fricker, G., Wunderli-Allenspach, H. (2004) *Biopharmazie*. Weinheim: Wiley-VCH.

Laue, T.M., Stafford, W.F. (1999) Modern application of analytical ultracentrifugation *Annu Rev Biophys Biomol Struct* 28: 75–100.

Lee, S.Y. (1996) High cell-density culture of *Escherichia coli*. *Trends Biotechnol* 14: 98–105.

LeHir, A. (2001) *Pharmacie galénique*. Paris: Edition Masson.

Li, R., Dowd, V., Stewart, D.J., Burton, S.J., Lowe, C.R. (1998) *Nat Biotechnol* 16: 190–195.

Liu, J., Nguyen, M.D.H., Andya, J.D., Shire, S.J. (2005) Reversible self-association increases the viscosity of a concentrated monoclonal antibody in aqueous solution. *J Pharm Sci* 94: 1928–1940.

Liu, T., Chu, B. (2002) Light-scattering by proteins. In: Somasundran P. (ed.) *Encyclopedia of Surface and Colloid Science*. New York: Marcel Dekker, pp. 3023–3043.

Manning, M.C. (2005) Use of infrared spectroscopy to monitor protein structure and stability. *Expert Rev Proteomics* 2: 731–743.

Mechref, Y., Novotny, M.V. (2002) Structural investigations of glycoconjugates at high sensitivity. *Chem Rev* 102: 321–369.

Middaugh, C.R., Tisel, W.A., Haire, R.N., Rosenberg, A. (1979) Determination of the apparent thermodynamic activities of saturated protein solutions. *J Biol Chem* 254: 367–370.

Murakami, S., Chiou, T.-W., Wang, D.I.C. (1991) A fiber-bed bioreactor for anchorage-dependent animal cell cultures: Part II. Scaleup potential. *Biotechnol Bioeng* 37: 762–769.

Nguyen, D.N. et al. (1995) Protein mass spectrometry: applications to analytical biotechnology. *J Chromatogr A* 705: 21–45.

Nicoud, R.M. (1998) In: Subramanian, G. (ed.) *Bioseparation and Bioprocessing.* Weinheim: Wiley-VCH, 3–38.

Patterson, S.D., Aebersold, R. (1995) Mass spectrometric approaches for the identification of gel-separated proteins. *Electrophoresis* 16: 1791–1814.

Perczel, A., Hollosi, M., Tusnady, G., Fasman, G.D. (1991) Convex constraint analysis: a natural deconvolution circular dichroism curves of proteins. *Protein Eng* 4: 669–679.

Perkins, M., Theiler, R., Lunte, S., Jeschke, M. (2000) Determination of the origin of charge heterogeneity in a murine monoclonal antibody. *Pharm Res* 17: 1110–1117.

Preissmann, A., Wiesmann, R., Buchholz, R., Werner, R.G., Noé, W. (1997) Investigations on oxygen limitations of adherent cells growing on macroporous microcarriers. *Cytotechnology* 24: 121–134.

Prigge, S.T. (1997) Amidation of bioactive peptides: the structure of peptidylglycine – hydroxylating monooxygenase. *Science* 278: 1300–1305.

Remmele, R.L. jr. et al. (2000) Differential scanning calorimetry: a practical tool for elucidating stability in liquid pharmaceuticals. *Biopharm Eur* 56: 58–60.

Rouessac, F., Rouessac, A. (2004) *Analyses chimique. Méthodes et techniques instrumentals modernes.* Paris: Edition Dunod.

Schein, C.H. (1990) Solubility as a function of protein structure and solvent components. *Bio/Technology* 8: 308–317.

Schimpf, M.E., Caldwell, K.D., Giddings, J.C. (eds) (2000) *Field-Flow Fractionation Handbook.* New York: Wiley-Interscience.

Schüle, S., Schultz-Fademrecht, T., Bassarab, S., Bechtold-Peters, K., Frieß, W., Garidel, P. (2004) Determination of the secondary protein structure of IgG1 antibody formulations for spray drying. *Resp Drug Delivery* 2: 377–380.

Schüle, S., Schultz-Fademrecht, T., Bassarab, S., Bechtold-Peters, K., Frieß, W., Garidel, P. (2005) Spray drying of IgG1 antibody formulations: a CD and FTIR investigation with respect to protein secondary structure. *Resp Drug Delivery* 1: 185–188.

Sheeley, D.M., Merrill, B.M., Taylor, L.C. (1997) Characterization of monoclonal antibody glycosylation: comparison of expression systems and identification of terminal α-linked galactose. *Anal Biochem* 247: 102–110.

Shields, R.L., Lai, J., Keck, R., O'Connell, L.Y., Hong, K., Meng, Y.G., Weikert, S.H., Presta, L.G. (2002) Lack of fucose on human IgG1 N-linked oligosaccharide improves binding to human FcγRIII and antibody-dependent cellular toxicity. *J Biol Chem* 277: 26733–26740.

Sine, J.P. (2003) *Séparation et analyse des biomolécules.* Paris: Edition Ellipses.

Stephenson, R.C., Clark, S. (1989) Succinimide formation from aspartyl and asparaginyl peptides as a model for spontaneous degradation of proteins. *J Biol Chem* 264: 6164–6170.

Stockert, R.J. (1995) The asialoglycoprotein receptor: relationships between structure, function, and expression. *Physiol Rev* 75: 591–609.

Storhas, W. (1994) *Bioreaktoren und periphere Einrichtungen.* Berlin: Springer Verlag, pp. 279–297.

Swartz, J.R. (2001) Advances in *Escherichia coli* production of therapeutic proteins. *Curr Opin Biotechnol* 12: 195–201.

Tatford, O.C., Gomme, P.T., Bertolini, J. (2004) Analytical techniques for the evaluation of liquid protein therapeutics. *Biotechnol Appl Biochem* 40: 67–81.

Tebbe, H., Lütkemeyer, D., Gudermann, F., Heidemann, R., Lehmann, J. (1996) Lysis-free separation of hybridoma cells by continuous disc stack centrifugation. *Cytotechnnology* 22: 119–127.

Teshima, G., Stults, J.T., Ling, V., Canova-Davis, E. (1991) Isolation and Characterization of a Succinimide Variant of Methionyl Human Growth Hormone. *J Biol Chem* 266: 13544–13547.

Van de Weert, M., Haris, P.I., Hennink, W.E., Crommelin, D.J. (2001) Fourier transform infrared spectrometric analysis of protein conformation: effect of sampling method and stress factors. *Anal Biochem* 297: 160–169.

Van Reis, R., Leonard, L.C., Hsu, C.C., Builder, S.E. (1991) Industrial scale harvest of proteins from mammalian cell culture by tangential flow filtration. *Biotechnol Bioeng* 38: 413–422.

Walter, J.K. (1998) In: Subramanian, G. (ed.) *Bioseparation and Bioprocessing*. Weinheim: Wiley-VCH, 1998: 447–460.

Wang, L., Amphlett, G., Blattler, W.A., Lambert, J.M., Zhang, W. (2005) Structural characterization of the maytansinoid–monoclonal antibody immunoconjugate, huN901–DM1: by mass spectrometry. *Protein Sci* 14: 2436–2446.

Waterman, K.C., Adami, R.C. (2005) Accelerated aging: prediction of chemical stability of pharmaceuticals. *Int J Pharm* 293: 101–125.

Werner, R.G. (2005) *BioProcess International* September 2005: 6–15.

Wildt, S., Gerngross, T.U. (2005) The humanization of N-glycosylation pathways in yeast. *Nat Rev Microbiol* 3: 119–128.

World Health Organization (1997) *WHO Epidemiological Records* 72: 143–145.

Wright, A., Morrison, S.L. (1998) Effect of C2-associated carbohydrate structure on IgG effector function: Studies with chimeric mouse-human IgG1 antibodies in glycosylation mutants of Chinese hamster ovary cells. *J Immunol* 160: 3393–3402.

Wurm, F.M. (2004) Production of recombinant protein therapeutics in cultivated mammalian cells. *Nat Biotechnol* 22: 1393–1398.

Zhang, W., Czupryn, J.M., Boyle, P.T. Jr, Amari, J. (2002) Characterization of asparagine deamidation and aspartate isomerization in recombinant human interleukin-11. *Pharm Res* 19: 1223–1231.

Zlokarnik, M. (2002) *Scale-up in Chemical Engineering*. Weinheim: Wiley VCH.

Zmak, P.M. et al. (2003) *J Chromatogr A* 1006: 195–205.

10
Pharmaceutical Formulation and Clinical Application

Gabriele Reich

10.1
Introduction

The therapeutic application of monoclonal antibodies (mAbs) has increased dramatically in recent years. More than 20% of all biopharmaceuticals currently being evaluated in clinical trials are full-length antibodies, antibody fragments, or conjugates (Pavlon and Belsey 2005). The key advantage of this class of therapeutics is their high level of specificity for the relevant disease targets. This pinpoint specificity serves to prevent harm to healthy cells and, hence, typically results in fewer side effects compared with traditional drugs.

Their clinical breakthrough as targeted effector molecules is a direct result of the successful engineering, large-scale production, and purification of tailor-made antibodies or fragments with optimal *in vivo* behavior. Chimeric, humanized, and fully human mAbs with reduced immunogenicity and improved half-life show high clinical efficacy for the treatment of a variety of diseases such as organ transplant rejections, cancer, and autoimmune, inflammatory, and infectious diseases. Bispecific antibodies are useful constructs to enhance immunological effector functions. In addition, Fab and single-chain Fv fragments have gained increasing interest for indications where the recruitment of Fc functions is not essential.

Besides their targeted effector function, antibodies are versatile molecules that may be used as carriers and vectors for the site-specific delivery of drugs, toxins, enzymes, radionuclides, and genes. Antibody-mediated targeting strategies hold promise for a wide number of clinical applications, especially since advances in recombinant antibody technology allow formats other than whole antibodies to be engineered. Modified antibodies with increased effector functions (e.g. bispecific mAbs) and fragments (e.g. Fabs, scFvs) of reduced size and thus improved cell penetrability are set to play a significant role in future product developments.

The huge clinical potential and broad variability of mAb designs are clearly reflected by a certain number of approved or nearly approved products being in

Handbook of Therapeutic Antibodies. Edited by Stefan Dübel

ISBN 978-3-527-31453-9

clinical use, and a rather large number of formulations being in the development stage. However, product development with mAbs is somewhat different from traditional low molecular weight drug substances. This is due to their large size and complex three-dimensional structure, which is mandatory for their functional performance. Major challenges are chemical and physical instability during manufacturing and storage, and delivery issues.

This article addresses pharmaceutical formulation, manufacturing, and delivery, as well as clinical applications of therapeutic antibodies.

10.2 Clinical Application

10.2.1 Therapeutic Areas of Antibody Drugs

The majority of antibody-based therapies in the clinic have exploited the activity of the antibody per se, namely to specifically bind to antigens and modulate several different pathways. Indeed, mAbs can function as targeted effector molecules for a wide number of indications. In the past, the pharmaceutical industry has heavily focused on oncology, arthritis, autoimmune, inflammatory and infectious diseases, and this trend is set to continue over the next years.

The major strategy for using antibodies in clinical oncology is a result of their ability to bind to tumor-associated antigens of primary and metastatic cancer cells, and to create antitumor effects by complement-mediated cytolysis, cell-mediated cytotoxicity, and/or signal-transduction leading to apoptosis or growth arrest (Ross et al. 2003). The clinical success of this approach is closely related to the selective binding of the antibody to tumor cells (i.e. a homogeneous overexpression of the target antigen on the tumor cell surface), and limited expression of the antigen by normal tissues. Appropriate tumor-associated target antigens are epidermal growth factor receptors (EGFR: HER1, HER2, HER3, and HER4) in carcinomas, CD19, CD20, and CD22 in B cell lymphomas, and CD25 or CD52 in T cell leukemia. Anticancer antibodies in clinical use include anti-CD20, anti-CD52, and anti-EGFR (HER1 and HER2) for the treatment of B cell lymphoma, lymphatic leukemia and carcinoma, respectively. Examples for approved commercial products are: rituxumab, a chimeric anti-CD20 B cell mAb for the treatment of relapsed or refractory non-Hodgkin lymphoma and related malignancies; alemtuzumab, a humanized anti-CD52 mAb for the treatment of refractory chronic lymphocytic leukemia; trastuzumab for the treatment of HER2-positive metastatic breast cancer; and cetuximab and bevazizumab for the treatment of HER1-positive colon cancer.

Another approach in solid tumor therapy is to inhibit new blood vessel formation and/or angiogenesis by targeting mAbs to antigens expressed on the tumor vasculature, rather than to tumor-associated antigens of solid tumors. Directing therapy to the vascular compartment reduces the impact of the physical barriers of solid tumors such as heterogeneous blood flow and elevated interstitial pressure. Novel advances in this field are the development of antibodies against angiogenesis-associated factors and receptors such as vascular endothelial growth factor (VEGF) and $\alpha_v\beta_3$ integrin. Examples are bevacizumab approved for the treatment of colorectal cancers, and vitaxin, which is being assessed in clinical trials (McNeel et al. 2005).

Antibody-based therapies have also gained increasing interest in chronic inflammatory diseases that are characterized by high levels of cytokines. Anti-tumor necrosis factor alpha (TNF-α) mAbs, in particular, have proven to be powerful therapeutics in Crohn's disease and rheumatoid arthritis (RA). These antibodies complex soluble TNF-α, a critical mediator in RA and Crohn's disease, thereby inhibiting its functional activity. Approved products are infliximab, adalimumab, and etanercept. Monoclonal antibodies targeting interleukins are another approach in this field. Two products – anakinra and daclazizumab – based on anti-interleukin (IL)-1 and anti-IL-2, have gained approval. Quite a number of anti-inflammatory mAb products targeting TNF-α or various interleukin subtypes are in preclinical and clinical trials (Sacre et al. 2005). To reduce immunological side effects triggered by the Fc part, and to increase circulating half-life, some of these are PEGylated fragments (see Section 10.2.3).

An example of the successful clinical use of a humanized anti-IgE antibody in allergic asthma therapy is omalizumab. When delivered systemically, the antibody targets the Fc portion of the human IgE, thus inhibiting IgE interaction with mast cells and basophils (i.e. removing it as a mediator of allergic asthma).

The potential of mAbs for the treatment of infectious diseases has not been fully exploited yet, although in theory, antibody-based therapies could be developed against any existing pathogen. Given the multitude of pathogens, the pathogen-specific nature of antibody therapies, and the high costs for product development, it seems reasonable that antibody-based therapies can only provide a therapeutic option for selected pathogens that affect primarily immunocompromised patients and/or for which there is no other antimicrobial therapy available. In view of this, it is not surprising that the only approved product yet is palivizumab, a humanized mAb against Rous sarcoma virus infection.

10.2.2
Antibody-Mediated Drug Delivery

In addition to their direct effector function, antibodies can be used as carriers or vectors for targeted drug delivery. The so-called immunoconjugates may be achieved by direct or indirect covalent linkage of an antibody (whole, fragment, or bispecific) to either a drug, toxin, enzyme, and/or radioisotope (Payne 2003; Cao and Lam 2004; Greenwald et al. 2004; Wu and Senter 2005), or to drug-containing sterically stabilized liposomes (Park et al. 2002; Schnyder and Huwyler 2005) or nanoparticles (Brannon-Peppas et al. 2004). Nanoparticles usually consist of biodegradable polymers, while liposomes are small vesicles composed of unilamellar or multilamellar phospholipid vesicles ranging from 20 to 10000 nm (Fig. 10.1). Immunoliposomes-nanoparticles are intended to improve the therapeutic efficacy by enabling a larger quantity of drug to be delivered per antibody molecule. Moreover, they protect the drug from metabolism and inactivation in the plasma. Attachment of high molecular weight poly(ethyleneglycols) (PEGs), leading to sterically stabilized so-called "stealth" liposomes or nanoparticles, may prevent recognition by phagocytotic cells, thus improving circulating half-life and distribution to peripheral tissues (see Section 10.2.3).

Immunoconjugates are an interesting approach for tumor cell and brain targeting. Antibody-mediated drug targeting to tumors is generated in the same manner as "naked" anticancer antibodies. Antibody conjugates with small cytotoxic drugs are intended to improve the therapeutic index of these drugs by prolonging the bioavailability of the drug, increasing drug uptake in the target cells, and reducing drug toxicity to nontarget cells. Successful approaches require internalization of the drug–antibody conjugate with subsequent intracellular drug release, or extracellular drug cleavage and subsequent cellular drug uptake by diffusion or active transport. To this end, pH- or enzyme-sensitive linkers are required. The use of antibody fragments (Fab, scFv), having a reduced size while retaining their target function, are well suited to increase tumor penetration of immunoconjugates. Various antineoplastic drugs such as methotrexate, 5-fluorouracil, doxorubicin, and maytansine have been developed as immunoconjugates. Gemtuzumab ozogamicin, a calicheamicin-conjugated humanized mouse anti-CD33 mAb, has been approved by the US Food and Drug Administration (FDA) for therapy of drug-refractory acute myloid leukemia.

The option of using liposomes or nanoparticles in targeted cancer therapy is based on the observation that discontinuities in the endothelium of the tumor vasculature favor extravasation and local accumulation at the tumor. Stealth immunoliposomes utilizing internalizing mAbs such as anti-HER2 or anti-CD19

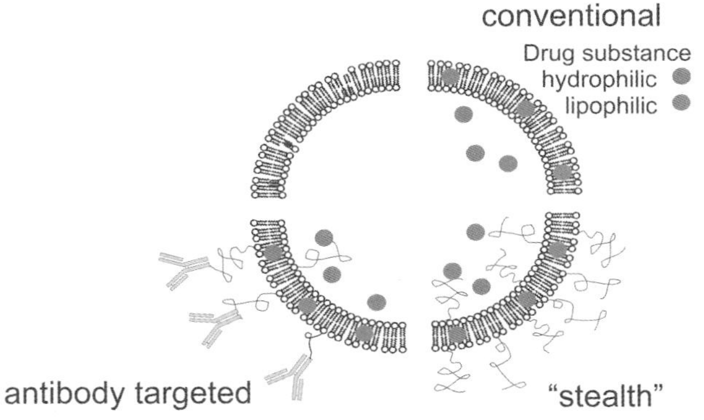

Fig. 10.1 Schematic representation of three different liposome types: conventional liposome; "stealth" liposome sterically stabilized by PEG attachment; "antibody targeted" stealth liposome (i.e. immunoliposome).

can be used to selectively deliver high drug concentrations into the cytoplasm of antigen-expressing tumor cells (Park et al. 2002). This approach has been shown to greatly enhance the therapeutic index of doxorubicin.

Radioimmunoconjugates are an interesting strategy for concentrating doses of radiation from isotopes such as iodine-131 or yttrium-90 to cancer tissues. It is supposed that even antigen-negative tumor cells may be eradicated, since these radionuclides are effective over a distance of several cell diameters. Two commercial products based on an anti-CD20 antibody are in clinical use for the treatment of patients with non-Hodgkin's lymphoma that have relapsed after chemotherapy and are no longer responding to rituximab.

Drug delivery to the brain is highly restricted by a tight vascular barrier, the blood–brain barrier (BBB). One option to enhance brain delivery is the use of an mAb to the extracellular domain of a BBB receptor which may trigger cellular uptake by the mechanism of receptor-mediated transcytosis. OX26, a anti-rat transferrin receptor mAb can be used for this purpose (Schnyder and Huwyler 2005). The mAb vector may be coupled to the drug via biotin–avidin conjugation. In case of sterically stabilized liposomes, the thiolated antibody or fragment is coupled to a thiol-reactive maleinimide-PEG-phospholipid. PEG-conjugated OX26 immunoliposomes are interesting carrier systems for brain targeting of small anticancer drugs or plasmids. Brain uptake of daunomycin was effectively increased in rats when it was entrapped in OX26 immunoliposomes. In a gene delivery approach based on OX26 immunoliposomes, gene expression in rat brain cells beyond the BBB was clearly demonstrated. Since alternative vectors, such as mAbs directed at the human insulin receptor, indicate the potential for higher targeting efficiency, it can be assumed that this approach will gain clinical relevance for human therapy in the near future.

10.2.3
PEGylated Antibodies and Antibody Fragments

Modification of proteins and colloidal carrier systems with PEG is a well-established technique (Caliceti and Veronese 2003) to

- reduce antigenicity, i.e. immunogenicity,
- increase circulating half-life by either evasion of renal clearance and/or cellular clearance mechanisms, and
- improve bioavailability and reduce overall drug toxicity.

Recent technological advances enabling humanized and fully human antibodies to be produced have largely overcome the problem of immunogenicity observed with murine antibodies. In addition, full-length antibodies generally have a long circulating half-life, since the Fc region binds to the neonatal Fc receptor (FcRn) providing a salvage mechanism (i.e. protection from *in vivo* degradation). In contrast, antibody fragments, lacking the Fc part and having a reduced size, usually suffer from relatively short half-lives (Weir et al. 2001). PEGylation of Fab or scFv fragments is therefore an attractive approach to increase the circulating half-life, thus improving their therapeutic potential (Chapman 2002; Greenwald et al. 2004). The pharmocokinetic effect is mainly attributed to an increased hydrodynamic size of the molecules to above the normal limit for glomerular filtration, resulting in reduced renal clearance. In addition, PEGylation can shield the molecules from immunological recognition and subsequent clearance from the circulation, and/or affect biodistribution including improved tumor targeting. Several studies have clearly shown greater accumulation of antibodies and antibody fragments in tumors without higher levels in normal tissues following PEGylation (references are given in Chapman 2002).

Additional benefits of PEGylation to date are related to the solubility of antibodies and antibody fragments. PEGylated antibody fragments in aqueous solution can be concentrated to >200 mg mL^{-1} without aggregation. This technological advantage over non-PEGylated fragments opens up new opportunities for chronic immunotherapy with high dosing regimens by subcutaneous injection (see Section 10.5.1). Moreover, the ability of PEG to dissolve in many different solvents, ranging from water to hydrophobic organic solvents, can be exploited for the development of improved slow-release formulations (see Section 10.5.4).

Numerous functionalized PEG molecules with different structures, chain lengths, and conjugation chemistries are available. Moreover, the conjugation site and number of PEG chains attached per antibody molecule can be varied. However, in the context of antibodies and antibody fragments, the choice of an appropriate PEGylation approach requires various aspects to be considered. To avoid significant loss of antigen binding and/or immunological effector functions mediated by the Fc region, site-specific PEGylation should be preferred over random PEG conjugation (Chapman 2002). In case of Fab' fragments, the method typically involves the use of PEG maleimide to react with thiol groups of free cysteine residues in the hinge region of specifically engineered antibody molecules. Site-specific PEGylation in this region ensures that the PEG is located away from the antigen-binding region, thus retaining full antigen-binding activity.

10.2.4
Routes of Administration

Monoclonal antibodies, fragments, and conjugates thereof are large molecules with a complex and sensitive three-dimensional structure. Because of poor *in vivo* stability and permeability (i.e. poor bioavailability by other routes), parenteral delivery has been the most conventional route of administration for mAb therapy in humans. Due to high dosing (mg kg^{-1} per day) and greater control in clinical settings, intravenous (i.v.) infusion is the preferred dosage form for mAb-based cancer therapy in hospitals. Indeed, the majority of approved mAb therapeutics used in oncology, whether naked or conjugated to toxins or radionuclides, are formulated for i.v. administration. Depending on the stability and solubility profile of the antibody molecule, ready-to-use solutions, concentrates and freeze-dried powders for reconstitution prior to administration are in place.

In contrast to cancer therapy, where clinical safety control aspects are extremely important, indications such as inflammatory diseases and immune disorders require outpatient administration and home use of antibody drugs, since frequent and chronic dosing is mandatory. For these settings, alternative delivery routes are more appealing than i.v. infusions to facilitate administration and improve patient compliance. However, as indicated above, systemic delivery by oral, transdermal, nasal, or pulmonary routes of administration has been difficult to achieve. Thus, a small volume (<1.5 mL) subcutaneous (s.c.) injection is currently the preferred dosage form for chronic administration of antibody therapeutics in the physician's office or by the patient at home. Few products have been developed as stabilized high-concentration solutions. However, stability and delivery problems frequently associated with highly concentrated, high-viscosity mAb solutions (see Section 10.5.1) have stimulated the search for alternative dosage forms. Low-viscosity suspensions based on highly stable crystalline antibody formulations are a novel approach that became feasible following recent advances in antibody batch-crystallization techniques (see Section 10.5.3). Due to reduced stability problems, crystalline antibodies may also provide a better platform for future developments of carrier-based mAb delivery systems (see Section 10.5.4).

Local delivery of whole antibodies or fragments to the site of disease is an alternative approach to enhance therapeutic efficacy and improve patient compliance. One such strategy is the pulmonary administration of anti-IgE antibodies for the treatment of allergic asthma. Spray-dried and spray freeze-dried powders yielding particle sizes that are appropriate for inhalation and lung deposition have been reported as promising dosage forms for this purpose (Maa et al. 1999). Formulation strategies to overcome mAb stability problems and achieve good aerosol performance upon drying will be addressed in Section 10.6.1.

Topical application of antibodies recognizing and neutralizing the oral pathogen *Streptococcus mutans* might be attractive for control and prevention of dental caries (Kupper et al. 2005). Ocular administration of antibody fragments via eyedrops has been proposed for local treatment of eye infections. Intestine- or colon-specific mAb delivery following oral administration could be a future option for targeted local therapy of infectious diseases in the gut.

10.3 Pharmaceutical Product Development

To gain regulatory approval for human use of a new drug product, the applicant has to perform a number of investigations that confirm pharmaceutical product quality. In classical "drug discovery and development processes" a new chemical entity (NCE) is usually examined in some sort of preformulation study before it is further progressed to formulation development and subject to an ascending series of preclinical and clinical trials to ascertain potency, efficacy, and safety as a pharmaceutical agent for human use.

Generally, antibody-based drug substances are treated no differently from their low molecular weight counterparts (i.e. an assessment of physicochemical and biological properties has to be made to fulfill the quality acceptance criteria defined in the ICH harmonized tripartite guideline). However, there are some important a priori differences between antibodies and conventional NCEs. The macromolecular character and the specific secondary, tertiary and/or quartary structure may have a controlling influence over the physicochemical and biological properties of an antibody in solid and solution state. This implies more extensive studies during the drug development stage, including assessment of purity, stability, molecular size and structure, *in vitro* and *in vivo* biological activity, pharmacokinetics, delivery, and immunogenicity.

Since stability problems may be crucial for product performance in many respects, the way from drug discovery to an approved, marketed antibody drug product is rather challenging. In view of this, the following chapter is dedicated to stability issues. Formulation challenges and strategies for the development of parenteral and local delivery systems are discussed in detail in Sections 10.5 and 10.6.

10.4 Stability Issues

Therapeutic antibodies may be defined as whole monoclonal antibodies, specifically designed fragments or constructs of various composition. Understanding the degradation pathways of each of these molecules and how they relate to drug potency, efficacy, and safety is the key issue for the development of stable antibody drug products. In Section 10.4.1, the current understanding of the main degradation pathways of therapeutic antibodies will be presented. Regulatory requirements for the design of stability studies including analytical and kinetic aspects will be addressed in Section 10.4.2.

10.4.1 Degradation Pathways

The degradation pathways of antibody molecules may be classified as chemical and physical instability (Wang 1999). Chemical instability involves modification of the antibody molecule via covalent bond formation or cleavage, while physical instability refers to conformational changes, adsorption, aggregation, and/or precipitation. Depending on the sites of degradation, the functional consequences may be different. Degradation sites in the variable region are likely to result in loss of antigen binding capability. Degradation pathways in the constant (Fc) regions may affect the *in vivo* effector function and/or the metabolism of the antibody. Unfolding and aggregation can lead to increased immunogenicity.

10.4.1.1 Chemical Degradation

Chemical instability, as distinct from proteolysis, is generally associated with specific amino acid residues or sequences. Chemical reactions affecting antibody molecules often involve deamidation, oxidation, and formation of incorrect disulfide bonds.

Deamidation Deamidation is a common reaction that may occur in aqueous antibody solutions under a variety of *in vitro* conditions. Asparagine (Asn) and glutamine (Gln) residues are both susceptible to deamidation, although at a different rate (Asn > Gln). The side chain amide linkage may either be simply hydrolyzed to form a free carboxylic acid, or proceed through a five- or six-membered cyclic imide intermediate formed by intramolecular attack of the succeeding peptide nitrogen at the side chain carbonyl carbon of the Asn or Gln residue. Since the formation of a cyclic imide involves participation of the neighboring amino acid, the size and physicochemical characteristics of the succeeding amino acid side chain as well as the conformational chain mobility play a significant role in the deamidation rate. Generally, the sequence -Asn-Gly- is most susceptible to deamidation, especially under neutral and alkaline conditions, increased temperature, and/or ionic strength. Asn and Gln accessibility within the overall three-dimensional structure is another important parameter to be considered. Typically, only few Asn and Gln residues are located in highly flexible hydrophilic domains or on the surface of the molecule, such as a small number of reactive Asn residues in the Fc region of the heavy chain. Since their degradation leads to changes in the charge pattern of the molecule, the overall structural stability may be compromised or additional chemical reactions induced. Inhibition of deamidation by adjusting solution conditions (temperature, pH, ionic strength) or freeze-drying is therefore an important issue, even if *in vivo* consequences are not likely to be expected.

Oxidation Oxidation is a second major degradation pathway of antibody molecules. Potential reaction sites are the side chains of methionine, cysteine, histidine, tryptophan, and tyrosine residues, several of which are typically in the heavy and light chains of IgG molecules. As with other side chain reactions, their susceptibility is closely related to their accessibility, which in turn may be affected by temperature, pH, and solvent composition. Methionine (Met) residues are selectively oxidized under acidic conditions. Two pathways have been described, namely a temperature-induced oxidation via formation of free radicals from hydrogen peroxides, or a light-induced reaction with singlet oxygen (Lam et al. 1997). Both pathways have been reported to potentially occur in aqueous solutions of therapeutic antibodies formulated with non-ionic polyether surfactants such as Tween 80 or 20, since these excipients can undergo autoxidation to form peroxides or dissipate light energy by reacting with molecular oxygen to generate singlet oxygen. Consequently, pH adjustment, nitrogen flushing, light protection, and low-temperature storage are effective means to reduce Met oxidation rate.

The thiol group of cysteine (RSH) can be oxidized in various steps to sulfenic acid (RSOH), disulfide (RSSH), sulfinic acid (RSO_2H), and finally, sulfonic acid (RSO_3H). Cysteine (Cys) oxidation takes place in the presence of oxidizing agents such as hydrogen peroxide, or spontaneously via autoxidation by oxygen from the air. The reaction rate strongly depends on temperature, pH, buffer salts, presence of molecular oxygen, and/or metal ion catalysts. Typically, Cys oxidation is accelerated with increasing pH. Under favorable steric conditions, it can lead to intra- and/or even intermolecular disulfide formation, which in turn may lead to irreversible aggregation. The intrinsic number and alignment of intra- and interchain disulfide bonds is an integral structural element of each antibody subclass. Interchange of disulfide bonds can result in incorrect pairings, which may lead to an altered threedimensional structure and consequently to an altered solubility, biological activity, and/or immunogeneity.

Disulfide exchange Disulfide exchange is favored in solution, but may also occur in powder products with a certain molecular chain mobility resulting from residual moisture. The reaction mechanism depends on the pH of the medium. Under neutral and alkaline conditions, the reaction is catalyzed by thiols, which, in the form of thiolate ions, carry out a nucleophilic attack on the sulfur atom of the disulfide. In acidic media, the reaction takes place through a sulfenium cation, which carries out an electrophilic displacement on a sulfur atom of the disulfide.

10.4.1.2 Physical Degradation

Antibody molecules, as other proteins, fold to a specific three-dimensional superstructure that is essential for their biological function. Since the thermodynamic stability of the native conformation, resulting from a unique balance between large stabilizing and large destabilizing forces, is only small, the molecules can undergo a variety of structural changes independent of chemical modifications. The process can lead to a stepwise unfolding of the molecule, and may be reversible or irreversible.

In solution, rate and extent of the conformational changes are strongly affected by temperature, shear forces, pH, salt type and concentration, surfactants, and the presence of organic solvents or cosolutes (see Section 10.5.1). Since unfolding is typically favored at both high and low temperatures, freeze-thawing and freeze-drying may damage the native structure of antibody molecules to the same extent as elevated temperatures. Even in the solid state, conformational changes may occur at temperatures near or above the glass transition temperature (see Sections 10.5.2 and 10.6.1).

Conformational perturbations typically promote irreversible adsorption and nonnative aggregation, which in some cases may lead to subsequent precipitation. Recent publications have clearly demonstrated that partially unfolded molecules rather than fully unfolded molecules are the reactive species that form physically and/or covalently linked aggregates (Chi et al. 2003; Minton 2005). Antibody aggregates often retain a large amount of their secondary structure, while losing their tertiary structure. Moreover, intermolecular disulfides are generally involved in the aggregation mechanism (Andya et al. 2003). Since aggregation of whole antibodies or fragments is inherently a nucleation and growth phenomenon, a lag phase resulting from an energy barrier to nucleation or assembly is often observed. Consequently, conditions that promote unfolding and/or the formation of soluble aggregates should be avoided, in order to obtain an acceptable long-term stability of the final antibody formulation.

10.4.2 Design of Stability Studies

Stability testing is intended to provide evidence on how the quality of a drug substance or drug product varies with time under the influence of a variety of environmental factors such as temperature, humidity, and light, and enable recommended storage conditions, retest periods and shelf-lives to be established. Consequently, stability studies are an integral part of drug preformulation and formulation development, and subsequent quality control of the marketed product. The design of stability testing programs within the different stages of the drug development process strongly depends on regulatory issues, practical needs, scientific knowledge, economic aspects, and accessibility of appropriate analytical methods.

10.4.2.1 Regulatory Aspects

From a regulatory point of view, stability is defined as "the capacity of a drug to remain within product specifications that have been established to assure its identity, potency, purity, efficacy and safety." The regulatory stability testing requirements for a Registration Application within the areas of Europe, Japan, and the USA are generally summarized in the ICH harmonized tripartite guideline entitled "Stability Testing of New Drug Substances and Products" (ICH Guideline Q1A). The guideline provides information on test procedures for drug substance and finished product including storage conditions for long-term and accelerated testing, test criteria, selection of batches, testing frequency, packaging, and labeling. The guideline emphasizes that expiration dating should be based on real-time, real-temperature data. The length of the studies and the storage conditions should be sufficient to cover storage, shipment, and subsequent use. Validated analytical methods must be applied.

For biopharmaceutical drug substances, including antibodies with physicochemical properties that differ from those of classical low molecular weight drugs, the general tripartite ICH guideline on stability was supplemented by Annex Q5C (ICH Guideline Q5C). Q5C is intended to provide information on how to consider these distinguishing properties in a well-defined testing program for the development of appropriate stability data. Typically, Q5C implies that "there is no single stability-indicating assay or parameter that profiles the characteristics of a biotechnological product." Consequently, various physicochemical, biochemical and immunochemical methodologies must be applied to comprehensively characterize drug substance and drug product, and accurately detect any changes resulting from chemical or physical degradation during storage. In addition, accelerated and stress conditions described in Q1A may not be appropriate for biotechnological products. Thus, it is recommended in Q5C that accelerated and stress conditions should be carefully selected on a case-by-case basis. This aspect will be addressed in detail in Section 10.4.2.3.

10.4.2.2 **Analytical Tools**

Several analytical techniques are usually applied to gain a comprehensive insight into the stability profile of antibody drugs. These include electrophoretic methods, chromatographic and spectroscopic techniques, laser light scattering, ultracentrifugation, calorimetry and peptide mapping (Wang 1999). Selecting the appropriate analytical tool for antibody stability testing requires a thorough understanding of the power and limitations of each of these tools. This section briefly emphasizes the potential of the most frequently used analytical techniques.

The presence of deamidated forms of asparagine or glutamine can easily be detected by an acidic shift in isoelectric focusing (IEF) bands or shifts in ion-exchange high-performance liquid chromatography (HPLC) retention times. Peptide mapping by reverse-phase HPLC is usually the tool employed to identify sites of deamidation. Peptide mapping may also be used for detection of methionine sulfoxidation sites. Low molecular weight degradation products resulting from antibody proteolysis can be determined by sodium dodecyl sulfate polyacrylamide gel electrophoresis (SDS-PAGE) and size-exclusion chromatography (SEC), respectively.

Various spectroscopic techniques are intended to evaluate conformational changes including secondary and tertiary structural features. UV and fluorescence spectroscopy both rely on the susceptibility of aromatic amino acid side chains (Phe, Tyr, Trp) to their microenvironment, thus allowing tertiary structural changes to be monitored in solution. Circular dichroism (CD), Fourier transform infrared spectroscopy (FTIR), and Raman spectroscopy are sensitive to secondary structural changes allowing α-helix and β-sheet content to be determined. Like UV and fluorescence spectroscopy, CD measurements can only be performed in solution, while FTIR and Raman spectroscopy are well suited in solid and solution state. The latter two have therefore been used as powerful tools for the determination of secondary structural perturbations of various mAbs upon changes in solution pH, lyophilization, or spray drying (Sane et al. 2004; Chang 2005a). Since the extent of drying-induced structural perturbations have been reported to exhibit good correlation with the aggregation rate of various mAbs upon long-term storage, spectroscopic methods may be considered as quick and reliable tools to screen excipients and excipient concentrations in dry powder formulations.

Qualification and quantification of mAb aggregates are rather challenging. Theoretically, a great number of analytical techniques, including HPLC/SEC, ion-exchange chromatography (IEX), SDS-PAGE, turbidimetry, laser light scattering, ultracentrifugation, and filtration are intended to provide information on the extent of aggregation. In practice, however, selective and reliable quantification of soluble and insoluble mAb aggregates requires advanced analytical methods such as asymmetric field flow fractionation (AFFF) combined with multiangle laser light scattering. Since operation is possible with dissolved and dispersed molecules, soluble and insoluble aggregates may be determined simultaneously. If only soluble aggregates are present, SEC and SDS-PAGE may provide a rough estimate of dimer, trimer etc. formation including information on the nature of aggregate formation (covalent versus noncovalent). Reliable quantification of soluble aggregates is feasible by sedimentation velocity in an ultracentrifuge.

Functional changes of "naked" whole antibodies or fragments can be analyzed by binding assays to purified antigens and/or defined regions of antigens. *In vitro* potency testing of drug–antibody conjugates is more sophisticated, since it requires an appropriate surrogate test that allows drug release under *in vivo* conditions to be considered.

10.4.2.3 **Practical Approach**

Stability studies during preformulation are intended to:

- Define stress parameters relevant for dosage form screening including manufacturing, shipping and storage conditions
- Provide information on degradation sites and pathways
- Evaluate degradation kinetics
- Screen potential stabilizers.

Since data generated at this stage serve as the basis for subsequent formulation development, efficient testing programs enabling fast assessment of stability characteristics with less expenditure of material are required. Procedures known as "accelerated stability testing" based on "classical isothermal" approaches require a large number of samples stored at different constant temperatures over a prolonged period of time. Moreover, their applicability to accelerated protein stability tests is still a matter of debate, since Arrhenius behavior of chemical protein degradation is only valid at temperatures where unfolding is not an issue. In view of this, accelerated "non-isothermal" kinetic methods in combination with analytical tools that allow conformational changes to be monitored, have recently been described as an alternative approach for the early-stage development phase of protein formulations (Reithmeier and Winter 2002). Degradation rate constants obtained from non-isothermal studies in the temperature range below T_m of the model protein showed good comparability to isothermal data. The resulting Arrhenius plots were linear and enabled shelf-life of various formulations to be estimated by extrapolating the stability data to the storage temperature. In addition, nonlinear behavior was detectable in a much shorter time. Since miniaturization and automation of non-isothermal experiments makes high-throughput screening of potential stress parameters and stabilizing excipients feasible, the non-isothermal approach is an interesting means to make preformulation stability testing of antibody drugs more efficient and less time consuming.

Contrary to preformulation studies, the design of long-term and accelerated stability testing programs of the final formulation (i.e. the finished product) is well described in the ICH tripartite guidelines Q1A and Q5C, allowing test conditions for antibody products to be defined on a case-by-case basis. Consequently, storage conditions for real-time/real-temperature stability studies are usually confined to the proposed storage temperature which is often precisely defined. In the case of freeze-dried products, the stability of the product after reconstitution should be demonstrated in addition to dry powder stability.

10.5 Formulation and Manufacturing of Parenteral Delivery Systems

10.5.1 Ready-to-Use Solutions and Concentrates

The formulation and manufacturing design of parenteral antibody solutions requires general quality attributes and more specific stability, delivery, and compliance aspects to be considered. Typically, aqueous antibody solutions intended for intravenous application have to be sterile, particle-free, and isotonic, and in case of high volume infusions, free of endotoxins. Osmolarity is not a major issue when formulating subcutaneous injections; however, isotonic solutions are desirable to allow painless injections. The pH value of the solution should provide sufficient solubility, optimal chemical and physical long-term stability, and biocompatibility. Viscosity should be as low as possible to allow economic processing and reduce pain upon injection. If the product is intended for multi-use, it is essential to include preservatives in the formulation.

To maintain chemical and physical stability during manufacturing, shipping, and/or storage, potential stress parameters such as high and low temperature, oxygen, light, pH changes, shear stress and adsorption to interfaces should be reduced to a minimum. Protection from oxygen and light to improve chemical stability are usually easy to perform by using nitrogen flushing during the manufacturing process, and selecting a packaging material that protects the solution from oxygen and light. Temperature control during manufacturing is also feasible. As an example, pathogen inactivation in antibody solutions is usually performed by sterile filtration at room temperature instead of using heat. The method is gentle provided that the filter material does not induce antibody adsorption and/or aggregation. Unlike the manufacturing process, precise control of storage conditions during shipping is not always feasible.

Additional concerns to be considered are shear stress and contact with interfaces. Adsorption is always an issue in syringes, filters, and packaging materials and can lead to unrecoverable product loss. Shear stress-induced mAb unfolding, adsorption, aggregation, and/or precipitation can occur in filter and pumping systems. For example, piston-driven pumps usually tend to generate more shear stress than rolling diaphragm pumps. This should be taken into consideration when filling shear sensitive antibody solutions, not only since unrecoverable product loss increases cost of goods and reduces potency, but also because non-native aggregates may have a major impact on the pharmacokinetic and safety profile of the drug product due to increased immunogenicity.

Viscosity implications are a major issue when developing small-volume, high-dose antibody solutions for subcutaneous injection (Liu et al. 2005). Reversible multivalent antibody self-association mediated by electrostatic interactions of charged residues on the molecule surface may result in unusually high solution viscosity. This nonlinear viscosity increase with increasing mAb concentration has been reported to complicate manufacturing processes, stabilization, and administration. High solution viscosity may result in reduced membrane flux and high pressure-drops during sterile filtration and tangential flow filtration (TFF), the main technology for large-scale buffer exchange (i.e. concentration and formulation of antibody solutions). Depending on the propensity of the antibody molecules to interact and unfold at the membrane surface, economically unacceptable losses may occur as a result of membrane clogging. Moreover, application is aggravated, since high-viscosity solutions require the use of large-bore needles, which may result in more painful injections. Furthermore, if the *in vivo* dissociation rate upon dilution is slow, reversible self-association may have an impact on antibody potency and overall immunogenicity.

In conclusion, formulation development requires an integrated approach whereby a stable formulation is developed that can be successfully administered and economically manufactured. Further to PEGylation, the proper choice of excipients such as buffers, tonicifiers, and various types of stabilizers is a primary concern (Wang 1999). Two examples of antibody formulations are described in Table 10.1.

10.5.1.1 Appropriate Excipients

Buffers Buffers serve to adjust the solution pH to optimal solubility, chemical and physical stability, and physiological compatibility. Various buffers such as

Table 10.1 Examples of excipients used in antibody solutions.

Bevacizumab concentrate (Avastin)	*Adalimumab concentrate (Humira)*
Na_2HPO_4/NaH_2PO_4	Na_2HPO_4/NaH_2PO_4
α,α-Trehalose $2H_2O$	Citric acid/sodium citrate
Polysorbate 20	Sodium chloride
Water for injection	Mannitol
	Sodium hydroxide
	Polysorbate 80
	Water for injection

citrate, histidine, Tris, etc. are in use. However, buffer choice can be critical, since salt type as well as ionic strength may have a major impact on the solution viscosity. Histidine is often the first choice in highly concentrated solutions, since it has been found to effectively reduce reversible self-association of monoclonal antibodies (Chen et al. 2003).

Tonicifiers Excipients that are often used as tonicifiers are mannitol, glycine, or sodium chloride. The latter is preferred when solution viscosity might be a concern, since sodium chloride has been observed to reduce electrostatic protein–protein interactions leading to antibody self-association. However, formulations containing sodium chloride should avoid being stored in stainless-steel vessels to prevent generation of iron ions that may catalyze oxidative degradation of antibody molecules (Lam et al. 1997).

Osmolytes To increase the thermodynamic stability of the native antibody structure (i.e. reduce temperature-, shear- or pH-induced antibody unfolding and nonnative aggregation), molar concentrations of osmolytes such as sugars and polyols have been applied. Since these compounds are preferentially excluded from the native antibody molecule surface, they increase the protein's chemical potential by preferential hydration of the native antibody structure. Despite this advantage, the use of osmolytes in antibody solutions may be limited, since they also add to the viscosity and osmolality of the formulation which in turn may render it impractical for subcutaneous delivery (Liu et al. 2005). High concentration of sugar can lead to hypertonic solutions and enhance self-association of native antibody molecules in high-concentration formulations, both of which will make injections painful.

Surfactants The use of surfactants is essential to reduce the intrinsic and agitation-induced adsorption tendency of mAbs to interfaces. Selection of the appropriate surfactant type and concentration may be critical, since surfactants usually have a certain tendency to bind to mAbs thereby compromising their conformation. In practise, stabilizing concentrations are very low and should be discerned experimentally. Regulatory approval for parenteral use is an issue to be considered when selecting the surfactant type. Nonionic polyether surfactants such as Tween 80 and 20 are usually applied.

Preservatives Multidose formulations must contain preservatives to protect them from microbial contamination upon multiple withdrawals. Selection of the optimal preservative depends on a number of factors including solubility, efficacy, and compatibility with the formulation and the route of administration (Gupta and Kaisheva 2003). In this respect, stability issues of the antibody are a major concern, since certain preservatives such as phenolic compounds are known to cause precipitation of humanized monoclonal antibodies in aqueous solution. Benzyl alcohol, one of the least toxic and most widely used parenteral preservatives, has also been reported to trigger mAb aggregation, although in a concentration-dependent manner. Combinations of benzyl alcohol/chlorobutanol and benzyl alcohol/methylparaben have been screened as potential candidates for the preservation of antibody solutions containing histidine buffer, Tween 80, and sodium chloride as surfactant and tonicifier.

10.5.2 Freeze-Dried Powders

If parenteral administration is intended, a ready-to-use solution is the most convenient dosage form for the end user. However, chemical and physical stability issues of therapeutic antibodies are generally more pronounced in solution than in solid state, since increased mobility of dissolved molecules facilitates conformational changes and accessibility of reactive sites. In fact, whenever preformulation studies indicate that sufficient stability cannot be achieved in solution, dry powder formulations for reconstitution prior to administration are an attractive alternative. Freeze-drying of sterile filtered solutions under aseptic conditions is the most conventional way to obtain antibody products with the following characteristics:

- Elegant cake structure without any collapse upon storage
- Long-term chemical and physical stability
- Sterility upon storage, reconstitution and optional multi-use
- Fast reconstitution
- Isotonicity upon reconstitution
- Maintainance of all chemical and physical characteristics of the original dosage form upon reconstitution.

To achieve these quality attributes, a comprehensive process and formulation understanding, including potential stress parameters upon freezing, drying, storage, and reconstitution, is essential (Carpenter et al. 1997).

The freeze-drying process consists of three stages, namely freezing, primary drying, and secondary drying. The freezing step is intended to form pure ice crystals and solidify the remainder of the solution in an amorphous state. The temperature that allows the latter to be achieved is usually between −20 °C and −60 °C, i.e. well below the collapse temperature (T_c) of the formulation. In the drying step, the product should be kept as well below T_c to remain in the solid amorphous state. During the primary drying step, ice is removed by sublimation at subambient temperatures (usually between −40 °C and −10 °C) under vacuum (40–400 mTorr). In the secondary drying stage, small amounts of bound water may be removed by increasing the temperature to between 5 °C and 20 °C (i.e. below the glass transition temperature (T_g) of the product, but high enough to promote adequate desorption rates). Precise control of cooling rate, drying temperature, and residual moisture content is essential for final product performance (e.g. cake structure and antibody stability).

Critical stability issues in the freezing process are: (1) exposure of the antibody to the ice/water interface, and (2) pH and/or salt effects in the highly concentrated undercooled solution. Antibody exposure to the ice/water interface is mainly affected by the cooling rate which determines the number and size of ice crystals, and the antibody concentration which determines the percentage of molecules being in contact with the ice/water interface. A fast cooling rate leads to a large number of small ice crystals (i.e. a large ice/water interface). Since ice crystals may be considered as solid interfaces with the potential risk of protein unfolding, adsorption and/or aggregation, fast cooling and a low overall protein concentration are unfavorable conditions in this respect. Fast cooling, however, can be desirable to reduce the exposure time of the antibody drug to unfavorable pH conditions that may occur as a result of gradual buffer salt crystallization upon freezing, and lead to chemical and/or physical instabilities. The most drastic changes have been observed with sodium phosphate and potassium phosphate buffer salts, leading to acidic or alkaline pH values after crystallization of one of the buffer salt components.

A potential stress parameter during dehydration is the removal of the protein hydration shell, which causes significant conformational changes in the absence of appropriate stabilizers. In many cases, the extent of antibody unfolding during freeze-drying was found to directly correlate with the subsequent rate of non-native aggregation and/or chemical degradation upon storage (Sane et al. 2004). In fact, the chemical and physical long-term stability of a lyophilized antibody formulation strongly depends on the extent of structural perturbations during freezing and dehydration, and the molecular mobility of the protein within the dry powder matrix (i.e. the formulation composition, the residual moisture content, and the storage conditions) (Chang et al. 2005a/b). Product storage below the glass transition temperature (T_g) is recommended to enhance long-term stability (e.g. improve shelf-life and retain reconstitution properties at ambient room conditions to allow 100% recovery of intact antibody molecules).

In conclusion, formulation design of freeze-dried antibody powder products aims to provide (1) minimal pH change upon freezing, (2) structural preservation upon freezing and dehydration, (3) product solidification in the amorphous state, (4) a high glass transition temperature of the final product, thus reducing the molecular mobility of the protein upon storage, and (5) minimal unfolding and/or aggregation upon reconstitution. Appropriate excipients (Table 10.2) may be classsified as buffers, tonicifiers, stabilizers, and bulking agents (Carpenter et al. 1997).

Table 10.2 Examples of excipients used in freeze-dried antibody products.

Infliximab (Remicade)	*Trastuzumab (Herceptin)*	*Omalizumab (Xolair)*
Na_2HPO_4/NaH_2PO_4	Histidine/histidine HCl	Histidine/histidine HCl
Sucrose	α,α-Trehalose $2H_2O$	Sucrose
Polysorbate 80	Polysorbate 20	Polysorbate 20

10.5.2.1 Appropriate Excipients

Buffer salts Selection of buffer salt type and concentration is an important issue, since drastic pH changes upon freezing may increase the risk of deamidation, unfolding, and/or aggregation of antibodies. Buffers that have minimal pH change upon freezing include citrate, histidine, and Tris. Quite a number of lyophilized antibody formulations have been formulated using noncrystallizing histidine as a buffer at pH from 5.5 to 6.5. Besides functioning as a buffer, histidine has also been reported to protect antibodies such as rhuMAb HER2 and ABX-IL8 upon freezing as evidenced by lower levels of aggregation after multiple freeze/thaw cycles (Chen et al. 2003).

Tonicity modifiers Tonicity modifiers such as mannitol, glycine, sucrose, glycerol, and/or sodium chloride can be included either in the reconstitution medium or in the powder formulation. In the latter case, their crystallization tendency and their effect on the glass transition temperature of the product should be considered.

Bulking agents Bulking agents are intended to increase powder mass and improve cake structure at low drug concentrations. Since the concentration of antibodies is usually high, bulking agents are not necessary.

Stabilizers Stabilizers may be characterized as cryoprotectants, lyoprotectants, and reconstitution aids. Cryoprotectants are excipients that can preserve the native antibody structure during the freezing process. Cryoprotection may be explained by the mechanism of preferential hydration of the native antibody structure, which can be achieved by the preferential exclusion of the cryoprotectant from the protein surface in both the liquid and frozen state. Polyols, monosaccharides, disaccharides and hydrophilic polymers such as polyethyleneglycol have all been reported to act as cryoprotectants thereby preserving the native antibody structure during the freezing process.

Lyoprotection refers to the prevention of drying-induced conformational changes of the protein and is best achieved by disaccharides. It may be explained by the "water replacement mechanism," which states that dissacharides protect proteins during dehydration by hydrogen bonding to polar and charged groups on the surface of the native conformation.

Reconstitution aids may help rehydration of the lyophilized powder prior to administration (i.e. they reduce the tendency of the antibody to form aggregates during reconstitution). Low concentrations of nonionic surfactants such as polysorbate in either the powder or the diluent are usually effective in preventing aggregate formation during rehydration. Although the exact mechanism is still unknown, it may be assumed that the surfactant interferes with the intermolecular protein interactions and/or serves as a wetting agent that hastens the dissolution of the freeze-dried cake (Webb et al. 2002).

The proper choice of stabilizers requires several product characteristics to be considered, namely (1) elegant and mechanically strong cake structure, (2) fast reconstitution, and (3) long-term chemical and physical antibody stability. In view of this, the excipient(s) should allow for preferential hydration of the native antibody structure in the frozen state, and hydrogen bonding to polar and charged groups upon dehydration. In addition, the formulation should provide an amorphous glassy matrix with a high glass transition temperature in the dried state to achieve long-term product stability at ambient conditions. Generally, disaccharides are superior to polyols and/or hydrophilic polymers, since they may function as cryo- and lyoprotectants. Their crystallization tendency is usually low, which is an additional advantage over mannitol, for instance. However, one group of compounds, namely reducing sugars should be avoided, since they have the propensity to degrade proteins via Maillard reaction between sugar carbonyls and protein free amino acid groups.

Numerous reports provide evidence that trehalose and sucrose are first choice for stabilizing antibodies and fragments thereof during freeze-drying and storage in the dried solid state (Andya et al. 2003). The mechanism of stabilization by these sugars has been proposed to occur by acting as water substitute and producing a glassy matrix that restricts mobility. Moreover, a diluent effect may not be excluded. Usually sucrose and trehalose provide equivalent protection against chemical and physical degradation of freeze-dried antibodies upon storage. Concentrations that are 3- to 4-fold below the iso-osmotic concentration and equivalent to a 360 : 1 molar ratio of sugar to antibody are generally sufficient to ensure product stability even at elevated temperature of 40 °C (Cleland et al. 2003).

10.5.3
Crystalline Suspensions

Antibody therapies generally require frequent delivery of between 100 mg and 1 g of protein per dose to achieve clinical efficacy. These doses are typically administered through large volume i.v. infusions in a hospital setting. Delivery of these large doses in a small volume appropriate for subcutaneous injection is likely to improve therapeutic opportunities and patient compliance of antibody treatments. However, highly concentrated solutions often result in very high viscosity, which may cause manufacturing problems, poor overall stability and delivery problems as indicated in Section 10.5.1. Crystallization of full-length antibodies or antibody fragments is a novel approach that holds great promise for the development of high-dose, low-viscosity antibody suspensions (Yang et al. 2003).

Recently, a large-scale batch crystallization process has been developed which allows full-length monoclonal antibodies to be crystallized efficiently in high yields (>90%) (Shenoy et al. 2002). Small crystals with excellent chemical and physical stability upon storage at room temperature including full retention of biological activity *in vivo* can be produced. Crystals and crystal formulations of different size, shape, and morphology, including spherical nanocrystalline composite particles with different dissolution properties, may be achieved by manipulation of the crystallization protocol (Yakovlevsky et al. 2005). In addition, the process can streamline the production of pharmaceutical antibody formulations by replacing some of the purification steps.

Once crystallized, the antibodies can be formulated into high-concentration suspensions that are biocompatible when injected subcutaneously. *In vivo* release rates may be controlled by varying the size and morphology of the crystals and/or the vehicle composition, thus allowing fast and carrier-free slow release dosage forms to be developed. PEG/ethanol mixtures have proven to be appropriate nonaqueous vehicles for providing low-viscosity formulations that maintain both crystallinity and integrity of various monoclonal antibodies such as rituximab, infliximab, and trastuzumab. Suspensions containing 200 mg mL^{-1} of crystalline trastuzumab in PEG/ethanol did not show any aggregation upon storage over a period of 20 weeks at 4 °C. Viscosity remained low and allowed injection of 1 mL suspension with a 26-gauge needle in less than 5 s. Efficacy in a preclinical mouse model of human breast cancer was clearly demonstrated. Histological analysis of the injection sites revealed rapid dissolution of the crystals after subcutaneous (s.c.) injection and high biocompatibility.

Crystalline infliximab suspensions injected subcutaneously in rats revealed slow release rates with an extended serum pharmacokinetic profile (i.e. a longer biological half-life compared with s.c. or i.v. injections of the commercially available solution). Furthermore, a higher area under the curve (AUC), indicating a higher bioavailability, was observed.

These examples clearly indicate that crystalline antibodies and suspensions thereof are an improved and versatile formulation approach for high-dose antibody delivery by the subcutaneous route. In addition, antibody crystals and stabilized crystal formulations may be advantageously encapsulated in a polymeric carrier to produce controlled-release microparticles.

10.5.4 Carrier-based Systems

In addition to crystalline suspensions, carrier-based delivery systems have gained increasing interest to provide parenteral controlled-release of whole antibodies or antibody fragments. One approach is the use of microparticles based on biodegradable polymers such as poly-D, L-lactide-co-glycolide (PLGA). PLGA polymers are available in a range of molecular weights and monomer ratios, thus providing a number of variables with which the antibody release rate can be adjusted. PLGA is legally approved for parenteral use, since it is nontoxic, nonimmunogenic and well tolerated after subcutaneous injection. A variety of microencapsulation techniques are available to produce PLGA microparticles (Benoit et al. 1996). They may be classified as phase separation or coacervation, emulsion solvent evaporation, and spray drying. The polymer is generally dissolved in an organic solvent, and the antibody drug may be encapsulated in either solid form or solution state.

When producing microencapsulated formulations of whole antibodies or antibody fragments, it is important that the chemical, physical, and biological properties of the antibody remain intact during encapsulation (Bilati et al. 2005). Since conditions such as exposure to water/organic solvent interfaces and homogenization may compromise the structural integrity of dissolved antibody molecules, formulation approaches based on PEGylated and/or solid crystalline antibodies are of particular interest. The solubility profile of PEG, including a certain shielding effect when attached to antibodies, opens up new opportunities for emulsion-based encapsulation techniques, which typically lead to adsorption and/or aggregation problems when applied to non-PEGylated antibodies or fragments. The use of stable nanocrystalline antibodies or antibody formulations is another option that may allow successful encapsulation with enhanced preservation of the native antibody structure during the encapsulation process (Yakovslevsky et al. 2005). Typically, the release rate from microparticles can be varied from days to months by adjusting the microencapsulation parameters, the polymer properties, the antibody loading, the crystal size, shape, and morphology, and the formulation used to prepare the nanocrystals. The potential of this challenging technology for controlled-release antibody formulations has to be verified in the future.

10.6 Formulation and Manufacturing of Local Delivery Systems

10.6.1 Inhalation Powders

Long-term stability of monoclonal antibodies is usually enhanced when stored in a dry solid rather than a liquid state. Contrary to parenteral delivery systems, dry powder formulations for local pulmonary delivery of antibody drugs require not only drug stability issues, but also aerosol performance (i.e. aerodynamic properties) to be addressed. Aerodynamic properties defining the fraction of aerosolized drug to be delivered to the lung are strongly affected by particle size and morphology. Since freeze-drying procedures usually generate cakes rather than powders, the process is not the method of choice when particles with good dispersibility and defined aerodynamic particle size distributions are required, such as for pulmonary delivery. Several drying techniques have been explored for their ability to produce fine inhalation powders. So far, spray drying is the most popular method, but supercritical fluid and spray freeze drying technologies have recently emerged as promising alternatives.

10.6.1.1 Spray Drying

Spray drying is a manufacturing process that typically yields powder particles small enough for aerosol delivery to the lower airways of the lung (1–7 μm). However, processes involved in spray drying impose potential stress to antibody drugs, including exposure to high temperature, distribution at air–water interfaces, and dehydration. In the absence of sugar or polyol stabilizers, any of these stress parameters may induce structural perturbations to antibody molecules, thus compromising their biological activity or immunogenicity. In fact, a significant decrease in β-sheet content and a corresponding increase in turn and unordered content upon spray drying was monitored with several antibodies using FTIR or Raman spectroscopy (Sane et al. 2004). A correlation between the extent of structural perturbations immediately after spray drying and the rate of aggregation upon long-term storage of the formulation was reported. This clearly indicates that preservation of the native mAb structure during spray drying is essential for long-term product stability, as was already pointed out for freeze-dried products.

To protect full-length mAbs and antibody fragments during spray drying, different excipients such as polyols, sugars, hydrophilic polymers, surfactants, amino acids, and/or proteins are generally included in the final formulation. The stabilizing mechanism of surfactants is thought to be a competitive adsorption at the air–water interface, thereby retaining the native antibody structure. The effect of polyols and sugars may be explained by preferential hydration and/or water substitution by hydrogen bonding to the antibody molecules. The stabilizing effects are concentration dependent (i.e. a certain excipient to mAb mass ratio is required to preserve the native structure). However, any excessive excipient can lead to a decrease in the physical stability of the formulation, which in turn may affect aerosol performance. This has been demonstrated for spray-dried rhuMAbE25 stabilized with mannitol, trehalose, or lactose, respectively (Andya et al. 1999). Trehalose was found to increase powder cohesiveness in a concentration-dependent manner. Lactose exhibited acceptable powder performance, but protein glycation was observed during storage. Mannitol at a molar ratio of higher than 200:1 resulted in crystallization, thus compromising aerosol performance. In conclusion, a balance must be achieved between addition of enough stabilizer to improve protein stability without compromising aerosol performance.

10.6.1.2 Spray Freeze-drying

Spray freeze-drying, as opposed to spray drying, produces particles with light and porous characteristics, which offer more favorable aerodynamic properties and thus a better aerosol performance (Maa et al. 1999). The process is highly efficient in terms of product recovery (>95%). However, selection of appropriate formulation excipients is crucial for product quality, since crystallization can deteriorate aerodynamic powder properties. Application of supercritical fluid (SCF) technologies is especially attractive for reasons of mild process conditions, cost-effectiveness, feasibility of scaling up, possible sterilizing properties of supercritical carbon dioxide, and capability of producing powder particles with defined physicochemical properties. Several concepts have been described in the literature using a SCF (usually carbon dioxide) either as antisolvent or propellant during a low-temperature spray-drying process. Since limited data are available on antibody stability during SCF processing, further research is required to identify stress parameters, and provide rationales for product development (Jovanovich et al. 2004).

10.6.2 Various Dosage Forms

As of today, limited data are available on the stability of antibodies or fragments in combination with oral and peroral delivery systems. Toothpaste, mouthwash, or chewing gum might be attractive formulations for local administration of antibodies in the mouth. To protect the antibody from pH- and/or enzyme-induced degradation in the upper gastrointestinal tract, controlled-release formulations based on gastroresistant polymers are a viable approach (Kälkert and Reich 2004). Delivery to the small intestine can be achieved using enteric-coating polymers such as Eudragit L or S that are insoluble at low pH, but soluble in the neutral environment of the small intestine. Direct compression of stable crystalline antibody powders with subsequent tablet coating, pellet formation, or microencapsulation based on stabilized antibody formulations are options to date. Colon targeting represents a greater technological challenge, as these dosage forms must pass through the whole upper gastrointestinal tract before delivering the antibody to the colon. Four different delivery principles based on pH-dependent dissolution, time-dependent erosion or dissolution, pressure-induced disintegration, or enzymatic degradation are available (Bauer 2001). Their clinical reliability for local antibody delivery to the colon has to be demonstrated.

10.7 Outlook

Site-specific drug delivery has been an ultimate goal of pharmaceutical product design. Antibodies and antibody fragments have the potential to realize this objective in various directions. To fully exploit their targeted effector and vector function, multidisciplinary collaborative efforts are key elements of future product development. Pharmaceutical research must focus on innovative formulation concepts for parenteral and local controlled delivery of tailor-made antibodies and antibody fragments. Major technological challenges are the design of high-dose, low-viscosity sustained-release systems, and antibody fragments with improved *in vitro* and *in vivo* stability. Moreover, optimization of biophysical properties of stealth immunoliposomes and antibody/drug conjugates are of major importance for further progress in tumor and brain targeting including gene delivery.

References

Andya, J.D., Maa, Y.-F., Costantino, H.R., Nguyen, P.-A., Dasovich, N., Sweeney, T.D., Hsu, C.C., Shire, S.J. (1999) The effect of formulation excipients on protein stability and aerosol performance of spray-dried powders of a recombinant humanized anti-IgE monoclonal antibody. *Pharm Res* 16: 350–358.

Andya, J.D., Hsu, C.C., Shire, S.J. (2003) Mechanisms of aggregate formation and carbohydrate excipient stabilization of lyophilized humanized monoclonal antibody formulations. *AAPS PharmSci* 5(2) Article 10.

Bauer, K.H. (2001) Colonic drug delivery: review of material trends. *Am Pharm Rev* 4: 8–16.

Benoit, J.P., Marchais, H., Rolland, H., Velde, V.V. (1996) Biodegradable microspheres: advances in production technology. In: S. Benita (ed.) *Microencapsulation: Methods and Industrial Applications.* New York: Marcel Dekker, pp. 35–72.

Bilati, U., Allemann, E., Doelker, E. (2005) Strategic approaches for overcoming peptide and protein instability within biodegradable nano- and microparticles. *Eur J Pharm Biopharm* 59: 372–388.

Brannon-Peppas, L., Blanchette, J.O. (2004) Nanoparticle and targeted systems for cancer therapy. *Adv Drug Deliv Rev* 56: 1649–1659.

Calicati, P., Veronese, F.M. (2003) Pharmacokinetic and biodistribution properties of poly(ethylene glycol)-protein conjugates. *Adv Drug Deliv Rev* 55: 1261–1277.

Cao, Y., Lam, L (2004) Bispecific antibody conjugates in therapeutics. *Adv Drug Deliv Rev* 55: 171–197.

Carpenter, J.F., Pikal, M.J., Chang, B.S., Randolph, T.W. (1997) Rational design of stable lyophilized protein formulations: some practical advice. *Pharm Res* 14: 969–975.

Chang, L., Shephard, D., Sun, J., Quellette, D., Grant, K.L., Tang, X., Pikal, M.J. (2005a) Mechanism of protein stabilization by sugars during freeze-drying and storage: native structure preservation, specific interaction, and/or immobilization in a glassy matrix. *J Pharm Sci* 94: 1427–1444.

Chang, L., Shephard, D., Sun, J., Xiahu, T., Pikal, M.J. (2005b) Effect of sorbitol and residual moisture on the stability of lyophilized antibodies: implications for the mechanism of protein stabilization in the solid state. *J Pharm Sci* 94: 1445–1455.

Chapman, A.P. (2002) PEGylated antibodies and antibody fragments for improved therapy: a review. *Adv Drug Deliv Rev* 54: 531–545.

Chen, B., Bautista, R., Yu, K., Zapata, G.A., Mulkerrin, M.G., Chamow, S.M. (2003) Influence of histidine on the stability of physical properties of a fully human antibody in aqueous and solid forms. *Pharm Res* 20: 1952–1960.

Chi, E.Y., Krishnan, S., Randolph, T.W., Carpenter J.F. (2003) Physical stability of proteins in aqueous solution: mechanism and driving forces in nonnativeprotein aggregation. *Pharm Res* 20: 1325–1336.

Cleland, J., Lam, X., Kendrik, B., Yang, J., Yang, T.-H., Overcashier, D., Brooks, D., Hsu, C., Carpenter, J.F. (2001) A specific molar ratio of stabilizer to protein is required for storage stability of a lyophilized monoclonal antibody. *J Pharm Sci* 90: 310–321.

Greenwald, R.B., Choe, Y.H., McGuire, J., Conover C.D. (2004) Effective drug delivery by PEGylated drug conjugates. *Adv Drug Deliv Rev* 55: 217–250.

Gupta, S., Kaisheva, E. (2003) Development of a multidose formulation for a humanized monoclonal antibody using experimental design techniques. *AAPS PharmSci* 5(2) Article 8.

ICH Guideline Q1A (1993) *Stability Testing Guidelines: Stability Testing of New Drug Substances and Products, International Conference on Harmonization of Technical Requirements for Registration of Pharmaceuticals for Human Use.* http://www.ich.org/cache/compo/363-272-1.html

ICH Guideline Q5C (1995) *Quality of Biotechnological Products: Stability Testing of Biotechnological/Biological Products, International Conference on Harmonization*

of Technical Requirements for Registration of Pharmaceuticals for Human Use. http://www.ich.org/cache/compo/363-272-1.html
Jovanovich, N., Bouchard, A., Hofland, G.W., Witkamp, G.-J., Crommelin, D.J.A., Jiskoot, W. (2004) Stabilization of protein dry powder formulations using supercritical fluid technology. *Pharm Res* 21: 1955–1969.
Kälkert, K., Reich, G. (2004) Spray dried gelatine/Eudragit microparticles as a gastroresistant release system for proteins. *Proc Int Meeting on Pharmaceutics, Biopharmaceutics and Pharmaceutical Technology*, 15–18 March, Nuremberg, pp. 269–270.
Kupper, M.B., Huhn, M., Spiegel, H., Ma, J. K.C., Barth, S., Fischer, R., Finnern, R. (2005) Generation of human antibody fragments against *Streptococcus mutans* using a phage display chain shuffling approach. *BMC Biotechnol* 5: 4–12.
Lam, X.M., Yang, J.Y., Cleland, J.L. (1997) Antioxidants for prevention of methionine oxidation in recombinant monoclonal antibody HER2. *J Pharm Sci* 86: 1250–1255.
Liu, J., Nguyen, M.D.H., Anya, J.D., Shire, S. J. (2005) Reversible self-association increases the viscosity of a concentrated monoclonal antibody in aqueous solution. *J Pharm Sci* 94: 1928–1940.
Maa, Y.-F., Nguyen, P.-A., Sweeney, T.D., Shire, S.J., Hsu, C.C. (1999) Protein inhalation powders: spray drying vs spray freeze drying. *Pharm Res* 16: 249–254.
McNeel, D.G., Eickhoff, J., Lee, F.T., King, D.M., Alberti, D., Thomas, J.P., Friedl, A., Kolesar, J., Marnocha, R., Volkman, J., Zhang, J., Hammershaimb, L., Zwiebel, J.A., Wildung, G. (2005) Phase I trial of a monoclonal antibody specific for $\alpha v\beta 3$ integrin (MEDI-522) in patients with advanced malignancies, including an assessment of effect on tumor perfusion. *Clin Cancer Res* 11: 7851–7860.
Minton, A.P. (2005) Influence of macromolecular crowding upon the stability and state of association of proteins: predictions and observations. *J Pharm Sci* 94: 1668–1675.
Park, J.W., Hong, K., Kirpotin, D.B., Colbern, G., Shalaby, R., Baselga, J., Shao, Y., Nielson U.B., Marks J.D., Moore, D., Papahadjopoulos, D., Benz, C.C. (2002) Anti-HER2 immunoliposomes: enhanced efficacy attributable to targeted delivery. *Clin Cancer Res* 8: 1172–1181.
Pavlon, A.K., Belsey M.J. (2005) The therapeutic antibodies market to 2008. *Eur J Pharm Biopharm* 59: 389–396.
Payne, G. (2003) Progress in immunoconjugate cancer therapeutics. *Cancer Cell* 3: 207–217.
Reithmeier, H., Winter, G. (2002) Development of an experimental setup for accelerated preformulation studies of liquid protein dosage forms. *Proc 4th World Meeting ADRITELF/APGI/APV*, Florence, 8/11 April, pp. 829–830.
Ross, J.S., Gray, K., Gray, G.S., Worland, P.J., Rolfe, M. (2003) Anticancer antibodies. *Am J Clin Pathol* 119: 472–485.
Sane, S.U., Wong, R., Hsu, C.C. (2004) Raman spectroscopic characterization of drying-induced structural changes in a therapeutic antibody: correlating structural changes with long term stability. *J Pharm Sci* 93: 1005–1018.
Sacre, S.M., Anreakos, E., Taylor, P., Feldmann, M., Foxwell, B.M. (2005) Summary of rheumatoid arthritis therapeutics in development. *Expert Rev Mol Med* 7: Issue 16.
Shenoy, B., Govardhan, C.P., Yang, M., Margolin, A.L. (2002) Crystals of whole antibodies and fragments thereof and methods for making and using them. International Patent WO 02/072636A2.
Schnyder, A., Huwyler, J. (2005) Drug transport to brain with targeted liposomes. *NeuroRx* 2: 99–107.
Wang, W. (1999) Instability, stabilization, and formulation of liquid protein pharmaceuticals. *Int J Pharm* 185: 129–188.
Webb, S.D., Cleland, J.L., Carpenter, J.F., Randolph, T.W. (2002) A new mechanism for decreasing aggregation of recombinant human interferon-γ by a surfactant: slowed dissolution of lyophilized formulations in a solution containing 0.03% polysorbate 20. *J Pharm Sci* 91: 543–558.
Weir, A.N.C., Nesbitt, A., Chapman, A.P., Popplewell, A.G., Antoniw, P., Lawson, A. D.G. (2001) Formatting antibody fragments to mediate specific therapeutic functions. *Biochem Soc Trans* 30: 512–516.

Wu, A.M., Senter, P.D. (2005) Arming antibodies: prospects and challenges for immunoconjugates. *Nat Biotechnol* 23: 1137–1146.

Yakovlevsky, K., Chamachkine, M., Khalaf, N., Govardhan, C.P., Jung, C.W. (2005) Spherical protein particels and methods of making and using them. European Patent EP 14925554.

Yang, M.X., Shenoy, B., Disttler, M. Patel, R., McGrath, M., Pechenov, S., Margolin, A.L. (2003) Crystalline monoclonal antibodies for subcutaneous delivery. *Appl Biol Sci* 100: 6934–6939.

11
Immunogenicity of Antibody Therapeutics

Huub Schellekens, Daan Crommelin, and Wim Jiskoot

11.1
Introduction

Since the first description of hybridoma technology in the 1970s (Kohler and Milstein 1975), a great deal of attention has been devoted to the development of monoclonal antibodies (mAbs) as therapeutic agents. However, it took 11 years before the first monoclonal antibody, OKT3, was approved for prevention of allograft rejection and another 7 years before the marketing authorization of Reopro to assist percutaneous coronary surgery was approved. There are many reasons for this slow development (Merluzzi et al. 2000), such as the difficulties of large-scale production, the high immunogenicity of the first generation of murine-derived monoclonal antibodies (Kuus-Reichel et al. 1994) and their lack of effector functions in humans, and the disappointing results in cancer trials, mainly caused by their bad penetration in cancer tissue (Reff and Heard 2001).

Several technical advances have been made over the years in both the development and the production of monoclonal antobodies. Recombinant DNA technology has made it possible to exchange the murine constant parts of the immunoglobulin chains with the human counterparts (chimeric mAbs) and later to graft murine complementarity determining regions (CDRs), which determine specificity, into a human immunoglobulin backbone, creating humanized mAbs. Today, transgenic animals, phage display technologies (Bradbury 1999) and other developments allow the production of completely human mAbs (Kellerman and Green 2002).

These technological advances have led to the introduction of an increasing number of therapeutics mAbs, some of which have provided major breakthroughs in the treatment of serious chronic diseases such as rheumatoid arthritis (Taylor 2003). Table 11.1 lists the monoclonals that have been allowed marketing authorization in the US and/or Europe to date. At least 400 others are in the pipeline and some of these will certainly reach the market in the near future.

However, the expectation that human mAbs would be devoid of immunogenicity proved to be naive. The scientists involved in the generation of completely

Handbook of Therapeutic Antibodies. Edited by Stefan Dübel

ISBN 978-3-527-31453-9

Table 11.1 Monoclonal antibodies registered in the EU and/or the US.

Trade name	*Generic name*	*Type of mAb*	*Ig type*	*% antibodies*[a]
Humira	adalimumab	Human	IgG1	12
Remicade	infliximab	Chimeric	IgG1	24
Reopro	abciximab	Chimeric	Fab	6
Herceptin	trastuzumab	Humanized	IgG1	1
Mabthera	rituximab	Chimeric	IgG1	1
Xolair	omalizumab	Humanized	IgG1	0
Simulect	basiliximab	Chimeric	IgG1	0
Synagis	palivizumab	Humanized	IgG1	1
Campath	alemtuzumab	Humanized	IgG1	2
Zenapax	daclizumab	Humanized	IgG1	9

a Frequency of antibody induction based on package inserts.

human antibodies generated from transgenic mice claimed: "Fully human mAbs are anticipated to be nonimmunogenic and thus to allow repeated administration without human anti-human antibody response" (Yang et al. 2001), but although humanization has reduced the immunogenicity of mAbs, completely human mAbs have been shown to induce antibodies (Table 11.1) as has been predicted (Clark 2000). As we know from other biopharmaceuticals, proteins that are considered completely identical to an endogenous protein may induce antibodies, sometimes in the majority of patients. The study of the immunogenicity of these therapeutic proteins has shown the dependence on many other factors, besides structural factors, such as the degree of nonself (Schellekens 2002). Importantly, because the antibodies induced by mAbs may interfere with efficacy and may enhance immune-mediated side effects, the issue of immunogenicity should be considered for every new therapeutic mAb.

In this chapter we will discuss the methods available to assess the immunogenicity of mAbs, consider the immunological mechanisms responsible for the induction of antibodies, and the factors that influence these mechanisms. We will discuss the biological and clinical consequences of immunogenicity and the methods to predict and prevent it.

11.2 Assays for Antibodies Induced by Monoclonal Antibodies

When comparing the immunogenicity reported for different mAbs or even different trials with the same mAb, it is important to realize that the assays for measuring antibodies have not been standardized. Standard ELISA type immunoassays are not appropriate for measuring these antibodies because of the high level of crossreactivity between the therapeutic mAb and the antibodies it may induce. The bridging assay has been advocated as the best assay (Buist et al. 1995; Pendley et al. 2003). In this assay the mAb is used to capture the antibodies present in the patient sera and the captured antibodies are detected by adding the labeled mAb as a probe (Thurmond et al. 1998). Such a bridging assay is independent of the type of antibodies to be detected. This enables the use of antisera induced in animals as positive control, although these sera will mainly contain antibodies directed to the constant part of the monoclonal antibodies, whereas human patients will mainly generate antibodies to the variable regions.

The bridging immune assay may miss a low-affinity IgM type of immune response because of the washing steps involved. Therefore, for the early immune response the use of surface plasmon resonance technology such as Biacore is advocated rather than the ELISA type of assay methodology (Ritter et al. 2001). On the other hand, the detection limit of ELISAs for high-affinity antibodies is generally lower than that for Biacore assays. In summary, the methods are complementary and should be used in parallel.

Both the bridging assay and the surface plasmon resonance technology determine the presence of binding antibodies and can be used as screening assays. In addition, it may be important to assay for the presence of neutralizing antibodies. These antibodies may interfere with the biological and clinical activity of the mAbs. Assays for neutralizing activity are based on the inhibition of a biological effect of the mAb *in vitro*. Because every mAb has its own specific biological effect, assays for neutralizing activity need to be designed for every individual mAb and are especially difficult to standardize because the basis is a bioassay.

Another aspect that needs to be considered in designing assays, sampling timing, and interpreting data is the relative long half-life (several weeks) of therapeutic mAbs, which may interfere with the detection of induced antibodies and therefore may lead to false negative results. Sampling sera up to 20 weeks after the patient has received the last injection avoids the interference of circulating mAbs. The presence of natural antibodies, receptors, and immune complexes may also interfere with assays and lead to either false positive or false negative results.

11.3 Mechanisms of Antibody Induction

As with other therapeutic proteins, there are two main mechanisms by which antibodies against mAbs are induced. If the mAbs are of foreign origin, like the first-generation mAbs derived from murine cells, the antibody response is comparable to a vaccination reaction. Often a single injection is sufficient to induce high levels of neutralizing antibodies which may persist for a considerable length of time. The other mechanism is based on breaking B-cell tolerance, which normally exists to self antigens, such as human immunoglobulins. To break B-cell tolerance prolonged exposure to proteins is necessary. In general it takes months before patients produce antibodies which are mainly binding and disappear when treatment is stopped.

It is likely that in the case of mAbs the induction of antibodies occurs by both mechanisms. B-cell tolerance may explain why patients do not make antibodies to constant regions of mAbs of human origin. It is, however, unlikely that tolerance exists for the full repertoire of variable regions an individual may produce.

To induce a classical immune reaction, a degree of non-self is necessary. The trigger for this type of immunogenicity is the difference between the human and murine immunoglobulin structure. The triggers for breaking tolerance are essentially different. The production of autoantibodies may occur when the self antigens are exposed to the immune system in combination with a T-cell stimulus or danger signal such as bacterial endotoxins, microbial DNA rich in CpG motifs, or denatured proteins (Goodnow 2001). This mechanism explains the immunogenicity of biopharmaceuticals containing impurities. When tolerance is broken via this mechanism, the response is often weak with low levels of low-affinity antibodies.

To induce high levels of IgG, the self antigens should be presented to the immune system in a regular array form with a spacing of 50–100 Å, a supramolecular structure resembling a viral capsid (Chakerian et al. 2002). Apparently the immune system has evolved to react vigorously to these types of structures, which normally are only found on viruses and bacteria. The most important factor in the immunogenicity of biopharmaceuticals is the presence of aggregates. Aggregates present the self antigens in a repeating form, which is such a potent inducer of autoantibodies.

11.4 Factors Influencing the Immunogenicity

Many factors influence the immunogenicity of mAbs (Table 11.2). The degree of non-self has been considered the main factor contributing to the immunogenicity of mAbs. Indeed, the exchange of the murine constant regions with human counterparts has resulted in a substantial reduction the induction of antibodies (Table 11.3). It is less clear whether further humanization has resulted in an

Table 11.2 Factors influencing the immunogenicity of monoclonal antibodies.

Product-related	*Other*
Presence of foreign sequences	Dose schedule
Complete/incomplete antibody	Patient characteristics
Specificity of the mAb	Concomitant treatment
Formation of immune complexes	
Fc functions of the mAb	
Purity and formulation	

Table 11.3 Immunogenicity of monoclonal antibodies related to murine sequences.[a]

Antibody response	*Marked*	*Tolerable*	*Negligible*	n
Murine mAbs	84%	7%	9%	44
Chimeric mAbs	40%	27%	33%	15
Humanized mAbs	9%	36%	55%	22

a Marked, >15% of patients; tolerable 2–15% of patients; negligible. <2% of patients. Data from Hwang and Foote (2005).

additional decrease in immunogenicity (Hwang and Foote 2005). The DNA sequence homology between the V regions of different species is higher than between the C regions. This explains why the V regions of chimeric mAbs sometimes show a higher homology with the V regions in the human germline than those of humanized mAbs (Clark 2000). Fully human mAbs have also been reported to induce antibodies, pointing to other factors that are responsible for antibody induction.

Although more injections and higher doses are associated with a higher immune response, this is not necessarily true for all mAbs. Rechallenge with mAbs is associated with a higher antibody response than the first treatment. However, in some cases chronic treatment and higher doses are less immunogenic than episodic treatment and lower dose (Hanauer 2003). The induction of tolerance by continuous treatment and higher doses has been used to explain the reduced induction of antibodies. These data should, however, be interpreted with caution because under these treatment conditions the level of circulating mAbs is higher and more persistent; the presence of circulating mAbs during the time of blood sampling may mask the detection of induced antibodies. The few studies that have compared subcutaneous and intravenous routes of administration of mAbs showed little difference in immunogenicity (Livingston et al. 1995).

Although it has been suggested that smaller proteins are less likely to be immunogenic, $F(ab')_2$ fragments of murine antibodies have been reported to be at least as immunogenic as complete antibodies. The use of Fab fragments, however, was associated with a substantial reduction of the induction of antibodies.

The immune status of the patients influences the antibody response. Cancer patients and transplantation patients are important categories receiving mAb therapy. These patients are usually immunocompromised by the disease or by immunosuppressive treatments. It has been shown with other therapeutic proteins that immunocompromised patients are less likely to produce antibodies than patients with a normal immune status. Sometimes immunosuppressive drugs such as methotrexate are given to patients on mAb therapy with the purpose of inhibiting an antibody response. Other immunosuppressive agents that have been reported to block the antibody response are cyclosporin and 15-deoxyspergualin. However, the use of these agents sometimes leads to severe toxicity.

The type of ligand also influences the immunogenicity of mAbs. In general, cell-bound antigens as target lead to a higher level of antibody induction than soluble targets. An exception to this rule is mAbs directed to antigens on immune cells with the purpose of inducing immunosuppression; these mAbs also suppress an antibody response.

In contrast with many other biopharmaceuticals, mAbs by definition have T-cell activation properties and may themselves act as the second signal to initiate an immune raction or break immune tolerance. Fc functions such as macrophage activation and complement activation may boost the antibody response. Removal of *N*-linked glycosylation at the Fc part of the immunoglobulin may reduce Fc function and thereby lead to a diminished immunogenicity.

It is known from the use of therapeutic proteins other than mAbs that purity, stability, and formulations are the most important risk factors for inducing antibodies. The presence of aggregates, denatured proteins, and other impurities has been identified as the main factor influencing immunogenicity. However, in the case of mAbs these factors have hardly been investigated as most attention has been directed to the degree of non-self. But, with the advent of fully human mAbs, properties such as purity and stability will increase in importance.

11.5 Consequences of the Immunogenicity of Monoclonal Antibodies

In patients with anti-mAb antibodies an increased incidence of immune complex syndrome, allergic reactions, and infusion reactions has been reported (Baert et al. 2003). There also seems to be a dependency of the type of antibody reaction on these side effects. Patients with a slow but steadily increasing antibody titer are reported to show more infusion-like reactions than patients with a short temporary response (Ritter et al. 2001).

Monoclonal antibodies are mainly present in the circulation and therefore their pharmacokinetic behavior is highly sensitive to the presence of induced antibodies. These antibodies may either increase or diminish their half-life, depending on the affinity of the antibodies and properties of the mAbs.

The presence of antibodies raised by the mAbs may decrease their efficacy either by decreasing their half-life or by neutralizing their antigen-binding capacity (Baert et al. 2003). The more specific the immune response, the lesser the chance that the antibodies interfere with the efficacy of a new mAb treatment because of the lack of crossreactivity.

There are some suggestions that an immune reaction to mAbs may in certain conditions increase their efficacy (Koprowski et al. 1984; Wagner et al. 1990). A correlation has been reported between the level of the immune response to mAbs and the prognosis of the patient. This may be explained by an anti-idiotypic response increasing the immunogenic response to tumor antigens. Alternatively, the level of antibody response may only be a reflection of the general immune status of the patients.

11.6
Prediction of the Anti-mAb Response

As shown by comparing murine and chimeric/humanized mAbs, the level of non-self is a predictor of an immune response. This is the classical response to a foreign protein. However, completely human mAbs still are capable of eliciting antibodies based on breaking immune tolerance. It is likely that the quality of the preparation and its formulation are important factors, although this has not been studied in detail. Aggregated and solubilized immunoglobulins have been used in the past to break or induce immune tolerance, respectively, so aggregation is likely to be a major factor in the induction of antibodies. The role of protein modification (e.g. oxidation, deamidation, etc.) needs further investigation, as does the optimal formulation to avoid immunogenicity.

The monoclonal system may provide a unique opportunity to study the factors influencing immunogenicity because of the availability of a complete range of mAbs ranging from completely murine, murine/human hybrid to completely human. In addition, transgenic animals with the nearly complete immunoglobulin repertoire, which were developed for the production of human mAbs, are available. These animals also have an immune tolerance to mAbs that is comparable to the immune tolerance in patients. Comparing the immune response of a human mAb in these transgenic mice with the immune response in the non-transgenic strain will offer the possibility of discriminating between the factors important for the classical immune response and those which contribute to the breaking of immune tolerance.

Immune tolerant transgenic mice may also provide an important model for the study of the mechanisms responsible for the induction of antibodies and possible prevention and treatment by immune suppressive drugs.

Monkeys also have the potential to predict the immune response in patients as their response is mainly anti-idiotypic. However, other important factors may be missing in monkeys, such as disease state and concomitant therapy. Indeed, in one of the few studies in which the responses in monkeys and patients were compared there were major differences in incidence and type of response (Stephens et al. 1995).

In theory, *in vitro* T-cell stimulation tests and computational models are also available to predict immunogenicity. However T-cell proliferation assays have the drawback that all antibodies are capable of inducing some level of T-cell activation. The computational alogarithms which predict binding of antigens to HLA class II only give limited information on the interaction of the mAbs with the immune system, and also underdetected epitopes.

Competition antibody assays have also been used to predict immunogenicity of mAbs. Sera of patients who were positive for antibodies to murine mAbs were tested in a surface plasmon resonance-based competition assay to variants of humanized antibodies (Gonzales et al. 2002). Lesser reactivity of the sera was interpreted as a sign of reduced immunogenicity of these variants. These assays, however, show the antigenicity of these variants, which is not necessarily predictive of their immunogenicity (El Kasmi et al. 2000). Although these assays help to define the immunogenic sites, they may miss new epitopes that may be present in the variants and they also ignore other factors important for immunogenicity such as aggregates and impurities. The same is true for studies in which the affinity of the grafted CDR was compared with its capacity to bind to patient sera in a competition radioimmunoassay to define the construct with the best clinical potential (Iwahashi et al. 1999).

11.7
Reduction of Immunogenicity of Monoclonal Antibodies

The immunogenicity of mAbs is associated with side effects and loss of efficacy and should be avoided. Various methods can be used to reduce the induction of antibodies, including:

- Replacing rodent sequences by human sequences
- Immunosuppressive treatment
- Altering Fc functions
- Reducing the size
- Attachment of polymers
- Improving product quality
- Optimizing formulation
- Inducing tolerance.

The main approach to reducing the immunogenicity of mAbs has been replacing the murine parts of the molecules by human-derived sequences as discussed earlier. The immunogenic response may be reduced by immunosuppressive treatment (Baert et al. 2003). This immunosuppressive effect seems to be dependent on treatment schedule (Hanauer 2003).

As with other therapeutic proteins, covalently linking polymers such as polyethylene glycol and low molecular weight dextran to mAbs reduces their immunogenicity (Fagnani et al. 1995; Trakas and Tzartos 2001; Chapman 2002). However, these modifications in general make the molecules less active, necessitating higher doses. This and the increased half-life of the proteins increases their exposure to the immune system, which may enhance the immunogenic potential.

Tolerance to mAbs has been induced by using soluble forms (Isaacs 2001). Tolerance induction to mAbs reacting with cell-associated targets may be restricted to the isotypic parts of the immunoglobulin. To achieve tolerance to the idiotypic parts of mAbs reacting with cell-bound protein, variants may be used which lack the affinity for cells. Tolerance may also be induced by pretreating with PEGylated mAbs.

Factors such as the presence of contaminants, impurities, and the effect of formulation, which have been shown to be important factors for the immunogenicity of other therapeutic proteins, have hardly been studied for mAbs. It can, however, be assumed that these factors are also important for the induction of antibodies by mAbs. Improving the quality of mAb products and optimization of production and formulation may reduce the immunogenicity further.

11.8 Conclusion

As with other therapeutic drugs, it is safe to assume that all mAbs will induce an immune response, although the incidence may differ widely between individual mAb products. An immune response to mAbs is associated with an increase in toxicity and a decrease of efficacy. Although the reduction of non-human sequences has reduced the induction of antibodies, complete human mAbs are still immunogenic. The immunological mechanisms which lead to an antibody response to are not completely understood. More research is required to prevent immunogenicity completely.

References

Baert, F., Noman, M., Vermeire, S., van Assche, G., D'Haens, G., Carbonez, A., Rutgeerts, P. (2003) Influence of immunogenicity on the long-term efficacy of infliximab in Crohn's disease. *N Engl J Med* 348: 601–608.

Bradbury, A. (1999) Display technologies expand their horizons. *Trends Biotechnol* 17: 137–138.

Buist, M.R., Kenemans, P., van Kamp, G.J., Haisma, H.J. (1995) Minor human antibody response to a mouse and chimeric monoclonal antibody after a single i.v. infusion in ovarian carcinoma patients: a comparison of five assays. *Cancer Immunol Immunother* 40: 24–30.

Chakerian, B., Lenz, P., Lowy, D.R., Schiller, J.T. (2002) Determinants of autoantibody induction by conjugated papillomavirus virus-like particles. *J Immunol* 169: 6120–6126.

Chapman, A.P. (2002) PEGylated antibodies and antibody fragments for improved therapy: a review. *Adv Drug Deliv Rev* 54: 531–545.

Clark, M. (2000) Antibody humanization: a case of the 'Emperor's new clothes'? *Immunol Today* 21: 397–402.

El Kasmi, K.C., Deroo, S., Theisen, D.M., Brons, N.H.C., Muller, C.P. (2000) Crossreactivity of mimotopes and peptide homologues of a sequential epitope with a monoclonal antibody does not predict cross reactive immunogenicity. *Vaccine* 18: 284–290.

Fagnani, R., Halpern, S., Hagan, M. (1995) Altered pharmacokinetic and tumour localization properties of Fab' fragments of a murine monoclonal anti-CEA antibody by covalent modification with low molecular weight dextran. *Nucl Med Commun* 16: 362–369.

Gonzales, N.R., Schuck, P., Schlom, J., Kashmiri, S.V.S. (2002) Surface plasmon resonance-based competition assay to assess the sera reactivity of variants of humanized antibodies. *J Immunol Methods* 268: 197–210.

Goodnow, C.C. (2001) Pathways for self-tolerance and the treatment of autoimmune diseases. *Lancet* 357: 2115–2121.

Hanauer, S.B. (2003) Immunogenicity of infliximab in Crohn's disease. *N Engl J Med* 348: 2155–2156.

Hwang, W.Y.K., Foote, J. (2005) Immunogenicity of engineered antibodies. *Methods* 36: 3–10.

Isaacs, J.D. (2001) From bench to bedside: discovering rules for antibody design, and improving serotherapy with monoclonal antibodies. *Rheumatology* 40: 724–738.

Iwahashi, M., Milenic, D.E., Padlan, E.A., Bei, R., Schlom, J., Kashmiri, S.V.S. (1999) CDR substitutions of a humanized monoclonal antibody (CC49): contributions of individual CDRs to antigen binding and immunogenicity. *Mol Immunol* 36: 1079–1091.

Kellerman, S.A., Green, L.L. (2002) Antibody discovery: the use of transgenic mice to generate human monoclonal antibodies for therapeutics. *Curr Opin Biotechnol* 13: 593–597.

Kohler, G., Milstein, C. (1975) Continuous cultures of fused cells secreting antibody of predefined specificity. *Nature* 256: 495–497.

Koprowski, H., Herlyn, D., Lubeck, M., DeFreitas, M., Sears, H.F. (1984) Human anti-idiotype antibodies in ancer patients: is the modulation of the immune response beneficial for the patient. *Proc Natl Acad Sci USA* 81: 216–219.

Kuus-Reichel, K., Grauer, L.S., Karavodin, L.M., Knott, C., Krusemeier, M., Kay, N.E. (1994) Will immunogenicity limit the use, efficacy and future development of therapeutic monoclonal antibodies? *Clin Diagn Lab Immunol* 1: 365–372.

Livingston, P.O., Adluri, S., Zhang, S., Chapman, P., Raychaudhuri, S., Merritt, J. A. (1995) Impact of immunological adjuvants and administration route on HAMA respones after immunication with murine monoclonal antibody MELIMMUNE-1 in melanoma patients. *Vac Res* 4: 87–94.

Merluzzi, S., Figini, M., Colombatti, A., Canevari, S., Pucillo, C. (2000) Humanized antibodies as potential drugs for therapeutic use. *Adv Clin Pathol* 4: 77–85.

Pendley, C., Schantz, A., Wagner, C. (2003) Immunogenicity of therpeutic monoclonal antibodies. *Curr Opin Mol Ther* 5: 172–179.

Reff, M.E., Heard, C. (2001) A review of modifications to recombinant antibodies: attempt to increase efficacy in oncology applications. *Oncol/Hematol* 40: 25–35.

Ritter, G., Cohen, L.S., Williams, C., Richards, E.C., Old, L.J., Welt, S. (2001) Serological analysis of human anti-human antibody responses in colon cancer patients treated with repeated doses of humanized monoclonal antibody A33. *Cancer Res* 61: 6851–6859.

Schellekens, H. (2002) Immunogenicity of therapeutic proteins: clinical implications and future prospects. *Clin Ther* 24: 1720–1740.

Stephens, S., Emtage, S., Vetterlein, O., Chaplin, L., Bebbington, C., Nesbitt, A., Sopwith, M., Athwall, D., Noval, C., Bodmer, M. (1995) Comprehensive pharmacokinetics of a humanized antibody and analysis of residual anti-idiotypic responses. *Immunology* 85: 668–674.

Taylor, P.C. (2003) Antibody therapy for rheumatoid arthritis. *Curr Opin Pharmacol* 3: 323–328.

Thurmond, L.M., Reese, M.J., Donaldson, R.J., Orban, B.S. (1998) A kinetic enzyme immunoassay for the quantitation of antibodies to a humanized monoclonal antibody in human serum. *J Pharm Biomed Anal* 16: 1317–1328.

Trakas, N., Tzartos, S.J. (2001) Conjugation of acetylcholine receptor-protecting Fab fragments with polyethylene glycol results in a prolonged half-life in the circulation and reduced immunogenicity. *J Neuroimmunol* 120: 42–49.

Wagner, V., Reinsberg, J., Oehr, P., Briele, B., Schmidt, S., Werner, A., Krebs, D., Biersack, H.J. (1990) Clinical course of patients with ovarian arcinoma after induction of anti-idiotypic antibodies against tumor-associated antigen. *Tumor Diagn Ther* 11: 1–4.

Yang, X., Jia, X., Corvalan, J.R.F., Wang, P., Davis, C.G. (2001) Development of ABX-EGF, a fully human anti-EGF receptor monoclonal antibody, for cancer therapy. *Crit Rev Oncol/Hematol* 38: 17–23.

12
Regulatory Considerations

Marjorie A. Shapiro, Patrick G. Swann, and Melanie Hartsough

12.1
Introduction

In the 30 years since the publication of George Kohler and Cesar Milstein's paper describing hybridoma technology (Kohler and Milstein 1975), therapeutic monoclonal antibodies (mAbs) and the concept of mAbs as "magic bullets" have gone from initial disappointment to great success. The first therapeutic mAb, OKT3, was licensed in 1986, but it was not until the late 1990s that the potential of therapeutic mAbs began to be realized. Between the licensure of OKT3 and 1996 only one additional therapeutic mAb and five diagnostic mAbs were licensed. Since 1997, 17 mAbs (16 therapeutic and one diagnostic) and two Fc-fusion proteins have been approved by the US Food and Drug Administration (FDA; Table 12.1).

The early failure of most mAbs to progress to phase III clinical trials has been attributed to the insufficient characterization of the mAb and its *in vivo* performance, incomplete preclinical testing, and inadequately designed clinical trials (Stein 1997). In addition, many of the early clinical trials employed murine mAbs, which have a short half-life in humans, are inefficient at eliciting effector functions, and frequently induced human anti-mouse antibodies (HAMA) (Glennie and Johnson 2000). Overall, only 3% of therapeutic murine mAbs evaluated in clinical trials have been successful and ultimately approved (Reichert et al. 2005).

The major factor contributing to the more recent successes of therapeutic mAbs has been the ability to genetically engineer chimeric mAbs (murine or other non-human variable regions expressed with human constant regions) or humanized mAbs (murine or other non-human complementarity determining regions grafted onto human framework regions expressed with human constant regions). In direct contrast to their murine counterparts, chimeric and humanized mAbs

Disclaimer: Opinions expressed in this chapter reflect the professional views of the authors and ought not to be viewed as official policy of the US Food and Drug Administration or the Government of the United States.

Handbook of Therapeutic Antibodies. Edited by Stefan Dübel

ISBN 978-3-527-31453-9

Table 12.1 Approved monoclonal antibodies and Fc-fusion proteins.

Trade name	*USAN name*	*Use*	*Indication*	*Year approved*	*Type*
Orthoclone OKT3	muromomab	Therapeutic	Immunologic	1986	Murine
OncoScint	satumomab pendetide	Diagnostic	Oncologic	1991	Murine
ReoPro	abciximab	Therapeutic	Cardiac	1994	Chimeric
CEA-Scan	arcitumomab	Diagnostic	oncologic	1996	Murine
Myoscint	imciromab pentetate	Diagnostic	Cardiac	1996	Murine
Verluma	nofetumomab	Diagnostic	Oncologic	1996	Murine
Prostascint	capromab pendetide	Diagnostic	Oncologic	1996	Murine
Rituxan	rituximab	Therapeutic	Oncologic	1997	Chimeric
Zenapax	daclizumab	Therapeutic	Immunologic	1997	Humanized
Simulect	basiliximab	Therapeutic	Immunologic	1998	Chimeric
Synagis	pavilizumab	Therapeutic	Infectious disease	1998	Humanized
Remicade	infliximab	Therapeutic	Immunologic	1998	Chimeric
Herceptin	trastuzumab	Therapeutic	Oncologic	1998	Humanized
Enbrel	etaneracept	Therapeutic	Immunologic	1998	Fc Fusion Protein
Mylotarg	gemtuzumab ozogomicin	Therapeutic	Oncologic	2000	Humanized
Campath	alemtuzumab	Therapeutic	Oncologic	2001	Humanized
Zevalin	ibritumomab tiuxetan	Therapeutic	Oncologic	2002	Murine
Humira	adilimumab	Therapeutic	Immunologic	2002	Human
Amevive	alefacept	Therapeutic	Immunologic	2003	Fc fusion protein
Xolair	omalizumab	Therapeutic	Immunologic	2003	Humanized
Bexxar	tositumomab	Therapeutic	Oncologic	2003	Murine
Raptiva	efalizumab	Therapeutic	Immunologic	2003	Humanized
Erbitux	cetuximab	Therapeutic	Oncologic	2004	Chimeric
Avastin	bevacizumab	Therapeutic	Oncologic	2004	Humanized
NeutroSpec	fanolesomab	Diagnostic	Immunologic	2004	Murine
Tysabri	natalizumab	Therapeutic	Immunologic	2004	Humanized

are less immunogenic, exhibit longer half-lives, and efficiently promote effector functions in humans. Indeed, the approval rates for chimeric and humanized mAbs at present are 21% and 18%, respectively (Reichert et al. 2005).

The expression of fully human mAbs through hybridoma technology is generally inefficient but development of appropriate fusion partners continues (Karpas et al. 2001). The generation of fully human mAbs through the expression of human immunoglobulin genes in transgenic animals or by phage display libraries however, has facilitated the development of human mAbs that share the advantages of chimeric and humanized mAbs and are also predicted to be even less immunogenic than chimeric and humanized mAbs. Such fully human mAbs, however, do not undergo selection on a human background so the potential exists to select mAbs with unusual structures that may be immunogenic or that cross-react with autoantigens. In spite of the effort to produce fully human mAbs, it is not expected that these technologies will provide a great advantage over chimeric mAbs in reducing immunogenicity (Clark 2000).

Several factors influence whether or not a mAb will be immunogenic in patients. These include; the patient population (immunosuppressed, autoimmune), intercurrent illnesses which may disrupt the distribution of the mAbs, the presence of pre-existing antibodies (rheumatoid factor may react with some IgGs), concomitant medications (chemotherapy or immunosuppressive drugs), increases in the dose and/or frequency of administration, and the route of administration. The subcutaneous and intramuscular routes of administration are generally found to be more immunogenic than the intravenous route.

Due to the multiple factors influencing immunogenicity, as well as inherent differences in the assays developed for the detection of HAMA, human antichimeric antibodies (HACA), and human anti-humanized or anti-human antibodies (HAHA) for each product, a direct comparison of the immunogenicity among products cannot be made. It can be seen from Table 12.2, however, that simply removing the Fc portion of a murine mAb reduces the incidence of HAMA to levels more consistently observed for chimeric and humanized mAbs.

To date only one human mAb generated by phage display has been approved. The majority of all mAbs entering clinical trials since 2001, however, have been generated by antibody phage display or in transgenic mice expressing human immunoglobulin genes (Reichert et al. 2005). It is anticipated that within the

Table 12.2 Immunogenicity of licensed monoclonal antibodies.[a]

Antibody type	*Total*	*% patients with HAMA, HACA, or HAHA*
Murine	8	Whole mAbs[b]: <3% to >80% (loss of effectiveness of OKT3 seen when titers were >1:1000) Fab or Fab′ fragments: <1% to 8%
Chimeric	5[c]	<1% to 13%
Humanized	8	<1% to 10%
Human	1	12%[d]

HAMA, human anti-mouse antibodies; HACA, human antichimeric antibodies; HAHA, human anti-humanized or anti-human antibodies.

a All immunogenicity data taken from package inserts.

b <3% of patients developed HAMA against Zevalin (at 90 days post treatment), which ablates B cells. All other murine mAbs induced HAMA in 55% to >80% of patients.

c Four are whole mAbs, one is a Fab.

d When used as a monotherapy, 12% patients made HAHA against Humira. When used in conjunction with methotrexate, <1% patients made HAHA.

next decade data will become available that demonstrate whether or not such fully human mAbs are indeed less immunogenic than chimeric or humanized mAbs.

The introduction of promising mAbs into the clinic is not only attributable to established biotechnology and pharmaceutical companies, but also to start-up companies, as well as academic researchers. This chapter is intended to assist small business and academic sponsors who have limited experience in preparing submissions for Investigational New Drug (IND) applications to the FDA. The primary focus will be on product and preclinical issues that should be addressed prior to the initiation of phase I clinical trials for both therapeutic and *in vivo* diagnostic mAbs. These product development issues should also be considered when a mAb is to be used with devices for enriching or purging specific cell populations or in conjunction with cell therapies. Issues arising relative to product or preclinical development as clinical trials progress and the necessity to provide additional information will also be addressed.

12.2 Regulatory Authority

The statutory authority for the regulation of biological products and drugs for human use in the USA are derived from the Public Health Service Act and the Food, Drug and Cosmetic Act, respectively (http://www.fda.gov/opacom/laws/lawtoc.htm). The implementing regulations can be found in Title 21 of the Code of Federal Regulations (CFR). The regulations for biological products are found in 21CFR Part 600. Other applicable regulations include 21CFR 210 and 21 CFR 211, which describe good manufacturing practices and 21CFR 312, which describes requirements for submission of an IND. Information and relevant forms for submitting an IND application can be found at http://www.fda.gov/cder/regulatory/applications/ind_page_1.htm. The statutory authority for the regulation of devices came under the Medical Device Amendments of the Food, Drug and Cosmetic Act with the implementing regulations for devices located in 21CFR Part 800. The website for the CFR is www.gpo.gov/nara/cfr/index.html.

The development program for an mAb (note that Fc-fusion proteins are grouped with mAbs) is dependent upon the intended use in humans. MAbs are developed as therapeutic or *in vivo* diagnostic agents and also as agents used in the manufacture of other products for *in vivo* use (ancillary mAbs). These ancillary mAbs can be used either alone or in conjunction with devices, such as for the *ex vivo* enrichment of specific cell populations for *in vivo* administration (e.g. hematopoietic stem cells) or for the *ex vivo* purging of unwanted cell types (e.g. tumor cells).

Most mAbs, including mAbs conjugated with toxins or radioisotopes, are regulated as biologics. As of October 1, 2003, the regulatory oversight of most mAbs was moved upon the transfer of the Division of Monoclonal Antibodies (DMA) as well as the Office of Therapeutic Research and Review (OTRR) from the Center for Biologics Evaluation and Research (CBER) to the Center for Drug Evaluation and Research (CDER). Currently, the CMC portion of an application is regulated by the DMA in the Office of Biotechnology Products (OBP), whereas the pharmacology/toxicology and clinical portions are regulated based on clinical indication by the divisions in the Office of New Drugs (OND). The Office of Cellular, Tissue, and Gene Therapies (OCTGT), CBER has oversight of anti-idiotype (Id) mAbs and Id-KLH products used as vaccines as well as ancillary mAbs used in cell therapies but the DMA provides collaborative reviews for the development of these mAbs.

All therapeutic and *in vivo* diagnostic mAbs and mAbs used *ex vivo* with devices or as ancillary reagents in cell therapy protocols, should be characterized and manufactured under current Good Manufacturing Practices (cGMP), regardless of the FDA Center with regulatory oversight for that product. Although drug-conjugated mAbs such as Mylotarg, in which the mAb is used as a mode of localization or used to affect the biodistribution of the drug, are regulated as drugs, the mAb itself should be manufactured using guidelines for biologics. MAbs included as part of *in vitro* diagnostic kits (regulated either by CBER or the Center for Devices and Radiological Health (CDRH)) will not be discussed in this chapter.

In addition to the regulations, the FDA publishes guidance documents that reflect the FDA's current thinking on a particular topic. Guidance documents clarify requirements imposed by Congress or promulgated by the FDA by explaining how IND Sponsors and the Agency should comply with those statutory and regulatory requirements. They often provide specific detail that is not included in the relevant statutes and regulations. The recommendations in these documents are not legal requirements and are therefore not binding on either the Sponsor or the FDA. The Sponsor, however, is required to provide an alternative scientifically based proposal for those recommendations they choose not to follow. All guidance documents, including those cited in the references in this chapter, can be found on the CDER web site at http://www.fda.gov/cder/guidance/index.htm. Several guidance documents are developed through the International Conference on Harmonisation (ICH). The goal of the ICH is to harmonize the interpretation and application of regulatory requirements for pharmaceuticals among the United States, the European Union, and Japan. ICH documents are also posted on the CDER web site or can be found at http://www.ich.org.

The guidance entitled "Points to Consider in the Manufacture and Testing of Monoclonal Antibody Products for Human Use" (mAb PTC 97, www.fda.gov/cber/gdlns/ptc_mab.pdf) (Food and Drug Administration 1997a) is a comprehensive document, which describes and recommends steps that should be taken in the manufacture, characterization, quality control, and product testing of mAbs. This document also describes considerations for preclinical studies and the design of phase I and phase II clinical trials. Therapeutic mAbs including drug–mAb conjugates, mAbs for *in vivo* diagnostic use, and those for use *ex vivo* with therapeutic devices or in cell therapy protocols, should be developed according to the guidance provided in this document.

12.3 Chemistry, Manufacturing, and Controls Considerations

This section contains a summary of quality control testing that should be performed at various stages of the manufacturing process. The mAb PTC 97 document should be carefully reviewed for specific details. This chapter, however, provides updates on some of these recommendations. In addition, the information that should be included in an IND for phase I studies can be found in the Guidance "Content and Format of Investigational New Drug Applications (INDs) for Phase 1 Studies of Drugs, Including Well-Characterized, Therapeutic Biotechnology-Derived Products" (Food and Drug Administration 1996a).

12.3.1 Cell Line Qualification

Most mAbs have been expressed as hybridoma proteins or recombinant proteins in rodent cell lines. Epstein–Barr virus (EBV)-transformed human or primate cell lines have also been used, but thorough studies for the detection of human pathogenic viruses and the demonstration of the removal EBV during purification of the product were necessary prior to use of these products in clinical trials. Because of these safety concerns, the use of EBV-infected primate or human cell substrates is not recommended. Products derived from such cell lines are not eligible for abbreviated safety testing allowed when products are intended for serious and life-threatening conditions (see Section 12.3.7). If mAbs derived from such cell lines show clinical potential, expression of the mAb as a recombinant protein in a non-primate, non-human cell line is a desirable alternative.

In general, for mammalian cell lines a Master Cell Bank (MCB) should be established and demonstrated to be free from bacterial, fungal, and mycoplasma contamination. The MCB should also be tested for the presence of adventitious and species-specific viruses. Murine hybridomas are considered to be inherently capable of producing infectious murine retrovirus and thus, it is not necessary to test these cell banks for the presence of endogenous retrovirus. All other cell substrates, however, including other rodent cells, should be tested for retrovirus. Authenticity testing should be performed to confirm the cell line species of origin, identity, and lack of cell line cross-contamination.

A Working Cell Bank (WCB) may be established to extend the lifetime of the MCB and requires less extensive testing than the MCB. The WCB should be free from bacterial, fungal, and mycoplasma contamination and tested for authenticity. It is not required to develop a WCB if the MCB is of sufficient size to last throughout product development. It is recommended, however, that a WCB be established, qualified, and used for production by the time of a Biologics License Application (BLA) submission.

Antibody fragments (Fab, sFv, and sFv fusion proteins) are usually produced in bacteria, which do not require adventitious virus testing. An MCB should be established and demonstrated to be free of other microbial, fungal, and bacteriophage contamination.

Cell substrates from other species, yeast, insect or plant, or transgenic animals and plants, have been used infrequently, but it is anticipated that their use for mAb production will increase. In addition to relevant information obtained from the mAb PTC 97, specific guidance on cell substrates or alternative sources can be found in the documents entitled, ICH Q5D: "Guidance on Quality of Biotechnological/Biological Products: Derivation and Characterization of Cell Substrates Used for the Production of Biotechnological/Biological Products" (Food and Drug Administration 1998a), "Points to Consider in the Manufacture and Testing of Therapeutic Products for Human Use Derived from Transgenic Animals (1995)" (Food and Drug Administration 1995) and in the draft guidance entitled "Draft Guidance for Industry: Drugs, Biologics and Medical Devices Derived from Bioengineered Plants for Use in Humans and Animals (2002)" (Food and Drug Administration 2002a).

12.3.2 Quality Control Testing

Lot-to-lot safety testing should be performed at three stages of the manufacturing process: (1) on the unprocessed bulk drug (nonsterile, filtered, harvested tissue culture supernatant); (2) on the drug substance (bulk purified product), and (3) on the drug product (final formulated and filled product). The unprocessed bulk drug should be assessed for bioburden and shown to be free of mycoplasma and adventitious viruses. Three lots should be quantitated for endogenous retrovirus in order to establish a target level for removal of retrovirus during purification (see Section 12.3.6). In recent years, polymerase chain reaction (PCR) methods have been developed as alternatives to traditional methodologies for determining levels of endogenous retrovirus (Brorson et al. 2001, 2002) and assessing the presence of mycoplasma (Eldering et al. 2004) or adventitious virus by MAP testing (Bauer et al. 2004). PCR methods that have been properly validated may be acceptable alternatives to the traditional, more expensive and cumbersome assays. Dialogue with the FDA to discuss the acceptability of such assays is encouraged.

Acceptable limits for bioburden should be established for all stages of the purification process, but the drug product should be sterile. Sterility testing for licensed products is described in 21CFR 610.12. For sterility testing of mAbs during clinical development, procedures described in 21CFR 610.12 or in the US Pharmacopeia or European Pharmacopoeia are acceptable. *In vitro* adventitious virus testing should be performed routinely on all unprocessed bulk drug production lots using multiple relevant cell lines, while *in vivo* testing is generally done once on the unprocessed bulk drug and repeated only when production methods change.

Other safety tests include assessing levels of endotoxin and polynucleotides. Testing for levels of residual host cell DNA is usually performed on the drug substance. Subsequent to the publication of the mAb PTC 97, the World Health Organization (WHO) Expert Committee on Biological Standardization revised its recommendation so that a maximum of 10 ng of residual DNA from continuous cell lines per dose of a purified product is acceptable (Griffiths 1987), rather than the 100 pg stated in the mAb PTC 97.

Endotoxin testing should be performed on the drug product and is typically performed on the drug substance as well. The endotoxin limit is defined as K/M, where $K = 5.0\,\mathrm{EU\,kg^{-1}}$ for parenteral drugs and M is the maximum human dose per kg of body weight administered in a single one-hour period. Thus, for a dose of $2\,\mathrm{mg\,kg^{-1}}$, the endotoxin limit would be $(5.0\,\mathrm{EU\,kg^{-1}})/(2\,\mathrm{mg\,kg^{-1}}) = 2.5\,\mathrm{EU\,mg^{-1}}$ (Food and Drug Administration 1987). Note that other routes of administration (e.g. intrathecal administration) have different acceptable limits of endotoxin.

CFR 610.13(b) requires that a rabbit pyrogen test be performed for commercial drug product. It is acceptable to assess levels of endotoxin rather than perform the rabbit pyrogen test during clinical development. When a BLA is submitted, the method for endotoxin detection should be validated against the rabbit pyrogen test. This study should demonstrate that at the maximum level of endotoxin per release specification, no positive result is observed in the rabbit pyrogen test (USP <151>). When this validation is successfully completed, it is acceptable to substitute endotoxin testing for the rabbit pyrogen test as a lot-release test for commercial product (21CFR 610.9).

21CFR 610.11 describes the General Safety test, which is used to detect extraneous toxic contaminants in biological products. Licensed mAbs as well as those under clinical development are exempted from the General Safety Test (21CFR 601.2(c)(1)).

In addition to the testing described above, routine analysis on the drug substance should include tests to establish biochemical purity, molecular integrity, identity, and potency. The drug product should be tested for protein concentration, potency, purity, identity, pH, and, when appropriate, moisture, preservative, and excipients.

Potency assays should be based on the proposed mechanism of action for the mAb. While ELISA tests or other binding assays are often employed as potency assays, unless the mAb works by blocking the binding of the antigen to its intended target, binding assays alone are not sufficient to establish potency. If the mAb is proposed to work through effector functions such as complement-dependent cytotoxicity or antibody-dependent cytotoxicity, by the induction of apoptosis, or other mechanisms, cellular-based assays that reflect these mechanisms should be developed. MAbs conjugated with drugs or toxins should employ a cytotoxicity assay to establish potency.

Tests to establish biochemical purity should include assays that demonstrate the reduction of process contaminants to levels below detection or, in some cases, to minimal acceptable levels. Such process contaminants include host cell proteins and DNA, materials that may be introduced during culture, such as bovine or human insulin or transferrin, bovine serum albumin, or immunoglobulin (from bovine serum), surfactants used to protect cells in agitated suspension cultures from shear and mechanical force, methotrexate or other agents used to maintain the antibody-expressing construct under selected pressure, inducing agents intended to maximize product expression, Protein A (or other proteins used in immunoaffinity columns), and solvents and detergents used in virus inactivation steps.

Assays to detect host cell proteins are usually developed for each product but commercial kits are now available. If commercial kits are used they should be demonstrated to be suitable for use for each product and its host cell substrate. Release specifications should be based on relevant product quality attributes. For early clinical development, release specifications are considered preliminary but should be quantitative when possible. Acceptance criteria may be broad unless constrained by safety considerations. Specifications such as "For information only" or "conforms to reference standard" are not usually acceptable as upper or lower limits of acceptability are not delineated or the reference standard itself, for a variety of reasons, may not always provide the same results. This is of particular importance for potency assays as potency should not vary significantly among lots of drug product as product development and clinical studies progress. "Conforms to reference standard" may be an acceptable release specification if the reference standard is fully characterized and the attribute for that assay is quantitated. For example, a reference standard may display a certain number of major and minor bands within a specific p*I* range on isoelectric focusing gels. Subsequent lots may "conform to reference standard" if they also have the specified number of bands within the specified p*I* range.

12.3.3 Transmissible Spongiform Encephalopathy (TSE)

TSEs are a concern for biotechnology products because bovine products (serum, transferrin, insulin, albumin) or human plasma-derived products (transferrin, albumin, IgG) may be used during manufacturing. Due to the mad-cow disease epidemic in Europe in the 1990s, the emergence of variant Creutzfeldt–Jakob disease (vCJD), and the inability to assess levels of prion protein in blood, cell substrates, raw materials, and unpurified bulk drug substance, the FDA issued a guidance in 2002 to reduce the possible risk of transmission of CJD and vCJD by human blood and blood products (Food and Drug Administration 2002b). This Guidance distinguishes the risks between CJD and vCJD and permits the use of donors for plasma derivatives from some affected countries because such products are highly processed materials. Although the use of plasma derivatives from donors who have resided in the UK, France, and on US military bases abroad may be permitted, such products may be subject to an immediate recall if any donor subsequently develops vCJD. Therefore, it is strongly recommended that the use of such plasma derivatives be avoided. Of particular concern for mAb production are Protein A preparations used in affinity chromatography that have been purified over human IgG immunoaffinity columns. Protein A resins are now available that do not use human IgG in the purification process, which addreses this particular risk.

Bovine-derived products used during manufacture should be from countries known to be free of bovine spongiform encephalopathy (BSE). The United States Department of Agriculture (USDA) keeps a list of countries (see www.aphis.usda.gov/NCIE/country.html) with verified or suspected cases of BSE and animal products from these countries should not be used.

Because of the safety issues surrounding TSEs, the FDA has encouraged sponsors to adapt cell lines to serum-free, protein-free media and to use Protein A resins where the Protein A was not purified over human IgG. It is anticipated that in the future, validated assays to detect TSEs will be developed. Until that time, however, sponsors should provide the FDA with information regarding the source and country of origin for every animal- and human-derived component used in manufacture. This list should include the bovine and human plasma-derived components discussed above as well as amino acids used in the tissue culture medium, enzymes used to make protein hydrolyzates, cholesterol, Tween, and any other reagent used in the manufacturing process that may be animal- or human-derived. It is important to keep track of the lot numbers for each animal or human-derived raw material, should new developments regarding TSEs and the spread of vCJD arise.

12.3.4 Product Stability

Expiration dates are not established prior to submission of a BLA. Stability studies, however, are required during clinical development to ensure product quality for the duration of the clinical study. Therefore, stability testing protocols for both drug substance and drug product should be developed and initiated prior to the phase I clinical trials. The stability protocols should include tests for physicochemical integrity, potency, sterility, and other specific assays as appropriate. Samples of drug substance or drug product are usually tested frequently during the first year of the protocol and then on a 6-month basis through the second year. Stability studies extending longer than 2 years usually involve testing on a yearly basis. Accelerated stability testing (i.e. testing of samples stored at temperatures exceeding the recommended storage temperature) are often useful for identifying which tests are stability-indicating. Tests should be performed in parallel with a properly qualified and stored reference standard. Refer to the ICH documents Q5C: "International Conference on Harmonisation; Final Guideline on Stability Testing of Biotechnological/Biological Products" and Q1A (R): "Stability Testing of New Drugs and Products (Revised guidelines)" for more detailed recommendations (Food and Drug Administration 1996b, 2002c).

12.3.5
Reference Standard

A reference standard should be developed and appropriately qualified using defined physicochemical characteristics, specificity and potency attributes. It should be stored under appropriate conditions and tested periodically to document its integrity. The reference standard should be used for lot-to-lot comparisons performed for both drug substance and drug product release as well as for stability studies. New reference standards should be qualified when major manufacturing changes are made. Typically, the reference standard is qualified using a more comprehensive panel of biochemical, biophysical, and immunological assays than those used for release or stability testing. A thorough characterization of the lot that is used for a reference standard provides the basis for future comparability studies (see Section 12.3.8).

12.3.6
Virus Clearance and Inactivation Studies

Prior to the initiation of phase I clinical trials, virus clearance and inactivation studies that demonstrate an adequate level of removal or inactivation of a relevant model virus should be completed. For murine hybridomas and other rodent cell lines, the relevant model virus is murine leukemia virus (MuLV). For primate or human cell lines, the relevant model virus would be any viruses known to be present in that particular cell line (e.g. EBV). Studies on the clearance and/or inactivation of additional model viruses should be performed on material that will be manufactured using the process anticipated for licensure and should ideally be completed prior to the pivotal clinical trial. These studies may need to be repeated when major manufacturing changes are made. The ICH Q5A: "Guidance on Viral Safety Evaluation of Biotechnology Products Derived From Cell Lines of Human or Animal Origin" (Food and Drug Administration 1998b) as well as the mAb PTC 97 documents describe the appropriate design, implementation, and interpretation of such studies.

Generic or modular virus clearance studies may be applied to subsequent mAbs manufactured at a given facility and are described in more detail in the mAb PTC 97. In general, a generic clearance study is one in which virus removal or inactivation has been demonstrated for several steps in the purification process of a model antibody. These data may then be applied to subsequent mAbs purified using the identical process, provided they are of the same species and class and are derived from the same cell substrate. A modular clearance study is one in which a single step in a purification process may differ from that of a model antibody. In such cases, only the unique module needs to undergo a virus clearance study, while the values obtained from the other modules of the model antibody may be applied to the new mAb.

Alternatively, bracketed virus reduction/inactivation studies may be performed. If more than one mAb will be manufactured at the same facility with similar, but not identical purification schemes (e.g. differences in ionic strength or pH of an elution buffer), studies that bracket the range of differences may be performed. Subsequent mAbs for which the purification parameters fall within the tested range may use the virus clearance values obtained from the bracketed study (Brorson et al. 2003).

12.3.7 Abbreviated Product Safety Testing for Feasibility Trials in Serious or Immediately Life-Threatening Conditions

Feasibility clinical trials are pilot studies to provide an early characterization of safety and an initial proof of concept in specific patient populations. They are limited in scope and are generally conducted at a single clinical site with a small number of patients. An immediately life-threatening condition is defined in 21CFR 312.34 as a "stage of disease in which there is a reasonable likelihood that death will occur in a matter of months or in which premature death is likely without early treatment." For such phase I clinical trials, the full battery of product safety tests is not required. Sterility (bacteria and fungi) should be performed and it is strongly recommended that mycoplasma and endotoxin testing be performed. If the purification scheme contains two orthogonal robust virus removal/inactivation steps (virus removal/inactivation based on different mechanisms), neither adventitious virus testing nor virus clearance studies need be performed. If clinical trials progress beyond these phase I feasibility trials, the full battery of safety testing, as well as virus clearance studies for a relevant model virus, should be performed prior to initiating phase II trials. Abbreviated testing does not apply to human or primate cell substrates. Clinical reviewers determine whether the indication in the IND application meets the criteria for a serious or life-threatening condition.

12.3.8 Comparability

Changes in the manufacturing of an mAb are expected during product development and, depending on the nature of the change, may necessitate an assessment of the comparability of the product pre and post change. The purpose of the assessment is to ensure that the manufacturing changes have not affected the safety, identity, purity, or efficacy of the mAb (Chirino and Mire-Sluis 2004). Demonstration of the comparability of a product made using two different manufacturing schemes or at different manufacturing facilities becomes more important during phase III clinical trials or after product approval, than when changes are implemented earlier in the course of product development. In IND applications for phase I studies, however, the drug product being proposed for use in the clinical trial should be appropriately compared to the drug product used in the animal toxicology studies in order to extrapolate the preclinical safety data to the clinical scenario (Food and Drug Administration 1996a). As clinical trials progress and product development matures, it is expected that changes will be introduced to improve the manufacturing process and thus, plans to demonstrate comparability between the product generated by the old and new manufacturing schemes should be devised. Major changes should be in place at the start of phase III trials but a scale-up of the process or additional manufacturing changes are occasionally introduced during phase III trials. Comparability studies during phase III may also include an analysis of key process intermediates (including, but not limited to, cell culture metrics and process contaminants), as appropriate.

In addition to maintaining a current reference standard, samples from several lots manufactured by each production scheme or scale should be properly retained. An early and thorough characterization of the mAb, including physical, chemical, biological, and immunological characteristics, determines the attributes that a mAb should retain after manufacturing changes or scale-up are introduced. The need for additional preclinical or clinical testing when manufacturing changes occur is discussed in Section 12.4.6 below. Sponsors are strongly encouraged to consult with the FDA regarding plans for demonstrating product comparability.

12.4 Considerations for Preclinical Testing

The primary goals of preclinical safety evaluation are: to determine a safe starting dose and subsequent dose escalation schemes in humans; to identify target organs and potential toxicities; to establish safety parameters for clinical monitoring and to provide a risk assessment for the intended human population (Serabian and Pilaro 1999). A summary of the types of studies utilized in the preclinical assessment of biotechnology products, including mAbs, can be found in two specific documents, mAb PTC 97 and ICHS6: “Preclinical Safety Evaluation of Biotechnology-Derived Pharmaceuticals” (Food and Drug Administration 1997b). In general, the unique properties of mAbs that make them desirable therapeutic agents (i.e. high specificity and affinity for their targets) limit the types of preclinical studies that can be performed, resulting in the need for flexibility in preclinical development strategies. Hence, some conventional “small molecule” drug approaches may not be useful or appropriate.

When an IND is submitted to the FDA for initiation of a first-in-human study, the pharmacology/toxicology data for an mAb should include the following:

- Tissue cross-reactivity studies
- Data justifying relevant species
- Pharmacodynamic studies

- Pharmacokinetic assessment
- Toxicology studies.

A generalized description of the pharmacology and toxicology information needed to support a phase I study is provided in the "Guidance for Industry: Content and Format of Investigational New Drug Applications (INDs) for Phase 1 Studies of Drugs, Including Well Characterized, Therapeutic, Biotechnology-Derived Products" (Food and Drug Administration 1996a). This section of the chapter will address and expand upon the principles and concepts that relate specifically to pharmacology and toxicology studies performed with mAb products.

12.4.1 Tissue Cross-Reactivity

MAbs, although developed for a specific epitope/antigen, may cross-react with different, but similar, epitopes and/or bind to the intended epitope in unexpected tissues. Since the binding of pharmacologically active antibodies to nontarget tissues may have serious toxicological consequences, both the mAb PTC 97 and ICH S6 documents stress the importance of characterizing epitope/antigen distribution across a panel of human tissues, prior to the initiation of a phase I clinical trial. In general, tissue cross-reactivity studies help define the desired and undesired binding properties of the mAb, provide a greater understanding of the potential *in vivo* toxicities, reveal potential target tissues and organs and provide confidence in the selection of a suitable animal species for preclinical studies.

Tissue cross-reactivity is assessed by immunohistochemistry. For these studies, a panel of 37 human tissues from at least three donors should be examined to ensure an accurate representation of possible *in vivo* tissue binding (Food and Drug Administration 1997b). A list of the suggested human tissues is provided in the mAb PTC 97. The tissues should be cryopreserved, rather than paraffin-embedded, due to the possibility for "false negatives" that may result from inconsistent or poor antigen retrieval, which often occurs with paraffin-embedded tissue sections. The mAb to be tested (test article) should be the one designated for the clinic and can be either unconjugated and detected by a labeled secondary antibody or conjugated directly to detection molecules such as peroxidase, biotin, or fluorescein. Since direct labeling of an mAb may alter its binding affinity and tissue-staining pattern, the affinity of a conjugated test article should be comparable to that of the unconjugated mAb. At least two concentrations (a low and high concentration) of test article should be investigated. The projected clinical serum concentration may be considered, but the test article concentrations should be optimized for the assay to produce strong, specific binding, with limited background staining. Positive and negative control tissues and a species- and isotype-matched negative control antibody should be incorporated into the study design. In unique cases where adequate assay development is impossible, alternative approaches (e.g. use of a chimeric or murine mAb test article) may be considered;

however, it is advantageous to discuss these alternatives with the FDA prior to implementation.

12.4.2 Relevant Species

Since the specificity of an mAb, developed against a human substrate, often limits its immunoreactivity to orthologous substrates in test species, it is extremely important to identify pharmacologically relevant animal species appropriate for pharmacodynamic, pharmacokinetic, and toxicology studies. With regard to mAbs, a relevant species is one in which the mAb (1) is pharmacologically active due to its specific cross-species binding to the orthologous antigen/epitope and (2) demonstrates a tissue cross-reactivity pattern similar to human (Food and Drug Administration 1997b). Relative affinity of the mAb to the non-human epitope, compared with the human epitope, should be assessed as well and may be useful in the interpretation of the toxicology data. In determining appropriate species, it may be helpful to first screen several species for mAb binding by immunohistochemistry on a limited panel of tissues or by flow cytometry on cells expressing the orthologous antigens or for mAb activity by functional assays (e.g. enzyme induction, cell signaling, and physiological changes). A subsequent, comprehensive comparison of mAb tissue cross-reactivity between non-human and human tissues will confirm the relevancy of the species choice. All data supporting the choice of a relevant species should be included in the initial IND application. Two studies performed by Boon et al., one with a chimeric anti-CD40 mAb in cynomolgus monkeys (Boon et al. 2002a) and the other with a humanized anti-CD4 mAb in rhesus monkeys (Boon et al. 2002b), provide excellent examples of the contribution of tissue cross-reactivity studies in establishing a preclinical safety assessment program. In neither study were the tissue distributions of the mAb binding identical for the human and non-human primate; however, patterns of tissue and cell-type binding were similar enough to suggest that a toxicology study in the particular non-human primate would provide adequate safety data.

12.4.3 Pharmacodynamic and Pharmacokinetic Studies

Pharmacodynamic studies, particularly those contributing to understanding the mechanism of action of the mAb and those providing "proof of concept" (i.e. potential human efficacy), should be included in the initial IND submission. *In vitro* cell culture experiments and *in vivo* studies with animals displaying the disease are commonly utilized to demonstrate efficacy. If cross-species reactivity of the mAb is limited, xenograft or transgenic models expressing the human antigen of interest may be useful tools. Overall, these types of studies are important because they aid in the estimation of the effective dose, dosing regimens and most appropriate plasma concentrations.

Pharmacokinetic (PK) parameters are most often calculated from the administration of the mAb to healthy animals of a relevant species. However, use of an animal model that shares common pathophysiology/symptoms of the clinical disease may not only better reflect the pharmacodynamic properties and clinical outcome, but may also provide a more accurate PK profile. For example, if the mAb is directed towards an antigen that is overexpressed in tumor tissue, presence of the tumor will likely alter the biodistribution of the mAb, thereby decreasing its blood concentration. As the tumor burden diminishes, more antibody will be available systemically, ultimately affecting overall animal exposure to the mAb. Assessing the PK profile in nonrelevant animal species (those lacking a high-affinity binding epitope) is discouraged, because the major mechanisms of elimination of the mAb from circulation (i.e. binding to antigen and internalization) are low or absent. Thus, changes in clearance attributable to saturation of antigen binding would not be observed.

PK studies assess the absorption, distribution, and excretion of a mAb and may be performed in independent studies or incorporated into general toxicology studies. Unlike "small molecule" drugs, proteins, including mAbs, are not metabolized by hepatic cytochrome P450 mechanisms, but are catabolized into their individual amino acids that can be reused for protein synthesis and energy production. Thus, metabolism assessments used for "small molecule" drugs are not warranted. When mAbs are administered intravenously, they most aften display a biphasic elimination profile, consisting of a rapid distribution phase and a long elimination phase in which antibody recycling and catabolism occur (Ghetie and Ward 2002). MAbs of various isotypes and subtypes have different elimination half-lives, following similar trends to that identified for endogenous antibodies. Endogenous IgG1, IgG2, and IgG4 have half-lives of approximately 23 days, whereas the half-lives of IgG3 and IgM average 7.5–9 days and 5 days, respectively (Trang 1992). In general, half-lives of mAbs, particularly as they become more "humanized", are shorter in animals than what is observed in humans; the projected exposure difference should be considered when extrapolating animal data to the human setting in the course of designing a clinical trial dosing regimen.

PK profiles of mAbs can be influenced by a variety of factors, including, but not limited to, affinity for specific and non-specific binding proteins, immunogenicity (see Section 12.4.5), posttranslational processing (see Section 12.4.6), concentration, antibody isotype, manipulation of the antibody (e.g. fragments, fusion proteins), formulation, and route of administration. It is therefore important to maintain consistency between the preclinical and clinical material and study designs.

12.4.4 Toxicology

All preclinical toxicology studies should be performed in relevant species, in order to prevent misleading safety interpretations. For example, Klingbeil and Hsu observed that administration of humanized mAb Hu1D10 to antigen-positive monkeys resulted in severe acute adverse effects (e.g. respiratory suppression, increased heart rate, and urticaria) that in some cases required life-sustaining intervention, while no such adverse effects were observed in antigen-negative and control monkeys (Klingbeil and Hsu 1999). In this case, if pharmacological relevancy was ignored, the Hu1D10-related toxicities would have gone undetected and a "false negative" safety interpretation would have been established. Conversely, the use of nonrelevant animals may result in "false positive" safety conclusions, most often reflective of toxicities arising from product immunogenicity (see Section 12.4.5).

In many cases, the only appropriate animal species for a mAb is a non-human primate. The most common non-human primate used for preclinical toxicology studies is the cynomolgus monkey; additional non-human primates that may be used are rhesus and marmoset monkeys and chimpanzees. However, because chimpanzees are a protected species many limitations to conducting nonclinical studies with these animals exist, including (1) difficulty in obtaining protein-naive animals, (2) limited numbers of animals available, (3) limits to dose frequency and level, (4) limited numbers of animals per treatment group, and (5) inability to sacrifice animals at the end of the study to obtain histopathology data. Thus, sponsors confronted with the chimpanzee as the only relevant species often consider alternative approaches. These approaches include analogous mAbs, transgenic models, and animal models of disease (Food and Drug Administration 1997b).

An analogous mAb is one developed against the antigen of another species (e.g. recognizes the ortholog of the original human target). While this approach allows for conducting the necessary nonclinical studies, it is not without disadvantages. The manufacturing of the analogous mAb differs from that being developed clinically; as a result, the product may contain different impurity and contaminant profiles and exhibit disparate potency and/or pharmacology. Conducting appropriate studies to define the pharmacology of the analogous mAb, to the greatest extent possible, can reduce the impact of these disadvantages. The pharmacology of the analogous mAb should be compared with that of the product intended for human use. An example of this approach is described for the evaluation of reproductive and chronic toxicities of infliximab (chimeric mAb to human TNFα), in which an analogous mAb was used to target mouse TNFα (Treacy 2000). A second alternative approach is the use of a transgenic mouse model expressing the human antigen.

The information gained from these models is most useful when the interaction of the product with the humanized antigen has similar physiological consequences to those expected in humans. For example, chronic and reproductive toxicity studies for keliximab, a chimeric mAb specific for human and chimpanzee CD4, were performed in a human CD4 transgenic mouse model (Bugelski et al. 2000).

A third option is one using an animal model of disease. These models are used rarely in practice due to complications of the underlying disease interfering with the interpretation of the toxicological data. They may, however, be useful in ascertaining whether a particular toxicity is dependent on the conditions inherent in the disease, such as the expression of a disease-associated receptor, or is an independent finding. Overall, these alternative approaches are not employed routinely, but are considered when the scientific need for more information arises.

Normally, safety assessments (toxicology and PK studies) are performed in two species, one rodent and one nonrodent (Food and Drug Adminstration 1997b, 1997c). However, due to the potential limitations presented by different relevant species, the majority of biological products are assessed in only one species. ICHS6 specifically states that this approach is acceptable "in certain justified cases (e.g., when only one relevant species can be identified or where the biological activity of the biopharmaceutical is well understood)." Two notable product exceptions are immunotoxins, mAbs that are fused to protein toxins such as diphtheria toxin, ricin, or pseudomonas toxin (produced as one protein), and immunoconjugates, mAbs conjugated through an organic chemical linker to "small molecule" drugs, radioisotopes, or toxins. Toxicology studies for these product types should be performed in two species when no relevant species is available and nonspecific factors play a predominant role in toxicity, such as when (1) the toxin, "small molecule" drug or radioisotope portion of the molecule results in activity and/or toxicity in non-mAb targeted tissues, and (2) immunoconjugates are unstable and degrade to the individual components, thereby negating the specificity provided by the antibody.

In cases where degradation is prominent, separate studies with the individual components may be necessary to identify target tissues and toxicities exerted by the different components. Additional special considerations for both the toxicology and PK studies exist for immunotoxins and immunoconjugates, but are beyond the scope of this chapter; a detailed discussion is found in mAb PTC 97.

Good preclinical study design is essential for identifying potential endpoints of toxicity to be monitored in clinical trials and for determining the human starting dose. In general, toxicity studies should be performed according to Good Laboratory Practices (GLP, part 58 of the Code of Federal Regulations), and both males and females should be included or justification provided if one sex is excluded. The duration of mAb exposure, and frequency and route of administration should mimic, as closely as possible, the proposed clinical use. ICHM3: "Nonclinical Safety Studies for the Conduct of Human Clinical Trials for Pharmaceuticals" outlines the minimum length of repeat dose toxicology studies and the timing of these studies in relation to clinic trial durations and phases (Food and Drug Administration 1997c). Various dose levels should be selected, including a no adverse effect level dose (known commonly as the NOAEL) and a toxic dose. If the mAb is expected to be relatively nontoxic, the highest dose should be (1) a scientifically reasonable multiple of the highest projected clinical dose, (2) a dose reflective of a pharmacodynamic marker (e.g. saturation of antigen) or (3) the maximum feasible dose in the animals. In addition to terminal sacrifice, usually performed within 1–3 days after the last dose, a treatment-free recovery period should be included in the study design to determine reversibility of effects and/or potential delayed toxic effects.

Although no guidance is provided by CDER on toxicology endpoints, it is expected that physical examination, body weight, food consumption, ophthalmologic evaluation, clinical pathology, gross pathology, organ weights, histopathology, and immunogenicity endpoints will be included in general toxicology studies. A good description of applicable endpoints can be found in *Principles and Methods of Toxicology* (Wilson and Hardisty 2001). Additionally, ICHS7A: "Safety Pharmacology Studies for Human Pharmaceuticals" (Food and Drug Administration 2001) describes a core battery of "safety pharmacology" tests designed to investigate potential toxicities on the cardiovascular, respiratory and central nervous systems. For highly targeted biologics, such as mAbs, these "safety pharmacology" endpoints can be incorporated into the general toxicology study.

In addition to the general toxicology studies, other specialized toxicology studies, such as reproductive toxicology and carcinogenicity, may be necessary depending upon the product, clinical indication, and intended patient population (Food and Drug Administration 1997b). Specifically, embryo–fetal developmental reproductive toxicology studies are conducted when the intended human population includes women of child-bearing potential. However, reproductive studies with mAbs that have only non-human primate as a relevant species pose many challenges, including obtaining sexually mature animals, low conception rate, high abortion rate, limited number of offspring, and immunogenicity. As a result, these studies often can be deferred until later stages of drug development, provided that the product does not produce an unacceptable reproductive risk and women of child-bearing potential are using suitable contraception. Guidance on reproductive study designs can be found in ICHS5A: "Detection of Toxicity to Reproduction for Medicinal Products" (Food and Drug Administration 1994). Two-year carcinogenicity studies are usually not conducted for mAb products because (1) mAbs are not expected to be translocated to the nucleus of an intact cell and interact with DNA or other chromosomal material to induce mutation and transform cells (i.e. genotoxicity); (2) species cross-reactivity may limit the usefulness of current rodent models; and (3) the long lifespan of non-human primates and the large number of animals required make the studies impractical. However, if the mechanism of action of the mAb suggests that it, directly or indirectly, might support or induce proliferation of transformed cells, tumor promotion studies may be required. These studies assess the ability of the product to stimulate the growth of cells already transformed, such as dormant tumors or micrometastases, and may consist of *in vitro* proliferation assays of normal and transformed cells and/or *in vivo* studies utilizing alternative approaches (e.g. transgenic mouse models). The need for and design of such studies should be discussed with the FDA.

12.4.5 Immunogenicity

Many biotechnology-derived pharmaceuticals, including mAbs, induce the formation of anti-product antibodies (anti-mAb) in animals and humans. An immune response in animals, however, does not reliably predict a similar response in humans. This is true particularly with the design of more "humanized" or fully human antibodies, because animals may recognize these proteins as "foreign," whereas humans may not. Nevertheless, it is important to measure antibody formation in the toxicology study to aid in its interpretation. Specifically, anti-product antibodies may affect the overall animal exposure to the active mAb by altering its rate of clearance or by neutralizing its function by inhibiting target binding. In addition, anti-product antibody formation may result in immune complex disease (or serum sickness), a condition caused by deposition of anti-mAb:mAb complexes in the vasculature with subsequent activation of inflammation pathways. Conversely, these anti-product antibodies may not affect product exposure or activity; thus, the presence of anti-product antibodies in the absence of PK effects or other toxicities is not sufficient to lead to the termination of a study. Instead, antibody responses should be characterized with respect to titer, number of animals, and function (i.e. neutralizing or non-neutralizing) and correlated with any pharmacological and/or toxicological changes in the animals. An in-depth discussion of this topic was published previously (Bussiere 2003).

The ability to assess the preclinical and clinical immunogenicity of biotechnology products is dependent upon the quality of the assay that is developed. Poorly designed assays most often impede product development and may result in post-marketing commitment studies upon licensure. Specifically for mAb products, assays detecting anti-product antibodies are complicated by the fact that antibodies are usually the detection reagents, the product, and the target of the assay. Thus, the sole presence of the product (mAb) in blood samples can interfere with the assay. Therefore, it is critical to either (1) collect blood samples at time points where the levels of mAb are negligible or (2) demonstrate that the presence of the mAb does not interfere with the specificity and sensitivity of the assay. Mire-Sluis and colleagues provide detailed recommendations for optimizing such immunoassays (Mire-Sluis et al. 2004).

12.4.6 Comparability

Since changes in the product-manufacturing scheme occur frequently during clinical development, the product used in nonclinical studies is not required to be "identical" to that going into the clinic, but it does have to be "comparable." The mAb PTC 97, ICH Q5E: "Comparability of Biotechnological/Biological Products Subject to Changes in their Manufacturing Process" (Food and Drug Administration 2005) and the document entitled "FDA Guidance Concerning Demonstration of Comparability of Human Biological Products, Including Therapeutic Biotechnology-Derived Products" (Food and Drug Administration 2002d) provide guidance regarding the demonstration of comparability in this setting.

Comparability of biotechnology products is usually determined by analytical and functional assays. Preclinical studies may be necessary if (1) the product quality is impacted adversely; (2) biochemical alterations in the active moiety are known to affect product exposure *in vivo* (e.g. glycosylation, charge); (3) it is not possible to determine if a potential biochemical change will be significant clinically because the analytical testing may be insufficiently sensitive, precise, or accurate; (4) the relationship of efficacy and/or toxicity to the product is not sufficiently established to determine the significance of the differences observed from analytical testing; or (5) a change in the formulation includes the addition of unknown excipients (Green 2002). Bridging PK studies are most often required; however, depending on the extent of the final product changes, more extensive toxicology studies may be required and necessitate discussion with the FDA.

12.5 Conclusions

The recent successes in approving mAbs for commercial use for oncologic, immunotherapeutic, and infectious disease indications encourage the further development of new mAbs directed against novel targets to treat or diagnose a broadening array of diseases. Furthermore, the approvals of the drug-conjugate Mylotarg and the radioimmunoconjugates Zevalin and Bexxar are likely to spur development of a variety of antibody conjugates. The sequencing of the human genome and mapping of human genetic variation (International HapMap Consortium 2005) provides many new targets for mAb therapy and it is anticipated that submission of INDs with mAbs recognizing novel targets will increase in the coming years. Regulators rely upon a flexible, case-by-case, science-based approach to safety evaluation needed to support clinical development and marketing authorization. Sponsors are encouraged to contact the appropriate review office at FDA for their mAb product to request a meeting before submitting an IND or IDE. Effective communication between the FDA and the sponsor is a crucial element of the pathway from drug discovery to the clinic. While the regulatory pathway is complex, an early understanding of the regulatory process and careful product and preclinical characterization will enhance the chances of success.

Acknowledgments

We would like to thank our colleagues Drs Kathleen A. Clouse, Christopher Ellis, M. David Green, Steven Kozlowski, and Andrea Weir for insightful comments on the manuscript.

References

Bauer, B.A., Besch-Williford, C.L., Riley, L.K. (2004) Comparison of the mouse antibody production (MAP) assay and polymerase chain reaction (PCR) assays for the detection of viral contaminants. *Biologicals* 32: 177–182.

Boon, L., Holland, B., Gordon, W., Liu, P., Shiau, F., Shanahan, W., Reimann, K.A.,

Fung, M. (2002a) Development of anti-CD4 MAb hu5A8 for treatment of HIV-1 infection: preclinical assessment in non-human primates. *Toxicology* 172: 191–203.

Boon, L., Laman, J.D., Ortiz-Buijsse, A., den Hartog, M.T., Hoffenberg, S., Liu, P., Shiau, F., de Boer, M. (2002b) Preclinical assessment of anti-CD40 Mab 5D12 in cynomolgus monkeys. *Toxicology* 174: 53–65.

Brorson, K., Swann, P.G., Lizzio, E., Maudru, T., Peden, K., Stein, K.E. (2001) Use of a quantitative product-enhanced reverse transcriptase assay to monitor retrovirus levels in mAb cell-culture and downstream processing. *Biotechnol Prog* 17: 188–196.

Brorson, K., Xu, Y., Swann, P.G., Hamilton, E., Mustafa, M., de Wit, C., Norling, L.A., Stein, K.E. (2002) Evaluation of a quantitative product-enhanced reverse transcriptase assay to monitor retrovirus in mAb cell-culture. *Biologicals* 30: 15–26.

Brorson, K., Krejci, S., Lee, K., Hamilton, E., Stein, K., Xu, Y. (2003) Bracketed generic inactivation of rodent retroviruses by low pH treatment for monoclonal antibodies and recombinant proteins. *Biotechnol Bioeng* 82: 321–329.

Bugelski, P.J., Herzyk, D.J., Rehm, S., Harmsen, A.G., Gore, E.V., Williams, D.M., Maleeff, B.E., Badger, A.M., Truneh, A., O'Brien, S.R., Macia, R.A., Wier, P.J., Morgan, D.G., Hart, T.K. (2000) Preclinical development of keliximab, a Primatized anti-CD4 monoclonal antibody, in human CD4 transgenic mice: characterization of the model and safety studies. *Hum Exp Toxicol* 19: 230–243.

Bussiere, J.L. (2003) Animal models as indicators of immunogenicity of therapeutic proteins in humans. *Dev Biol (Basel)* 112: 135–139.

Chirino, A.J., Mire-Sluis, A. (2004) Characterizing biological products and assessing comparability following manufacturing changes. *Nat Biotechnol* 22: 1383–1391.

Clark, M. (2000) Antibody humanization: a case of the 'Emperor's new clothes'? *Immunol Today* 21: 397–402.

Eldering, J.A., Felten, C., Veilleux, C.A., Potts, B.J. (2004) Development of a PCR method for mycoplasma testing of Chinese hamster ovary cell cultures used in the manufacture of recombinant therapeutic proteins. *Biologicals* 32: 183–193.

Food and Drug Administration (1987) *Guideline on Validation of the Limulus Amebocyte Lysate Test as an End-Product Endotoxin Test For Human and Animal Parenteral Drugs, Biological Products and Medical Devices.* FDA.

Food and Drug Administration (1994) International Conference on Harmonisation: Guideline on detection of toxicity to reproduction for medicinal products. [Docket No. 93D-0140]. *Federal Register* 59: 48746–48752.

Food and Drug Administration (1995) Points to consider in the manufacture and testing of therapeutic products for human use derived from transgenic animals. [Docket No. 95D-0131]. *Federal Register* 60: 44036.

Food and Drug Administration (1996a) Guidance for industry; content and format of investigational new drug applications (INDs) for phase 1 studies of drugs, including well-characterized, therapeutic, biotechnology-derived products. [Docket No. 95D-0386]. *Federal Register* 61: 1939–1940.

Food and Drug Administration (1996b) International Conference on Harmonisation: Final guideline on stability testing of biotechnological/biological products [Docket No. 93D-0139]. *Federal Register* 61: 36466–36469.

Food and Drug Administration (1997a) Points to consider in the manufacture and testing of monoclonal antibody products for human use. [Docket No. 94D-0259]. *Federal Register* 62: 9196–9197.

Food and Drug Administration (1997b) International Conference on Harmonisation: S6 Guidance on preclinical safety evaluation of biotechnology-derived products. [Docket No. 97D-0113]. *Federal Register* 62: 6515–61519.

Food and Drug Administration (1997c) International Conference on Harmonisation: Guidance on nonclinical safety studies for the conduct of human clinical trials for pharmaceuticals. [Docket No. 97D-0147]. *Federal Register* 62: 62922.

Food and Drug Administration (1998a) International Conference on Harmonisation: Guidance on quality of biotechnological/biolcgical products Q5D: Derivation and characterization of cell substrates used for the production of biotechnological/biolcgical products. [Docket No. 97D-0159]. *Federal Register* 63: 50244–50249.

Food and Drug Administration (1998b) International Conference on Harmonisation: Guidance on viral safety evaluation of biotechnology products derived from cell lines of human or animal origin. [Docket No. 96D-0058]. *Federal Register* 63: 51074–51084.

Food and Drug Administration (2001) International Conference on Harmonisation: Guidance on S7A safety pharmacology studies for human pharmaceuticals. [Docket No. D-1407]. *Federal Register* 66: 36791–36792.

Food and Drug Administration (2002a) Draft Guidance for Industry: Drugs, biologics, and medical devices derived from bioengineered plants for use in humans and animals. [Docket No. 02D-324]. *Federal Register* 67: 57828–57829.

Food and Drug Administration (2002b) Guidance for Industry: Revised preventative measures to reduce the possible risk of transmission of Creutzfeldt-Jakob disease (CJD) and variant Creutzfeldt-Jakob disease (vCJD) by blood and blood products. [Docket No. 97D-0318]. *Federal Register* 67: 2226–2227.

Food and Drug Administration (2002c) International Conference on Harmonisation: Draft revised guidance on Q1A(R) stability testing of new drug substances and products. [Docket No. 93D-0139]. *Federal Register* 65: 21446–21453.

Food and Drug Administration (2002d) Guidance concerning demonstration of comparability of human biological products. [Docket No. 96D-0132]. *Federal Register* 70: 37861–37862.

Food and Drug Administration (2005) International Conference on Harmonisation: Guidance on Q5 comparability of biotechnological/biological products subject to changes in their manufacturing process. [Docket No. 2004D-0118]. *Federal Register* 59: 48746–48752.

Ghetie, V., Ward, E.S. (2002) Transcytosis and catabolism of antibody. *Immunol Res* 25: 97–113.

Glennie, M.J., Johnson, P.W. (2000) Clinical trials of antibody therapy. *Immunol Today* 21: 403–410.

Green, M. (2002) Comparability studies for human therapeutics, preclinical and pharmacokinetic aspects. In *Comparability Studies for Human Plasma-Derived Therapeutics – Workshop*. http://www.fda.gov/cbr/minutes/plasma053002.htm.

Griffiths, E. (1987) WHO Expert Committee on Biological Standardization. Highlights of the Meeting of October 1996. *Biologicals* 25: 339–362.

Karpas, A., Dremucheva, A., Czepulkowski, B.H. (2001) A human myeloma cell line suitable for the generation of human monoclonal antibodies. *Proc Natl Acad Sci USA* 98: 1799–1804.

Klingbeil, C., Hsu, D.H. (1999) Pharmacology and safety assessment of humanized monoclonal antibodies for therapeutic use. *Toxicol Pathol* 27: 1–3.

Kohler, G., Milstein, C. (1975) Continuous cultures of fused cells secreting antibody of predefined specificity. *Nature* 256: 495–497.

Mire-Sluis, A.R., Barrett, Y.C., Devanarayan, V., Koren, E., Liu, H., Maia, M., Parish, T., Scott, G., Shankar, G., Shores, E., Swanson, S.J., Taniguchi, G., Wierda, D., Zuckerman, L.A. (2004) Recommendations for the design and optimization of immunoassays used in the detection of host antibodies against biotechnology products. *J Immunol Methods* 289: 1–16.

Reichert, J.M., Rosensweig, C.J., Faden, L.B., Dewitz, M.C. (2005) Monoclonal antibody successes in the clinic. *Nat Biotechnol* 23: 1073–1078.

Serabian, M.A., Pilaro, A.M. (1999) Safety assessment of biotechnology-derived pharmaceuticals: ICH and beyond. *Toxicol Pathol* 27: 27–31.

Stein, K.E. (1997) Overcoming obstacles to monoclonal antibody product development and approval. *Trends Biotechnol* 15: 88–90.

International HapMap Consortium (2005) A haplotype map of the human genome. *Nature* 437: 1299–1320.

Trang, J.M. (1992) Pharmacokinetics and metabolism of therapeutic and diagnostic antibodies. In: Ferraiolo, B., Mohler, M.A., Gloff, C.A. (eds) *Protein Pharmacokinetics and Metabolism.* New York: Plenum Press, pp. 223–270.

Treacy, G. (2000) Using an analogous monoclonal antibody to evaluate the reproductive and chronic toxicity potential for a humanized anti-TNFalpha monoclonal antibody. *Hum Exp Toxicol* 19: 226–228.

Wilson, N., Hardisty, J.H. (2001) Short-term, subchronic and chronic toxicology studies. In: Hayes, W. (ed.) *Principles and Methods of Toxicology.* Philadelphia, PA: Taylor and Francis, pp. 917–957.

13 Intellectual Property Issues

Michael Braunagel and Rathin C. Das

13.1 Introduction

Intellectual property (IP) is a core aspect of any commercialization effort in the pharmaceutical industry. While protecting the ownership of substances of therapeutic value has always been a central concern for the pharma companies, recent developments in IP laws, the steadily increasing number of patents on both methods and compounds, and not least a fair number of high-profile litigation cases have put the IP area into the spotlight. The court battles especially have given the field a somewhat mixed reputation. While everyone agrees that protecting one's inventions is necessary, there are also some concerns that extensive IP protection on every aspect of research may inhibit research and drug development in the long run. Before analyzing the specific intellectual property right (IPR) issues on therapeutic antibodies, it may therefore be helpful to summarize two central aspects of IP protection.

13.2 Why Intellectual Property Rights are Important

While trademarks and copyrights do play a role in product development and sales and marketing of a product, the most relevant aspect of the property law for technologies and products in the pharmaceutical industry pertains to patents. Patents are not at all a new invention. The patent system dates back to the seventeenth century, and terms such as "international priority dates" were introduced more than 100 years ago. A patent is a deal between an inventor and the society. If an inventor is willing to disclose his or her invention to the public in enough

Certain portions of this article were previously published in a market research report, titled: "Antibody Therapeutics: Product Development, Market Trends, and Strategic Issues," by Rathin C. Das and K. John Morrow, Jr., 2004, D & MD Publications, West Borough, MA, USA.

Handbook of Therapeutic Antibodies. Edited by Stefan Dübel

ISBN 978-3-527-31453-9

detail that anyone could copy it, then the inventor should be allowed for a certain time period to have the sole right of commercializing the invention. After this time period, however, all rights to the invention would belong to the public, and anyone would be free to use it. At its very core, a patent is one of the very few monopolies accepted in today's world.

The major importance of patents for the pharma industry is based on the huge costs of drug development and the associated risks. A company involved in R&D of drugs needs to cover the costs for their activities, both for work done on the compounds introduced into the market, as well as for the many projects that do not come to fruition. The time span given by the patent on a compound, where the drug developer is essentially free of competition, is the only time in which those investments can be recuperated. As soon as the patent expires, other companies that have not invested in the drug development are free to compete. This is the reason why patenting is central not only for large pharma companies, but also for universities and small companies, which research on new compounds or methods.

Most players in the field will not have the knowledge or the financial capacity to follow a drug development project from researching the lead to market introduction; most will need to involve other players sooner or later. A large company is not likely to join any project on a compound with unclear IP ownership, as this may mean that its investments might not bring any future monetary value.

The second crucial issue with patents is to understand that a patent is by its very nature a negative right. A patent and the monopoly associated with it does not allow its owner to do whatever is claimed in the patent. However, it gives the owner the right to stop others from doing what is claimed in the patent. To give an example, a patent might cover a specific antibody and its use as therapeutic agent against cancer. In this case, anyone who wants to market this antibody for this purpose will need a license from the owner of this patent. However, the owner of the patent might have used a method to find the antibody covered by IP from a third party. To produce that antibody, he or she might need a production method, which is patent protected, too. It may also be that the antigen recognized by the antibody is patent protected, together with all ligands binding to it for therapeutic purposes, including antibodies in general. In such a scenario, the owner of the patent on that specific antibody will need to negotiate three licenses to be entitled to commercialize the antibody – one each for the research tool, the production method, and the antigen. In each case a license might bear payment of significant royalty by the owner of the antibody. This phenomenon is known as "royalty stacking," as royalties are a very common condition in licenses. Given the large number of patents filed or granted, royalty stacking is a significant issue in the pharma field in general, and the recombinant antibody field, in particular.

A company having central patents on key technologies or compounds is in a strong position in a particular field of product development. Indeed, any company having a proprietary position beginning with the development tools to production, which are covered largely by its own patents without the need of paying large amounts of license fee for any reach-through IPs, will have a larger share on potential revenues and would enjoy a dominant position in a specific drug development sector. Consequently, from the reverse perspective, entering a field in which several aspects are already covered by IP might be less attractive for a commercial organization, unless the licenses can be readily obtained and/or the potential revenue generation prospects are very high.

We have highlighted below the status of major IP positions in the recombinant antibody area and some of the interesting developments that have taken place in the field in recent years. We have also considered providing a roadmap for the newcomers in the field to facilitate solving some of the puzzles of the recombinant antibody IPs.

13.3 Recombinant Antibody Technologies

As incredible as it may sound from the perspective of current biotechnology commercialization, it is a fact that Georges Kohler and Cesar Milstein, inventors of mouse hybridoma, never attempted to patent their technology. Although many companies since established have utilized freely available mouse monoclonal antibody technology, only one completely mouse antibody is currently in the market, primarily because of the deleterious effect of mouse antibodies in humans, commonly known as the HAMA (human antibody mouse antibody) response. Of course, hybridoma technology has significantly improved during the past decades to generate hunanization technology, chimeric antibody technology, and completely human antibody technology. This has also led to the generation of patents covering every single aspect of such technologies and any improvements thereon, and production of antibodies generated by such technologies.

Because of this plethora of patents, many companies interested in entering the antibody field have been navigating the IP landscape with significant caution, while some of the IP holders in this field have been embroiled in lengthy legal battles to sort out their exact stake about the freedom of operation. Indeed, a great many elements involving discovery, engineering, and manufacturing of an antibody need to be considered in the case of production and marketing of a monoclonal antibody (mAb). Consequently, both dominant and reach-through IP issues need to be sorted out by an antibody developer and company financiers since these factors influence directly many facets of conducting a business in the mAb area, such as permissibility of development, freedom of operation as well as the royalty stacking issues. Most important IP issues in the mAb field deal with the following areas: (1) humanization of mouse monoclonals, (2) human antibody isolation by transgenic mouse system, (3) application of phage display in generating human antibodies and antibody fragments *in vitro*, and (4) production and manufacturing of humanized and/or human antibodies.

13.4 Antibody Humanization

Celltech Group plc (currently UCB SA) and Protein Design Labs (Currently PDL Biopharma) are the two companies that have the proprietary rights in the area of humanization of antibodies. Celltech's process is covered under the Adair patent (US 5,859,205), while Protein Design Labs has broad-ranging intellectual property coverage in the humanization process, covered under the Queens patents (US 5,585,089, US 5,693,761, US 5,693,762, US 6,180,370). PDL appears to have the dominant position in humanization technology, which it utilizes along with its in-house expertise to develop its own antibodies as well as to humanize promising murine antibodies of its partner companies. For a protracted period of about 6 months in 2003, Genentech and Protein Design Labs had developed significant disputes regarding PDL's antibody humanization patents and certain of Genentech's humanized antibodies. However, in late 2003, the two companies announced the resolution of such disputes and agreed that Genentech would exercise licenses under the patent licensing master agreement between the parties for various humanized products. Currently, there are seven marketed products that utilize PDL's humanization technology and more than 40 humanized antibodies in late stage clinical trials.

Companies currently holding patent licenses for technologies responsible for the discovery of various marketed antibodies are shown in Table 13.1.

13.5 Human Antibody Technology

Two primary technologies used for the discovery of recombinant human antibodies are transgenic mouse system and phage display method. Two leading companies that have strong intellectual patent position in the transgenic mouse technology area are Abgenix and Medarex. Abgenix's technology is known as XenoMouse technology, while Medarex's transgenic mouse system is called HuMAb-Mouse. Abgenix was launched as a wholly-owned subsidiary by Cell Genesys in 1996 by incorporating the XenoMouse technology as its primary technology platform. Medarex obtained its strong IP in its transgenic mouse technology by the acquisition of GenPharm in October of 1997. Although both companies were initially at bitter odds with each other, their cross-licensing agreement in March 1997 of their respective IP rights has proved to be a successful strategy for both companies since each has generated well over 50 collaborations and partnerships since conclusion of this agreement. Both technologies have successfully produced an exceedingly increased number of human antibodies, of which at least 33 are in the clinic with no apparent patient immune response to the products (*Nature Biotechnology*, September 2005).

Unlike the transgenic mouse system, there are several players in the *in vitro* human antibody discovery area who have certain proprietary rights applicable to

Table 13.1 Patented technologies applied to the FDA approved therapeutic monoclonal antibodies.

Product name/type	*Developer/marketer*	*Indication*	*Mechanism of action*	*Year approved*	*Patented technologies of organizations*
Orthoclone OKT3 Murine	Ortho Biotech	Acute kidney transplant rejection	Anti-CD3	1986	–
Zevalin Murine	IDEC Pharma (currently Biogen IDEC), Schering AG	Relapsed or refractory low-grade, follicular, or transformed B-cell non-Hodgkin's lymphoma (NHL)	Anti-CD20	2002	Corixa (GlaxoSmith Kline)
Bexxar Murine	Corixa, Glaxo Smith Kline	NHL, CLL	Anti-CD20	2003	University of Michigan
ReoPro Chimeric		Coronary intervention and angioplasty	GpIIa/IIIb antagonist	1994	Centocor, Malvern, PA
MAbThera/ Rituxan Chimeric	IDEC Pharma (Biogen IDEC), Genentech, Roche	CD20-positive non-Hodgkin's lymphoma	Anti-CD20	1997	Celltech
ReoPro Chimeric	Centocor (J & J), Eli Lilly	Refractory unstable angina	GpIIa/IIIb antagonist	1997	Celltech
Synagis Chimeric	MedImmune, Gaithersburg, MD/ Abbott, Chicago, IL	Respiratory syncytial virus disease	Anti-RSV F protein	1998	PDL, Celltech, Genentech, Centocor
Simulect Chimeric	Novartis Pharma	Kidney transplant rejection	Inhibits interleukin-2 activation of T cells	1998	Celltech
Remicade Chimeric	Centocor (J & J), Schering-Plough	Rheumatoid arthritis	Anti-TNF-α	1999	Celltech, Genentech
Erbitux Chimeric	ImClone Systems, Bristol Meyers Squibb, Merck KgaA	Colorectal cancer	Anti-EGFR	2004	Genentech

Table 13.1 *Continued*

Product name/type	***Developer/marketer***	***Indication***	***Mechanism of action***	***Year approved***	***Patented technologies of organizations***
Zenapax Humanized	PDL, Roche	Kidney transplant rejection	Inhibits interleukin-2 activation of T cells	1997	Celltech
Herceptin Humanized	Genentech, Roche	Metastatic breast cancer	Anti-Her-2/neu receptor	1998	PDL, Celltech
Remicade Humanized	Centocor	Crohn's disease	Anti-TNF-α	1998	
Mylotarg Humanized	Celltech, Wyeth	Chemotherapeutic mAb for the treatment of CD33-positive acute myeloid leukemia in patients 60 and older in first relapse	Anti-CD33	2000	PDL
Campath Humanized	Genzyme (Ilex Oncology), Berlex (Schering AG)	B-cell chronic lymphocytic leukemia (B-CLL)	Anti-CD52	2001	Cambridge University, BTG
Xolair Humanized	Genentech,Novartis Pharma, Tanox	Allergic asthma	Anti-IgE	2003	PDL
Raptiva Humanized	Genentech, XOMA	Psoriasis	Prevention of the activation of T cells and their migration to the site of inflammation	2003	PDL
Avastin Humanized	Genentech	Angiogenesis in colorectal cancer	Anti-VEGF	2004	PDL
Humira Human	Cambridge Antibody Technology, Abbott	Rheumatoid arthritis	Anti-TNF-α	2002	CAT, MRC, Scripps, Stratagene, Genentech

antibody discovery. Some of the forerunners are Cambridge Antibody Technologies (CAT), Morphosys, Dyax, Biosite, BioInvent, Affitech, and Xoma. However, all of these companies realized that they need other IPs from within their peer groups to be able to practice antibody discovery and development for commercialization purposes. Consequently, during the past several years significant cross-licensing of IPs within these group of companies have taken place.

CAT originally acquired certain aspects of its core technology platform from the Medical Research Council of the UK and was issued six patents in the US (Table 13.2) as well as one in Europe (EP0368684). In addition, CAT was recently issued several continuations in part and divisional patents. CAT's core technologies are covered by three main families of patents, commonly known as Winter II, Griffiths and McCafferty patents. Winter II patent covers antibody expression libraries, McCafferty patent is for phage display, and the Griffiths patent is for the isolation of human antibodies to human proteins by phage display.

The US Winter II patent (US 6,248,516) is entitled "Single domain ligands, receptors comprising said ligands, methods for their production, and use of said ligands and receptors," and was issued on June 19, 2001. It is generally directed to antibody variable domain expression libraries carrying a diversity of CDR3 sequences and methods of making such libraries. This patent is co-owned by the UK Medical Research Council, Scripps Research Institute, La Jolla, CA, USA and

Table 13.2 Key antibody phage display patents of Cambridge Antibody Technology.

Patent	*Description*	*Patent status United States*	*Europe*
Griffiths	Methods for obtaining of anti self antibodies from antibody phage display libraries (covers isolation of human antibodies against all human proteins except for those that generate natural antibodies in humans)	Granted in March, 1999 (US 5,885,793)	Pending
McCafferty	Covers phage display of antibody fragments	Granted in October, 1999 (US 5,969,108)	Granted in November, 1996 (EP 589 877)
Winter II	The use of antibody genes for constructing a library	Granted in June, 2001 (US 6,248,516)	Granted in April 1994 (EO-O-368-684)[a]
Huse/Lerner/Winter	The use of antibody genes for constructing a library	Granted in September 2001 (US 6,291,158; US 6,291,159; US 6,291,160 and US 6,291,161)	Pending

a Claim 32 of the Winter II patent was subsequently amended following opposition by Morphosys.

Stratagene Corporation. CAT obtained an exclusive commercialization rights to the IP. CAT obtained four US patents covering antibody expression libraries – the "Huse/Lerner/Winter" patents entitled: "Method for Producing Polymers Having a Preselected Activity" (US 6,291,158 and US 6,291,159) and "Method for Tapping the Immunological Repertoire" (US 6,291,160 and US 6,291,161). The patents are generally directed to nucleic acid libraries for expression of functional immunoglobulin variable domains (or portions thereof) and methods of making such libraries. All of these patents are co-owned by the Medical Research Council, The Scripps Research Institute and Stratagene. Following the settlement of interference proceedings in June 1999, CAT is the sole exploiter of the intellectual property rights arising under the patents, subject to certain rights reserved by the co-owners and their pre-existing licensees.

With its dominant IP position in antibody libraries and antibody phage display, CAT pursued vigorously for several years both execution of its IP rights as well as maintaining those rights through litigation in some cases, such as those with Morphosys and Crucell, and business negotiations with others, such as those with Xoma and Dyax.

Morphosys owns the HuCAL antibody library system comprising synthetically prepared genes. The company has since expanded its technology to include HuCAL-GOLD, Cys Display, HuCAL-EST etc. Table 13.3 lists Morphosys' most important patents.

Table 13.3 Key antibody phage display patents of Morphosys Technology.

Patent (primary authors)	***Description***	***Patent status***	
		United States	***Europe***
Pack, Lupas	Mini antibodies, kept together by heterodimerization	Granted in Sept 2001 (US 6,294,353)	Pending
Pack, Lupas	Mini antibodies, kept together by heterodimerization of complementary association domains	Granted in Feb 2004 (US 6,692,935)	Pending
Knappik et al.	HuCal libraries, construction of modular synthetic antibody libraries, methods for using them.	Granted in Oct 2001 (US 6,300,064)	Granted in June 2002 (EP 859841)
Knappik et al.	HuCal libraries, antibody libraries based on a consensus framework, methods for using the libraries	Granted in Feb 2004 (US 6,696,248) and Mar 2004 (US 6,706,484)	Pending
Frisch et al.	HuCal libraries, displaying ESTs	Granted in Nov 2003 (US 6,300,064)	Pending
Lohning	HuCal Gold libraries, CysDisplay	Granted in Jun 2004 (US 6,753,136)	Pending

Dyax owns the Ladner patent family (US 5,223,409, US 5,403,484, US 5,571,698, US 5,837,500), which is a dominant patent family for practicing basic phage display technology. Additionally, Dyax has been practicing its Fab antibody library, the basic technology that it acquired through the purchase of Target Quest NV, a Dutch company. Some of the most important patents of Dyax in the phage display area has been noted in Table 13.4.

Another leading phage display company is San Diego, CA-based Biosite, which obtained patent rights to US Patents US 5,427,908 and US 5,580,717 and European Patent EP 527839B1, through acquisition from Affymax Technologies N.V. in 1998. The patents are directed to phage display of Fab and multichain antibodies wherein one of the polypeptide chains comprises a heavy-chain variable domain or region and a light-chain variable domain or region. Additionally, Affimed, based in Heidelberg, Germany, cross-licensed its IgM display library against the Ladner IP estate of Dyax, as well as obtaining a license from CAT to carry out its antibody discovery activities.

The Swedish antibody company BioInvent utilizes a phage display library system called n-CoDeR for the fast and efficient discovery of human antibodies. It is utilized in the development of proprietary drug candidates in diseases areas as HIV, atherosclerosis, cancer, and osteoarthritis as well as in development partnerships with pharmaceutical and biotech companies. The n-CoDeR library

Table 13.4 Key antibody phage display patents of Dyax Technology.

Patent (primary author)	***Description***	***Patent status***	
		United States	***Europe***
Ladner	Displaying proteins other than scFv on the surface of viruses/phages	Granted in June 1993 (US 5,223,409)	Pending
Ladner	A virus, presenting a protein other than scFv fused to a coat protein of said virus on the surface of the virus	Granted in April, 1995 (US 5,403,484)	Pending
Ladner	Making display libraries of proteins other than scFv, and selecting a member with desired binding properties from that library	Granted in May, 1996 (US 5,571,698)	Pending
Ladner	Making display libraries of proteins other than by mutating a protein via randomization, and selecting a member with desired binding properties from that library	Granted in November 1998 (US 5,837,500)	Pending
Ladner	Making display libraries of antibodies, and selecting a member with desired binding properties from that library	Granted in Dec 2005 (US6,979,538)	Pending

is covered primarily by the EP1352959, EP0988378B1, US 6 989 250. Bioinvent has also finalized a cross-licensing agreement with Xoma, and obtained licenses from CAT, Micromet, and Dyax. BioInvent's more than 160 patents and pending applications cover its core technology to discover antibody candidates and various aspects thereof, as well as antibody product candidates in development and their use in therapy.

Another approach is followed by Domantis, which is located in both Cambridge, UK and Cambridge, MA, USA. The Company's Domain antibodies, about 13 kDa in size, correspond to either the variable heavy or the variable light region of a human antibody. Some features of this format allow for specific selection methods, which was developed by Domantis. Domain antibody libraries, their selection methods, and therapeutic formats with extended serum half-life are covered by WO05/035572, WO04/101790, WO04/081026, WO04/058821, WO04/003019, and WO03/002609 and their equivalents, amongst others. Furthermore, Domantis has licenses of Winter II, Huse/Lerner/Winter, McCafferty and Griffiths from the UK Medical Research Council. The relevance of those patents has been explained above in more detail.

Affitech, based in Oslo, Norway has the IP for a phagemid system of antibody display that contains full-length pIII protein. It has been granted the European rights (EP 547201) to its already approved US patent (US 5,849,500) and several divisionals (US 5,985,588, US 6,127,132, US 6,387,627, and US 6,730,483) on the use of phagemid display of antibodies and antibody fragments that it exclusively licensed from Deutsches Krebsforschungszentrum Stiftung des Offentlichen Rechts (German Cancer Research Center), Heidelberg, Germany. It has another European divisional application currently pending (EP 1065271), however, both the European patent and the divisional is being opposed by CAT at the European patent office. Recently, the original patent has been revoked in the first court hearing but the ruling is under consideration for an appeal. It should be emphasized here that since Norway is not covered by most of the dominant antibody display patents, Affitech's business strategy allows the company to practice all aspects of phage display-based antibody selection without any license from the primary IP holders, and to export non-infringing affinity-matured antibody products for the international market. Additional to that Affitech has recently entered into cross-licensing agreements with Dyax for Ladner IP and with XOMA for its antibody expression technology. Affitech's patent portfolio is summarized in Table 13.5.

CAT and Morphosys were entangled in litigation for several years over their respective IPs. However, since the ruling by the US District Court in March 2002 that Morphosys' HuCAL antibody library technology is distinct and independent from CAT's patent coverage by Griffith's IP, the two companies have ended their disputes after signing cross-licensing agreements. Additionally, CAT has also cross-licensed its phage display patents with (1) Dyax's Ladner phage display IP, (2) Xoma's antibody expression technology, (3) Micromet and Enzon Pharmaceutical's single chain antibody technology, and (4) Crucell's Mabstract phage display technology.

Table 13.5 Phage display library intellectual properties of Affitech.

Inventors:	Breitling, Little, Dübel, Braunagel and Klewinghaus
Priority date:	08 July 1991
Owner:	German Cancer Research Centre (DKFZ)
Licensee:	Affitech AS
License terms:	Exclusive with right to sublicense

Patent/application	***Priority date***	***Pat./appl. no.***	***Status***
European parental	8 July 1991	EP 0547201	Granted*
European divisional	8 July 1991	EP 1065271	Pending**
US parental	8 July 1991	US 5,849,500	Granted
US divisional 1	8 July 1991	US 5,985,588	Granted
US divisional 2 ***	8 July 1991	US 6,127,132	Granted
US divisional 3	8 July 1991	US 6,387,627	Granted
US divisional 4	8 July 1991	US 6,730,483	Granted

* Revoked, revocation under appeal
** Under opposition
*** Title: "Phagemid library for antibody screening"

13.6 Antibody Production

"Boss" patent of Celltech and old "Cabilly" of Genentech were the earliest patents for making monoclonal antibodies. Boss patent covered the production of engineered antibodies and antibody fragments together with vectors and host cells related to these processes. The old Cabilly patent (US 4,816,567) covered the production of altered and native immunoglobulins in recombinant cell culture. In 2001 Genentech obtained New Cabilly patent, the circumstances leading up to the issuance of this patent and the impact of the New Cabilly is detailed below. The other dominant player in the antibody production area is Xoma Corporation of Berkeley, CA, which holds several IPs involving expression of antibodies in bacterial systems.

13.6.1 Genesis of New Cabilly

Celltech has had a broad ranging intellectual property position in recombinant antibody production in various cell types under the Boss patent (US 4,816,397), which was valid until 2006. The Boss patent covered production of engineered antibodies but not murine antibodies produced from hybridomas and may not cover production of antibodies from transgenic mice. Consequently, companies producing chimeric and humanized antibodies and those produced by phage display were paying 1–3% royalty to Celltech. Celltech is also in possession of Bodner patent (US 5,219,996) for the production of antibody fragments in bacteria. This patent runs up to 2008 and is of significance to Celltech for the development of its own product line as well as for generating royalties from other companies. Genentech, which owns the Cabilly patent (US 4,816,567, now known as the old Cabilly patent) for the production of altered and native immunoglobulins in recombinant cell culture, has had a cross-licensing deal with Celltech for payment of royalties to each other on sales of antibody products. Both the Boss and Cabilly patents were issued on the same date, March 28, 1989, even though Celltech's patent application was filed on March 25, 1983 while Genentech's was done on April 8, 1983. Prior to the grant of the old Cabilly patent, Genentech filed a continuation application, which was further amended in March 1990 and contained copied claims of the Boss patent. Genentech also applied to the US patent office for declaring an interference between Celltech's patent and Genentech's application. After a lengthy proceedings which lasted for 7 years, the Board of Interferences favored the Celltech patent. Genentech appealed the decision for a summary judgment motion in the US District Court in San Francisco and provided new evidence claiming to predate Celltech's original filing date. However, Genentech's appeal was denied, and a mediation process was ordered by the US District Court of the Northern District of California in San Francisco. This resulted in the following: (1) a settlement between the two companies was reached and Celltech did not challenge Genentech's priority of invention; (2) a US Patent No. US 6,331,415 ("New Cabilly patent") was granted to Genentech on December 18, 2001, and the US Patent and Trademark Office revoked the Boss patent of Celltech. This ruling resulted in the highly unusual situation that a patent essentially got a lifetime of 29 years granted, from 1989 to 2018. The reason is a quirk in the US law. In 1989 the lifetime of a patent was 17 years from the grant. However, before the mediation leading to Cabilly II was initiated the law changed, and the patent lifetime is now 20 years from filing; Cabilly II in its current form was technically filed during the mediation in 1998.

The newly issued patent has 36 claims for recombinant methods and vectors to produce immunoglobulin (Ig) molecules and functional Ig fragments in transformed host cells. The new Cabilly patent extends patent coverage on antibody manufacture until 2018, and Celltech will receive compensation from Genentech in terms of income from sales of products, which would otherwise have been covered by the Boss patent until its normal expiry date in 2006. Genentech also granted license of new Cabilly patent to Celltech for the production of its products.

The New Cabilly patent covers one of the principal processes used in the manufacturing of therapeutic and diagnostic antibody drugs, and could potentially block the production of antibody products by rival companies or increase their royalty burden.

Indeed, Gaithersburg, MD-based MedImmune has filed a lawsuit on April 11, 2003 in the US District court in Los Angeles charging that the settlement of Celltech and Genentech was illegal and anticompetitive under both Federal and State anti-trust laws and California's unfair competion law.The Plaintiff stated that Celltech and Genentech essentially received an exclusive and dominant position until New Cabilly patent expires in 2018, essentially a 29-year monopoly. While the expiry of Boss and old Cabilly patents in 2006 would have allowed free access to the antibody manufacturing processes, various antibody manufacturers would have to consider now obtaining a license from Genentech. The case was resolved in late 2003 (MedImmune v. Genentech, Inc. et al., 2003 U.S. Dist. LEXIS 23443 (C.D. Cal. 2003)), amended Jan 14, 2004. Judge Pfaelzer of the District Court held that MedImmune is a licensee in good standing, and can therefore not bring forth a declaratory action for formal reasons, as there was no actual case of controversy.

But more importantly, MedImmune's alleged anti-trust and unfair competitions violations were rejected. Judge Pfaelzer held that the settlement between Genentech and Celltech was protected by the Noerr-Pennington doctrine, and granted a summary judgment for the defendants. The Noerr-Pennington doctrine permits collaboration between competitors to petition the government to take action that may restrain competition without incurring anti-trust liability by the act of collaborating. In the given case, the government action was the mediation between Genentech and Celltech with respect to Boss, Cabilly I and Cabilly II. The District Court found that the mediation of Judge Chesney was entitled to be protected under the Noerr-Pennington doctrine.

After an appeal, the Federal Circuit affirmed the District Court in a recent opinion (October 18, 2005). The Federal Circuit also dismissed the two further arguments of MedImmune: first, MedImmune had stated that permitting them to attack Cabilly served public interest, and second, according to MedImmune, if the Federal Circuit follows the District Court's argumentation that a licensee in good standing should not have the right to attack the rights the license is based upon (which it did), than the anti-trust and unfair competition counts should not be judged by the Federal Circuit but the United States Court of Appeals. The Federal Court did not follow MedImmune's argumentation in those cases.

In consequence, Cabilly II is still valid, and stronger than ever. There are some indications that the case may be brought to the Supreme Court. However, there are reasons to doubt that the Supreme Court will respond to the case.

13.6.2 Xoma Patents

Xoma Limited of Berkeley, California also holds a strong IP position in the expression of antibodies, but in bacterial systems. Xoma has received nine US patents to date relating to aspects of its bacterial cell expression system. The expression system includes an *araB* promoter, which allows controlled expression of the desired protein in bacterial host cells, and a *pelB* signal sequence, which allows protein secretion from the host cell. Xoma has also developed a genetically engineered *E. coli* host strain and an easy-to-use fermentation process that complements the *araB* and *pelB* technologies.

Out of the nine patents, six patents broadly cover key methods for the secretion of functional antibody molecules from bacteria, including antibody fragments

Table 13.6 List of Xoma's antibody expression intellectual properties.

Title	*Inventors*	*US Patents*
Modular Assembly of Antibody Genes, Antibodies Prepared Thereby and Use	Robinson, Liu, Horwitz, Wall, Better	5,618,920
		5,595,898
		5,576,195
		5,846,818
Novel Plasmid Vector with Pectate Lyase Signal Sequence	Lei, Wilcox	6,204,023
		5,698,435
		5,693,493
		5,698,417
		5,576,195
AraB Promoters and Method of Producing Polypeptides, Including Cecropins, by Microbiological Techniques	Lai, Lee, Lin, Ray, Wilcox	5,846,818

such as Fab and single-chain antibodies. On March 20, 2001, Xoma was issued its sixth patent in its antibody expression family of patents and the third patent in this family that broadly covers methods for the secretion of functional immunoglobulins from bacteria, including single-chain antibodies and antibody fragments (US 6,204,023, "Modular Assembly of Antibody Genes, Antibodies Prepared Thereby and Use"). Bacterial antibody expression is a key enabling technology for the discovery and selection, as well as the development and manufacture, of many recombinant antibody-based pharmaceuticals. Antibody discovery by phage display technology, for example, depends upon the expression of antibody domains in bacteria as properly folded, functional, secreted proteins, as described in Xoma's patent claims.

Xoma's other patents include one that relates to improved methods and genetic constructs for process control utilizing the *araB* promoter. Two patents relate to the technology for *pelB* signal sequence secretion. A number of foreign patents also have been granted to Xoma, which, along with pending applications, correspond to the issued US patents. Xoma's primary US-granted patents are listed in Table 13.6.

13.7 Litigations and Cross-licensing

From 1999, CAT was embroiled in patent opposition or infringement suits with Morphosys until December of 2002. The litigations between the companies were fought in the US courts regarding CAT's Griffiths, McCafferty, Winter II, and Winter/Lerner/Huse patents. Additionally, MorphoSys also launched opposition at the European Patent Office against CAT's Winter II and McCafferty patents. However, since the ruling by the US District Court in March 2002 that Morphosys' HuCAL antibody library technology is distinct and independent from CAT's patent coverage by Griffith's IP, the two companies have ended their disputes after signing cross-licensing agreements.

There was, however, a substantial amount of payment from Morphosys to CAT, which included an annual payment of 1 million euros over the next 5 years. CAT will also receive other financial consideration from MorphoSys' activities related to its HuCAL GOLD libraries for a defined period of time in addition to milestone and royalty payments under the license for products developed using previous HuCAL libraries. Furthermore, CAT will receive an equity stake of 588 160 ordinary shares in MorphoSys under the license agreement. MorphoSys retains the option to buy out its obligations to CAT for a predefined fixed amount at any time during the duration of the agreement.

Both the ruling and the subsequent cross-licensing agreement should allow Morphosys to indemnify its several collaborators about the sole proprietorship of its HuCAL library and any potential patent infringement issues. Additionally, CAT has also cross-licensed its phage display patents with (1) Dyax's Ladner phage display IP, (2) Xoma's antibody expression technology, (3) Micromet and Enzon Pharmaceutical's single chain antibody technology and Crucell's Mabstract phage display technology.

CAT and Dyax expanded their licensing agreement of 1997 under which Dyax licensed its Ladner phage display patents to CAT in exchange for receiving a worldwide license for research and to develop therapeutic and diagnostic antibody products under all the antibody phage display patents controlled by CAT. In return, CAT will receive milestone and royalty payments on antibody products advanced into clinical trials by Dyax and Dyax's customers. CAT also gains the option to co-fund and co-develop with Dyax antibodies discovered by Dyax, as well as the right to share in Dyax's revenues from certain other applications of antibody phage display. In a further expansion of the agreement, in September 2003, Dyax realized from CAT, among other benefits, an increased number of options for licenses to develop therapeutic and diagnostic antibody products under CAT's patents for Dyax's own use and on behalf of its partners. CAT and Dyax have further agreed that, as a result of various co-licensing agreements, CAT shall not have to pay royalties to Dyax in respect of any antibody products CAT develops including Humira.

CAT also finalized a cross-licensing agreement with XOMA under which CAT and its collaborators receive license to use the Xoma antibody expression technology for developing products using CAT's phage-based antibody technology. In return, Xoma receives a license payment, and in addition receives the right to use CAT's phage antibody libraries for its target discovery and research programmes, with an option to develop antibodies into therapeutics. Xoma will pay license payments to CAT if the company identifies and develops any therapeutic antibodies using CAT's libraries.

CAT and Crucell also formed a worldwide license agreement under which Crucell receives access to all CAT's antibody phage display technology patents, both pending and granted. This provided Crucell freedom under the CAT patents to fully exploit its proprietary MAbstract technology and other phage display technology. In return, CAT will receive an initial license fee from Crucell, and obtains an option to develop certain antibodies, opening the way for further collaboration between the companies. CAT will also receive milestone and royalty payments for any antibody products that Crucell or its partners develop that are derived from Crucell's MAbstract technology or other technology involving phage display.

13.8 Other Cross-licensing

Other notable cross-licensing agreements of recent years involved the antibody expression technology owned by XOMA. Among others, XOMA has cross-licensed with Morphosys, Dyax, BioInvent, and Affitech.

Indeed, such cross-licensing arrangements have provided a freedom-to-operate position for practicing human antibody discovery *in vitro* to several companies, akin to the transgenic mouse system of Abgenix and Medarex. Hopefully this will help to increase the interest level of many of the companies contemplating entering into the human antibody therapeutic area, since the constraints of entry to a greater extent has been minimized by the co-licensing process.

13.9 Litigation between CAT and Abbott

A recently resolved litigation case involved the distribution of royalties on Humira between CAT and Abbott. Humira is the first fully human antibody approved by the FDA for marketing on December 31, 2002. It is currently used for the treatment of early and late rheumatoid arthritis as well as psioratic arthritis. The antibody targets tumor necrosis factor alpha (TNF-α) and is in phase III trials for, amongst others, Crohn's disease and juvenile arthritis. Humira sales in 2005 were at \$1.4 billion, and are expected to rise to \$1.9 billion in 2006.

Humira was developed by CAT in collaboration with Knoll, which was later acquired by Abbott. The basic structure of the deal was that CAT was using their phage display technology to identify antibodies of clinical interest against targets provided by Knoll/Abbott. As a payment, CAT was to receive royalties at a little bit over 5%. However, under certain circumstances, the contract allowed for an offset of royalties due to CAT, if third-party rights were necessary to utilize CAT technology. Under that situation, the minimum royalty level to be received by

CAT was set at 2%. It should be mentioned that CAT was obliged under the initial contract to give a certain part of their royalty to the Medical Research Council (MRC), Scripps Institute, and Stratagene, where the CAT technology originated.

Abbott has disclosed that it is paying royalties on Humira to Serono and Peptech, and it is likely that royalties are also paid to Genentech. Consequently, Abbott reduced CAT's share of the royalties. CAT filed for legal proceedings, as they did not believe that these third-party rights are necessary to utilize CAT's technology and hence such a third-party royalty should not be deducted from the royalties paid to CAT. The case was brought to the High Court in London in November/December 2004. The Court ruled strongly in favor of CAT. The argumentation of Abbott was discarded in full. Abbott appealed, but before the proceedings at the Court of Appeal were initiated, an agreement between the parties was reached. This agreement was not only between CAT and Abbott but included also the licensors of CAT. The details were disclosed by CAT as follows:

- Abbott will pay CAT the sum of US$255 million, which CAT will pay to its licensors, the MRC, Scripps Institute and Stratagene, in lieu of their entitlement to royalties arising on sales of Humira from 1 January 2005 onwards.
- Abbott will also pay to CAT five annual payments of US$9.375 million commencing in January 2006, contingent on the continued sale of Humira. US$2 million from each of these payments will be payable to CAT's licensors.
- Abbott will pay CAT a reduced royalty of 2.688% from approximately 5.1% on sales of Humira from January 1, 2005. CAT will retain all of these royalties. CAT will also retain royalties received from Abbott in respect of sales of Humira up to December 31, 2004, net of approximately £1187.6 million of which will be paid to its licensors.
- CAT will refund to Abbott approximately £9.2 million for royalties paid from January 1, 2005 through June 30, 2005. Abbott will also pay CAT a reduced royalty of 4.75% on any future sales of ABT-874, from which CAT will pay a portion to the MRC and other licensors (according to CAT's 1997 agreement with the MRC).
- Abbott will capitalize and amortize the upfront payment, net of the refund, and annual payments to CAT through the term of the agreement. When this amortization is combined with the revised royalty rate of 2.688%, the blended effective royalty rate is reduced from the approximate 5.1% as previously instructed by the Court.

13.10 Importation of Data

For several years companies have wondered whether data or drugs generated by molecular screening methods outside of USA could be imported into the US without infringing a US patent, and if such activities constitute a "Product made" under the Process Patent Amendments Act of 1988. This Act was passed by the US Congress to protect holders of US process patents in response to concerns that competitors can avoid infringement of method patents simply by carrying out the methods in other countries and importing the products into the USA. The amended patent statute included the section 35 U.S.C.271(g) which states that importing, using, selling, or offering for sale in the US a product that was made abroad by a process protected by a US patent would constitute infringement of that patent.

A verdict on August 22, 2003, by the US Court of Appeals for the Federal Circuit in the case of Bayer AG vs Housey Pharmaceuticals has clarified the scope of patent protection and ruled that drug discovery data are not "product" under the Act.

In this case Housey, the patent holder, sued Bayer, alleging that Bayer's importation of drug discovery data to the US violated the Act. However, Bayer contended that it was importing only information, and since 271(g) applies only to manufacturing processes in which a product is physically made, it could not infringe the Housey patent claims. Housey argued that 271(g) should be interpreted broadly to cover any patented method irrespective of whether it produces a physical product or not. The lower court disagreed with Housey and ruled that drug discovery data are not "product" covered by the Act. This ruling was subsequently upheld by the Federal Appeals Court, and by so doing determined that Bayer's importation of screening data was not an infringement of the Housey patents.

As a result of this decision many international companies would be able to carry out their drug discovery screening efforts outside of the US. Here it should be stated that the US Senate has been considering changing laws in order to close the possibility of importing data, which might trigger a new round of licenses and litigation.

13.11 The Single-Chain Antibody Technology

Single-chain antibodies (SCAs) are composed of the antigen-binding regions of antibodies on a single polypeptide chain. Micromet AG of Munich, Germany has recently consolidated the SCA patent estate, first by acquiring on June 29, 2001 from Curis, Inc. (Cambridge, MA, USA) all of its patent rights and patent applications directed to SCAs and SCA fusion proteins, and following to that by forming an exclusive marketing partnership with Enzon (Piscataway, NJ, USA) for its patents, patent applications, and technology pertaining to SCAs. Consequently, the comprehensive IP portfolio includes patents and patent applications claiming broad aspects of SCA technology. Table 13.7 lists the EU and US rights of the combined SCA IP portfolio of Micromet/Enzon.

Table 13.7 Single-chain antibody intellectual property portfolio of Micromet/Enzon.

Country	*Title*	*Patent number*
USA	Single Polypeptide Chain Binding Molecules	4,946,778
USA	Single Polypeptide Chain Binding Molecules	5,260,203
USA	Single Polypeptide Chain Binding Molecules	5,455,030
USA	Single Polypeptide Chain Binding Molecules	5,518,889
USA	Single Polypeptide Chain Binding Molecules	5,534,621
CANADA	Single Polypeptide Chain Binding Molecules	1,341,364
EP	Single Polypeptide Chain Binding Molecules	0281604
USA	Computer Based System And Method For Determining And Displaying Possible Chemical Structures For Converting Double or Multiple Chain Polypeptides to Single Chain Polypeptides	4,704,692
EP	Method For The Preparation of Binding Molecules	0349578
US	Multivalent Antigen Binding Proteins	5,869,620
US	Multivalent Antigen Binding Proteins	6,121,424
US	Multivalent Antigen Binding Proteins	6,027,725
US	Methods for Producing Multivalent Antigen Binding Proteins (As Amended)	6,025,165
US	Linker For Linked Fusion Polypeptides	5,856,456
US	Linker For Linked Fusion Polypeptides	5,990,275
US	Antigen-Binding Fusion ProteinsPolyalkylene	5,763,733
US	Antigen-Binding Fusion ProteinsPolyalkylene	5,767,260
US	Stabilized Monomeric Protein Compositions	5,656,730
US	Stabilized Monomeric Protein Compositions	5,917,021
US	Single-Chain Antigen-Binding Proteins Capable of Glycosylation, Production And Uses Thereof	6,323,322
US	Method For Targeted Delivery of Nucleic Acids	6,333,396
US	Nucleic Acid Molecules Encoding Single-Chain Antigen-Binding Proteins (As Amended)	6,103,889
US	Polypeptide linkers for production of biosynthetic proteins	5,482,858
US	Polypeptide linkers for production of biosynthetic proteins	5,258,498
CAN	Targeted Multifunctional Proteins	1,341,415
AU	Targeted Multifunctional Proteins	612 370
AU	Targeted Multifunctional Proteins	648 591
EP	Targeted Multifunctional Proteins	0318554
EP	Targeted Multifunctional Proteins	0623679
US	Biosynthetic Antibody Binding Sites	5,132,405
US	Biosynthetic Antibody Binding Sites	5,091,513
US	Biosynthetic Antibody Binding Sites	5,476,786
US	Genetically Engineered Antibody Analogues and Fusion Proteins Thereof	6,207,804
US	Biossynthetic Binding Protein for Immuno-targeting	5,534,254
US	Biossynthetic Binding Protein for Immuno-targeting	5,837,846
US	Biosynthetic Binding Protein for Immuno-targeting	5,753,204
US	Serine-rich Peptide Linkers	5,525,491
CAN	Serine-rich Peptide Linkers	2,100,671
AU	Serine-rich Peptide Linkers	664 030
EP	Serine-rich Peptide Linkers	0573551
US	Methods and Compositions for High Protein Production from Non-native DNA	5,658,763
US	Methods and Compositions for High Protein Production from Non-native DNA	5,631,158
US	Methods and Compositions for High Protein Production from Non-native DNA	5,733,782

Source: Micromet.

13.12
US Patent Issued on Polyclonal Antibody Libraries

In January 2002, it was announced that a US patent (US 6,335,163 B1) has been issued to Boston University, research partner of Symphogen, which holds a worldwide exclusive license to this newly issued patent and to the previously issued methods patent (US 5,789,208). The patent claims relate to the methods for the creation and use of libraries of proteins which comprise polyclonal antibodies to a common antigen or group of antigens, receptor proteins with related variable regions, or other immune related proteins with variable regions. The company believes that this patent gives it a unique position in the development of recombinant polyclonal antibodies.

Some of the important US patents relevant to the discovery, development, and production of recombinant antibodies, mentioned in the text and their assignee names are listed in Table 13.8.

Table 13.8 Selected list of relevant US patents.

Patent number	*Commonly used name (where known)*	*Assignee/company*
US 4,816,397	Boss patent	Celltech Group
US 4,816,567	Old Cabilly patent	Genentech Inc.
US 5,219,996	Bodner patent	Celltech Group
US 5,223,409	Ladner patent family member	Dyax Corporation
US 5,403,484	Ladner patent family member	Dyax Corporation
US 5,427,908	Not known	Biosite Inc.
US 5,571,698	Ladner patent family member	Dyax Corporation
US 5,580,717	Not known	Biosite Inc
US 5,585,089	Queen's patent family member	Protein Design Labs
US 5,595,721	Not known	Corixa Corporation
US 5,693,761	Queen's patent family member	Protein Design Labs
US 5,693,762	Queen's patent family member	Protein Design Labs
US 5,723,323	Kauffman patent family member	Applied Molecular Evolution, Inc.
US 5,763,192	Kauffman patent family member	Applied Molecular Evolution, Inc.
US 5,789,208	Not known	Symphogen
US 5,814,476	Kauffman patent family member	Applied Molecular Evolution, Inc.
US 5,817,483	Kauffman patent family member	Applied Molecular Evolution, Inc.
US 5,824,514	Kauffman patent family member	Applied Molecular Evolution, Inc.
US 5,837,500	Ladner patent family member	Dyax Corporation
US 5,849,500	Breitling patent family member	Affitech AS
US 5,859,205	Adair patent	Celltech Group
US 5,885,793	Griffiths patent	Cambridge Antibody Technology

Table 13.8 *Continued*

Patent number	*Commonly used name (where known)*	*Assignee/company*
US 5,967,862	Kauffman patent family member	Applied Molecular Evolution, Inc.
US 5,969,108	McCafferty patent	Cambridge Antibody Technology
US 5,985,588	Breitling patent family member	Affitech AS
US 6,015,542	Not known	Corixa Corporation
US 6,051,230	Not known	Peregrine Pharmaceutical
US 6,054,561	Not known	Chiron corporation
US 6,090,365	Not known	Corixa Corporation
US 6,127,132	Breitling patent family member	Affitech AS
US 6,180,370	Queen's patent family member	Protein Design Labs
US 6,204,023	Antibody Expression patent family member	Xoma Ltd.
US 6,248,516	Winter II patent	Cambridge Antibody Technology
US 6,287,537	Not known	Corixa Corporation
US 6,291,158	Huse/Lerner/Winter patent	Cambridge Antibody Technology
US 6,291,159	Huse/Lerner/Winter patent	Cambridge Antibody Technology
US 6,291,160	Huse/Lerner/Winter patent	Cambridge Antibody Technology
US 6,291,161	Huse/Lerner/Winter patent	Cambridge Antibody Technology
US 6,300,064	HuCAL patent family member	MorphoSys AG
US 6,331,415	New Cabilly patent	Genentech Inc.
US 6,335,163		Symphogen
US 6,387,629	Breitling patent family member	Affitech AS
US 6,451,312	Not known	Peregrine Pharmaceutical
US 6,692,935	Not known	MorphoSys AG
US 6,696,248	HuCAL patent family member	MorphoSys AG
US 6,730,483	Breitling patent family member	Affitech AS

13.13 Conclusion

In essence, a company venturing into the mAb therapeutic area must consider at the outset which discovery tool is the company's preferred system. Is it (1) humanization of mouse or rabbit antibodies, (2) human antibodies from transgenic mouse, or (3) *in vitro* antibody display technology? Following that comes the decision of mAb formating: should it be a full-length IgG or antibody fragments. This decision, to a large extent, is dependent upon the cellular functions, targeting, and efficacy of the antibody. This will also lead to the selection of manufacturing

Table 13.9 Essential technologies in recombinant antibody and primary intellectual property holders.

Technologies	*IP holders*
Antibody discovery	
Humanization	PDL
Transgenic mouse system	Abgenix
	Medarex
In vitro display technology	Affitech
	CAT
	Dyax
	Morphosys
Antibody production	
Mammalian	Genentech
Bacterial	Xoma
Plants	Various companies
Aquatic	Biolex
Yeast	GlycoFi
Antibody formats	
Single chain	Micromet/Enzon
Fab	Dyax
Domain antibodies	Domantis
Antibody modifications	
Glycosylation	Glycart (a Roche company)
	GlycoFi
	BioWa
Antibody variants	AME (an Eli Lilly company)
	Xencor

system since a full-length IgG must be made from an eukaryotic expression system because of the presence of oligosaccharides in the Fc region, while scFv and Fab formats could be produced from a prokaryotic system because of the lack of glycosylation moieties in such molecules. There are also plants, yeast, and aquatic organisms available for production that several companies are attempting in lieu of mammalian cell systems.

Although we have not considered here IP issues relating to the effect of modified glycosylation or antibody engineering for generating mAbs with superior efficacy or higher production yield or improved binding to the Fc receptor, several companies, such as Applied Molecular Evolution, Xencor, Glycart, GlycoFi, and BioWa in recent years have built important IP estates in such areas.

While the advice of a good patent attorney is imperative for an organization interested in entering or dealing with the labyrinth of mAb IPs, Table 13.9 depicting various technologies important for mAb discovery and production and the companies holding the most important IPs in respective areas should facilitate initial assessment of the recombinant mAb landscape and evaluation of the business perspectives of an mAb product.